DIGITAL FORMATIONS

DIGITAL FORMATIONS

IT and New Architectures
in the Global Realm

Edited by
Robert Latham and Saskia Sassen

PRINCETON UNIVERSITY PRESS   PRINCETON AND OXFORD

Published by Princeton University Press, 41 William Street,
Princeton, New Jersey 08540
In the United Kingdom: Princeton University Press, 3 Market Place,
Woodstock, Oxfordshire OX20 1SY

Library of Congress Cataloging-in-Publication Data

Digital formations: IT and new architectures in the global realm / edited by Robert
Latham and Saskia Sassen.
p. cm.
Includes bibliographical references and index.
ISBN 0-691-11986-4 (cloth : alk. paper) — ISBN 0-691-11987-2 (pbk. : alk. paper)
1. Information technology. 2. Computer networks. 3. Communication, International. I. Latham, Robert. II. Sassen, Saskia.
HM851.D53 2005

303.48′33—dc22 2004062467

British Library Cataloging-in-Publication Data is available

This book has been composed in Sabon

Printed on acid-free paper. ∞
pup.princeton.edu

Printed in the United States of America

10 9 8 7 6 5 4 3 2 1

Contents

List of Illustrations vii

Acknowledgments ix

Introduction
Digital Formations: Constructing an Object of Study 1
ROBERT LATHAM AND SASKIA SASSEN

SPACES OF KNOWLEDGE 35

Recombinant Technology and New Geographies of Association 37
JONATHAN BACH AND DAVID STARK

Electronic Markets and Activist Networks:
The Weight of Social Logics in Digital Formations 54
SASKIA SASSEN

The New Mobility of Knowledge: Digital Information Systems
and Global Flagship Networks 89
DIETER ERNST

NETWORKS OF COOPERATION 115

Cooperative Networks and the Rural-Urban Divide 117
D. LINDA GARCIA

Networks, Information, and the Rise of the Global Internet 146
ROBERT LATHAM

The Political Economy of Open Source Software and
Why It Matters 178
STEVEN WEBER

DESIGNS AND INSTITUTIONS 213

Designing Information Resources for Transboundary Conflict
Early Warning Networks 215
HAYWARD R. ALKER

Discourse Architecture and Very Large-scale Conversation 242
WARREN SACK

Transnational Communication and the European Demos 283
LARS-ERIK CEDERMAN AND PETER A. KRAUS

Information Technology and State Capacity in China 312
DOUG GUTHRIE

List of Contributors 339

Index 341

List of Illustrations

Sassen: Electronic Markets and Online Activist Networks
TABLE 1. Financial Assets of Institutional Investors, 1990 to 2001 60
TABLE 2. Cross-border Transactions in Bonds and Equities, 1975 to 2002 63
TABLE 3. The Twelve Biggest Stock Markets in the World, 2000 and 2003 66
TABLE 4. Foreign Listings in Major Stock Exchanges, 2000 and 2003 67

Ernst: The New Mobility of Knowledge
FIGURE 1. GFNs, DIS, and Knowledge Diffusion 91

Latham: Networks, Information, and the Rise of the Global Internet
FIGURE 1. Transboundary Internetworking Styles 153
FIGURE 2 172

Alker: Designing Information Resources for Transboundary Conflict Early Warning Networks
FIGURE 1. The Case of Chiapas in the CEWS Explorer 229

Sack: Discourse Architecture and Very Large-scale Conversation
FIGURE 1. Mozilla News 245
FIGURE 2. soc.culture.palestine during the period August 1–7, 2001 261
FIGURE 3. soc.culture.palestine during the period August 4–11, 2001 263
FIGURE 4. soc.culture.afghanistan during the period September 24–28, 2001 264
FIGURE 5. alt.politics.election for the week prior to the 2000 U.S. election 270
FIGURE 6. The Same Conversation Map as Shown in Figure 5 with the Discussion Theme "Gore" Selected 271
FIGURE 7. A Partial List of the Terms Associated with Bush and/or Gore 272
FIGURE 8. A Sentence from the Newsgroup Associating a Term (Bush) with a Verb 273
FIGURE 9. Close Examination of the Structure of a Thread 273

FIGURE 10. alt.politics.election for the Week after the 2000 U.S. Election 275

Cederman and Kraus: Transnational Communications and the European Demos
FIGURE 1. The Logic of National Substantialism 288
FIGURE 2. The Logic of Civic Voluntarism 290
FIGURE 3. The Logic of Bounded Institutionalism 292

Guthrie: Information Technology and State Sovereignty in China
TABLE 1. Access to Media of Information in China, 2000 318
TABLE 2. Growth of Information Technology in China, 2001 319
FIGURE 1. Number of Foreign Joint Ventures in Selected Industries 321
TABLE 3. Foreign Capital Invested in China in the Reform Era, 1985 to 1999 325
FIGURE 2. Gross Industrial Output by Ownership Type 331
TABLE 4. Vital Statistics on Higher Education in China, 1980 to 1999 334

Acknowledgments

THIS VOLUME IS the end product of a research working group on "Cooperation and Conflict in a Connected World," organized through the Social Science Research Council Program on IT and International Cooperation. The program was established to help advance the interdisciplinary, social scientific study of the international dimensions of IT. Building a social science of IT means that research and theory-building around IT should be well integrated into each of the social science disciplines. This will better enable researchers to understand not only IT itself, but also how it intersects with other conditions and challenges existing theories and empirical generalizations. The aim is to avoid a segregated domain of "IT studies"; instead, information and its technologies should become a category of analysis for the social sciences in much the same way as "institutions." We hope that the chapters that follow live up to these aims.

The Ford Foundation funded the program and pushed it to take risks and be innovative. Mahnaz Ispahani at Ford was a key partner whose initial support was essential, as is the continued support of both Lisa Jordan and Becky Lentz.

The aid of the original program Steering Committee, some of whom became part of the working group and contributed chapters, was indispensable to the project. They included Hayward Alker, John Seely Brown, Dorothy E. Denning, Dieter Ernst, Jane Fountain, D. Linda Garcia, Dina Iordanova, Robert Keohane, Rohan Samarajiva, Saskia Sassen (chair), David Stark, Nigel Thrift, Steven Weber, Barry Wellman, and Ernest Wilson. We owe special thanks to former committee member Robert Keohane, who was an active and engaged participant in the working group. His critical comments and suggestions along the way shaped the book in innumerable ways.

The program and this volume owe special thanks to Michael Watts and David Stark for organizing summer institutes at, respectively, Berkeley and Columbia. All of the many participants, too numerous to name here in those institutes provided important analyses and ideas that have become part of the collective wisdom this volume represents. Additionally, a meeting led by Bill Drake and Ernie Wilson in Budapest, co-organized and co-funded by the Open Society Institute, contributed further intellectual capital to this program and volume. Key to making that meeting possible were Darius Cuplinskus and Vera Franz of OSI. (See www.ssrc.org/programs/itic for a list of participants and profiles of the various meetings.)

Michael Chwe also participated in the working group and contributed to the intellectual agenda. Paul Price helped formulate some of the initial purposes of the working group. Deborah Matzner assisted ably in the initial organization of the working group. Marcela Sabino and Gretchen Schwarz helped bring the chapters to final form. Without them both, this volume would not have been possible.

DIGITAL FORMATIONS

Introduction

Digital Formations: Constructing an Object of Study

ROBERT LATHAM AND SASKIA SASSEN

COMPUTER-CENTERED NETWORKS and technologies are reshaping social relations and constituting new social domains. These transformations assume multiple forms and involve diverse actors. In this volume we focus on a particular set of instances: communication and information structures largely constituted in electronic space. Examples are electronic markets, Internet-based large-scale conversations, knowledge spaces arising out of networks of nongovernmental organizations (NGOs), and early conflict warning systems, among others. Such structures result from various mixes of computer-centered technologies and the broad range of social contexts that provide the utility logics, substantive rationalities, and cultural meanings for much of what happens in these electronic spaces. In this regard, the electronic spaces that concern us in this volume are social. Digital formation is the construct we use to designate these specific types of information and communication structures. Digital formations are to be distinguished from digital technology tout court; not all digital networks are digital formations.

This volume seeks, then, to advance research that is at the intersection of what we might simplify as technology and society. We do not assume that technology and society are actually separate entities, and we accept many of the propositions in the critical social science literature that posit that technology is one particular instantiation of society—society frozen, that is, one moment in a trajectory that once might have been experienced as simply social (Latour 1991). Without losing this critical stance, we want, nonetheless, to capture the distinctiveness and variable weight of "technology" and to develop analytic categories that allow us to examine the complex imbrications between the outcome of society that we call technology and the social, economic, political, and cultural dynamics through which relations and domains are constituted. Much rides in social analyses of IT on the category of "newness," and this volume is no exception. We believe we are looking at formations that have not existed before, and we mean this to imply two things: that the forms were not present in a given social context before, and that the formations in question are novel social forms.

That these are novel forms implies that we are looking at entities that are likely in the early—if not initial—stages of formation. We are not claiming this status for IT itself. Beniger (1986) underscores that the reflexive development and organization of complex IT-based formations is discernible as early as the nineteenth century.[1] Rather, we attach this status to the emergence of a wide range of formations of varying scales that depend on digital technologies, cross a variety of borders (national or otherwise), and engender a diverse array of spatial, organizational, and interactive practices.

The set of cases explored in the chapters that follow is meant to give readers a sense of that range and to cover topics that have been considered important to the social analysis of IT, especially as it bears on transboundary phenomena, including transnational civil society, transboundary public spheres, global finance, transnational corporate networks, global technological diffusion, regional integration, and international economic development. There has been no attempt to be comprehensive, however.[2] What joins the chapters is not only the effort to capture constitutive and transformative processes, but also concerns with design and social purpose.

Locating a New Field of Inquiry

One of the distinct capabilities of these technologies when it comes to the communication and information structures that concern us in this volume is the rescaling of social relations and domains. What has tended to operate or be nested at local scales can now move to global scales, and global relations and domains can now, in turn, more easily become directly articulated with thick local settings. In both types of dynamics, the rescaling can bypass the administrative and institutional apparatus of the national level, still the most developed scalar condition. As a result of the growing presence and use of these technologies, an increasing range of social relations and domains have become de facto transboundary. It need not be this way, and indeed many of these digital formations are not, but the trend is definitely toward expanding the world of transboundary re-

[1] Another significant historical analysis that is U.S. focused is Chandler and Cortada (2000).

[2] One noticeable omission is the security sphere. But see the related SSRC-sponsored volume, *Bombs and Bandwidth* (Latham 2003), which focuses exclusively on this realm. Further, a new SSRC volume on global civil society and the Internet is in progress (edited by Jon Anderson, Jodi Dean, and Geert Lovink).

lations and domains. This trend is evident in this volume, where even digital formations that need not be transboundary, such as large-scale conversations or knowledge spaces, wind up being so directly or indirectly.

We are, then, seeing the transnationalizing of a growing range of local or national relations and domains, as well as the formation of new ones. Such transformations enable nonstate actors to enter international arenas once exclusive to states and the formal interstate system. This is well illustrated by specific features of the growing numbers and types of international nongovernmental organizations, global business alliances, and diasporic networks. These transformations have also furthered the formation of new types of spaces constituted partly through cross-border actors and transactions. All of this partly reconstitutes the world of cross-border relations and takes this world beyond formulations common in the specialized literature on international relations.

To some extent these transformations in the world of cross-border relations are overdetermined in that they entail multiple causalities and contingencies. This volume's focus on computer-based interactive technologies and networks does not presume to posit a single causality. What we refer to below for short as sociodigitization is deeply imbricated with other dynamics.[3] In some cases sociodigitization is "derivative"—a mere instrumentality of these dynamics—but in others it is "transformative"—by reshaping social relations—and even "constitutive"—by producing new social domains of action. Yet even when derivative, sociodigitization is contributing to the rescaling of a variety of processes with the resulting implications for territorial boundaries, national regulatory frames, and cross-border relations. The outcome is a set of changes in the scope, exclusivity, and competence of state authority over its territory, and, more generally, the place of interstate relations in the expanding world of cross-border relations.

An organizing assumption in this volume and in the larger Social Science Research Council (SSRC) project on information technologies to which it contributes is that these new conditions have implications for theory and for politics.[4] The social sciences are not well prepared to take on these developments. The discipline that has had cross-border relations at its core, international relations, remains mostly focused on the logic of relations between states and has not generally treated communication and information as essential to analysis. Exceptions to the state-centric focus in IR include work on transnational relations (Nye and Keohane 1971),

[3] Sociodigitization, as defined below, is the process whereby activities and their histories in a social domain are drawn up into digital codes, databases, images, and text.

[4] In particular, the SSRC program, IT and International Cooperation. See www.ssrc.org/programs/itic.

which assumes new relevance under current conditions.[5] Also warranting greater attention is pioneering work incorporating information and communications (Deutsch 1953, 1957; Jervis 1976) and more recent research and analysis that focuses on information technologies.[6] However, this work cannot quite fully encompass today's multiplication of nonstate actors and new conditions in transboundary cooperation and conflict.

An alternative line of scholarship is centered on the technical properties of the new technologies and their capacities for producing change.[7] These technologies increasingly dominate explanations of contemporary change and development, with technology seen as the impetus for the most fundamental social trends and transformations.[8] Such explanations also tend to understand these technologies exclusively in terms of technical properties and to construct the relation to the social world as one of applications and impacts.

Neither theorizations centered on the state nor those centered on technology as the key explanatory variable can adequately capture the transformations in the world of cross-border relations that concern us in this volume. Understanding the place of these new computer-centered networks and technologies from a social science perspective requires avoiding a purely technological interpretation and recognizing the embeddedness of these technologies and their variable outcomes for different economic, political, and social orders.

Confining interpretation to the properties of these technologies neutralizes or renders invisible the social conditions and practices, placeboundedness, and thick environments within and through which these technologies operate. Such readings also lead, ironically, to a continuing reliance on analytic categorizations that were developed under other spatial and temporal conditions, that is, conditions preceding the current digital era. Thus the tendency is to conceive of the digital as simply and exclusively digital, and the nondigital (whether represented in terms of the physical/material or the actual, all problematic though common concep-

[5] Of note is the special issue of *Millenium: Journal of International Relations* on Territorialities, Identities, and Movement in International Relations (1999).

[6] See, for example, Choucri (2000), Deibert (2000), Der Derian (2001), Laguerre (2000), and Wilson (2004).

[7] Latham (2002) offers a fuller discussion of ways that newness has figured into analyses of IT and social change.

[8] For critical examinations that reveal particular shortcomings of technology-driven explanations see, e.g. Wajcman (2002), Loader (1998), Nettime (1997), Hargittai (1998), and more generally Latour (1987), Munker and Roesler (1997), Mackenzie (1999), and Mackenzie and Wajcman (1999). For a critique by "technologists" of such technology-driven explanations, see Brown and Duguid (2000).

tions) as simply and exclusively nondigital. These either/or categorizations filter out alternative conceptualizations, thereby precluding a more complex reading of the intersection and interaction of digitization with social, other material, and place-bound conditions. Another consequence of this type of reading is to assume that a new technology will ipso facto replace all older technologies that are less efficient, or slower, at executing the tasks the new technology is best at. We know that historically this is not the case.

Nonetheless, it is important for our effort to recognize the specific capacities of digital technologies.[9] They are central to the emergence of new information and communication structures and the transformation of existing ones.[10] In their digitized form, these structures exhibit dynamics of their own that derive from technological capacities that enable specific patterns of interaction. These technology-driven patterns are, then, endogenous to these digitized structures rather than the product of an exogenous context such as the interstate system. Among such patterns are the simultaneity of information exchange, capacity for electronic storage and memory, in combination with the new possibilities for access and dissemination that characterize the Internet and other computer-centered information systems.[11]

These technical capacities can change the relationship between information and a broad range of entities and conditions. For instance, new resources and capabilities are being created for NGOs and other private

[9] There are important types of computer technology that we are not addressing in this volume, notably robotics, data processing, and virtual reality.

[10] Studies of new or transforming structures have typically focused on various dimensions of social life, including individual identity, community, social development, work, politics, and economic organization. Illustrative are Webster (1995), May (2002), and of course Castells (1996), the latter being mainly focused on socioeconomic change. Note that much of this literature is anchored in the notion that modern societies are transforming into information societies driven by an information revolution. This sort of thinking caught on in the early 1970s, and a particularly notable statement is Bell (1973). Among the structures that are seen as developing through and around the use of these technologies are "virtual communities," "virtual corporations," and multi-user-domains (MUDS). On communities, see Smith and Kollock (1998); on virtual corporations, see the journal at www.virtual-organization.net; and on MUDs, among other virtual social forms, see Turkle (1995).

[11] For most producers and consumers of research on IT, knowledge begins and ends with the Internet. While the Internet is crucial to the development of digital formations, in and of itself it is not a formation but, as conveyed in the chapter by Latham, a global communication system that comprises myriad electronic networks. These networks, in turn, are the underlying platforms for digital formations. But a digital network need not be part of the open Internet tied to e-mail and the World Wide Web if it is a private network as considered by Ernst and Sassen.

associations via web pages and document storage (Garcia, this volume). This matters because groups, particularly when involved in contestational politics, can use these information resources to challenge certain kinds of interpretations of developments, events, or policies. Such challenges lead to new knowledge spaces (Bach and Stark). Groups, such as diasporas connected to zones of conflict, can construct their histories and make them accessible to insiders and outsiders. These possibilities, in turn, prompt a reexamination of assumptions about the role of "knowledge" circulating within and across groups in the shaping of intergroup cooperation or conflict (Alker). Technology here makes it easier to trace the history of interactions and events, which in turn has implications for reciprocity and repeated strategic interaction. When it comes to major economic actors such as transnational corporations, the typically private information systems offer whole new organizational and managerial capabilities, such as the global flagship networks examined by Ernst.

From a social science perspective, as compared to a purely engineering one, such digitized information and communication structures and dynamics—what we call digital formations—filter and are given meaning by social logics. By social logics we intend to refer to a broad range of conditions, actors, and projects, including specific utility logics of users as well as the substantive rationalities of institutional and ideational orders. The distinctiveness of digital formations can contribute to the rise of social relations and domains that would otherwise be absent. Examples of such distinctive structurations in our volume are open source software communities (Weber), the formation of digitally based large-scale conversations (Sack), new types of public spheres (Cederman and Kraus), certain types of early warning systems (Alker), and electronic markets for capital (Sassen).

The presence of social logics in the structuring of these formations means, from a social science perspective, that the technical capacities of these new technologies get deployed or used in ways that are uneven and contradictory within diverse digital formations. They unfold in particular contexts and evince both variability and specificity. Digital formations, as we define them here, do not exist as purely technological events. This, in turn, makes it difficult to generalize their transformative and constitutive outcomes. Variability and specificity are crucial dimensions emerging from the diverse foci of analysis in the volume. The choice of chapters seeks to address this as each focuses in great detail on a different subject. While variability and specificity make generalization difficult, detailed study can illuminate patterns and structures helpful in hypothesizing future trends and in developing agendas for research and analysis as IT continues to evolve.

The uneven and often contradictory character of these technologies and their associated information and communication structures also lead us to posit that these technologies should not be viewed simply as factor endowments. This type of view is present in much of the literature, often implicitly, and represents these technologies as a function of the attributes of a region such as Asia or an actor such as an NGO—ranging from regions and actors fully endowed, or with full access, to those without access. Rather, we recognize that any given region or actor can be associated with uneven or inconsistent technological capacities. Cederman and Kraus make clear that even in wired Europe, attempts to construct a rich communicative space confront the limits of online public engagement.

Variability also emerges because the deployment and diffusion of these technologies is shaped by the diverse operational logics of social forms, including prominently states and markets. For instance, technologies relating to the Internet, satellite surveillance, and data banks can be strongly associated with cooperative policies and practices (e.g., transborder access to IT infrastructures, data, and human capital, greater transparency, the formation or strengthening of transboundary public spheres) or they can be linked to conflict (e.g., applications of IT in the military, the identity politics of ethnic groups involved in violent conflicts, the confrontational politics of activists, and the competition for sectoral economic dominance among large transnational corporations).

Variability is also a function of unintended consequences. Guthrie shows us how the state-controlled development of an IT industrial sector in China had the effect of setting in motion processes of change not foreseen by any of the players involved, most importantly a trend toward reducing some aspects of state authority as networked individuals could gain access to information about foreign models of economic development. Developing the industrial side of these technologies had the perhaps ironic effect of altering—if ever so minimally—the position of individuals toward the state.

The concepts that have been central to work on cooperation and conflict—such as alliances, regimes, and institutions—may not analytically capture what some of these types of communication systems are. The Internet illustrates this well. For instance, it has some of the features through which we specify institutions—in this case a transnational institution. It is so in the sense that there is a set of rules, compliance procedures, and norms that shape human action. But with its varied uses and forms of information, the Internet is also more than an institution: it is worthy of study as a global phenomenon in its own right, with interesting implications for cross-border relations (Latham).[12]

[12] The uniqueness of the Internet (compared to the telephone, telegraph, or television)

In brief, there is considerable diversity in the types of actors and logics that constitute communication and information structures. Their endogenous technical properties vary as do their endogenized social logics. Recapitulating the above, we identify at least three sets of implications for their study from a social science perspective. One is the difficulty of prediction in a domain of contradictory and uneven patterns and processes, a fact that may help undermine various types of regimes for control and governance. A second implication is that these systems have endogenous capabilities that may enable them to escape partly the conditioning of existing systems, such as the interstate system, and transform these or constitute whole new domains. A third implication is that communication and information structures need to be treated as distinct from information technology. That is, the first are human "habitats" or ecologies anchored in the social relations associated with public spheres, networks, organizations, and markets.[13] They are therefore not subsumed by or reducible to the technology that helps make them possible.

Digital Formations: Constituting an Object of Study

Methodologically, the types of concerns present in this volume require us to go beyond the notion that understanding these technologies can be reduced to the question of impacts.[14] That is, impacts are only one of several forms of intersection of society and technology—understood in the qualified sense discussed above. Other forms of intersection have to do with the constitution of whole new sociotechnical relations and domains—digital formations—that in turn need to be constructed as objects of study. This means examining the specific ways in which these technologies are embedded in often very specialized and distinct contexts. And it requires examining the mediating cultures that organize the relation between these technologies and users, where we might think of matters as diverse as gendering or the utility logics that organize use. Because they are specific, these mediating cultures can be highly diverse; for example, when the objective is control and surveillance, the practices and disposi-

rests on a combination of (1) ready-at-hand storage capacity for documents; (2) diffuse networks of communication and interactivity (including many-to-many rather than just one-to-one or one-to-many); (3) simultaneous access and interactivity produced by 1 and 2. The first factor may seem trivial at first, but it should be noted that the capacity to store data and documents of political import to wide bodies of actors was a virtual monopoly of the state (government archives, libraries, data bases such as tax rolls, etc).

[13] For an exploration of the concept of "information ecologies" see Nardi and O'Day (1999).

[14] We see this as consistent with the analytical frame in Castells (1996).

tions involved are likely to be different from those involved in using electronic markets or engaging in large-scale computer-based conversations.

The search for impacts means framing analysis in terms of independent and dependent variables, which is by far the most common approach in the social sciences. Our understanding that these technologies are part of transformative and even constitutive processes means we cannot confine the analytic development of this field of inquiry to that type of framing. We also need to develop analytic categories able to capture formations that incorporate what would be conceived of as mutually exclusive conditions or attributes in the independent-dependent variable framing. This is what we intend for the construct, digital formation.

The construct obviously builds on the concept of social form and the process of formation. The term "social form" is meant to convey that digital formations have ontological status as social "things" (with coherence and endurance), but not as fixed units whose attributes are pregiven to analysis.[15] We are adopting a relational perspective that emphasizes that forms emerge in and through complex social processes and changing relations.[16] By formation we mean to imply four things. These forms are, as mentioned above, in the early stages of development. Second, their emergence is not likely to be signaled by some sort of founding event, formal constitution, or charter, but by a mix of informal elements ranging from network blueprints (see Latham) to manifestos (Weber). Third, they will tend toward a developing and variable structure and nature because any social form is subject to changes in relevant contexts, agents, relations, and logics from one time to the next or one instantiation to the next (across different times and places). Finally, our understanding of digital formations is nascent and will change considerably as analyses of existing and newly emerging formations cumulate.

As that understanding begins to develop, we will need to think through strategies for delineating, however contingently, general categories of formations and their corresponding instantiations. How will we know we have the right categories in place? Are research networks, knowledge communities, and electronic markets, for instance, the right categories? How far up in generality or far down in specificity does one go? How will we identify the trajectories of change in categories? On what terms and with what basis of confidence should we generalize from individual cases and categories? These are important questions because their pursuit will open the way for comparisons across types and cases and for the identification of overarching logics and patterns relevant beyond digital formations.

[15] Coherence and endurance as important qualities for marking the existence of a social form is mentioned by Abbott (1995).

[16] See Tilly (1995), Emirbayer (1997), and Cederman (2002) for discussions of this perspective.

How would you recognize a digital formation if you encountered one? As we emphasized in the discussion above, you should be able to identify a coherent configuration of organization, space, and interaction. By organization we mean the ordering of practices (e.g., via rules and roles), content (data, images), and relations among actors (individual, collective, and even machine).[17] By interaction we mean the flow of exchange and transmission among actors.[18] And by space we mean the electronic staging of the substance (or content) and social relations at play in a digital formation.

These three dimensions of formation (organizing/interacting/spatializing) are of course overlapping and mutually constitutive: space is organized; organization is spatial and interactive; interaction requires organization; and interaction produces spaces. This imbrication among dimensions brings coherence and identity to a formation.

Of the three dimensions, space is likely to be the most troubling to readers. Organization and interaction are common conceptual tools in social analysis. Space is less familiar, and the electronic space associated with digital formations is even more so because it is not primarily geo-corporeal in nature. In thinking about electronic space, we can build on the two main ways the broader category of social space is understood: as the lived environment of social artifacts (homes, factories, schools, etc.) and as the expansive range of realized and potential relations and actions that can unfold in and across such environments.[19] Instead of geo-corporeal social artifacts, electronic space is composed of picto-textual social artifacts embodied in electronic stagings of texts, images, and graphics through software and hardware.[20] A range of realized and potential relations and actions is opened up to produce electronic space. Manifestations of such relations are found in the linking, searching, and interacting described by Bach and Stark.[21]

[17] Bach and Stark employ the argument associated with Bruno Latour (1987) that machines can be nodes in a network.

[18] By interaction we do not mean to imply parity and symmetry in flows and exchange. We need not go as far as Lev Manovich (2001: 55–58) in rejecting use of the term because it can be taken to denote symmetry.

[19] The most developed work on social space is Lefebvre (1974, see esp. pp. 33–59). Obviously, the issues at stake in the concept are far more complex than we can give justice to in this meager context. See here also the work of Poster (1997).

[20] This is a departure from the usual practice of describing electronic space as either virtual or cyber. We would save the term virtual to describe a type of picto-textual spatialization where geo-corporeality is staged electronically. To explore this form of picto-textual space, see Barfield and Furness (1995). While "cyber" is a popular adjective, it does not help us here gain a sense of the character of space. It also refers back to control via feedback schema—as in cybernetics. This does not mean the term should not be applied to artifacts where control is paramount, such as in virtual reality. See Benedikt (1991) for some thoughtful essays united under this term.

[21] Bach and Stark pick up on the contingent aspects of social space in their chapter.

The picto-textual dimension of electronic space emphasizes the materialization and visualization of the digital that depends on a mix of screens, logics of sequencing, and graphic presentations of text and images. The use of the term staging—borrowed from the theater and the military—is meant to convey the putting into order and motion of semantic configurations. Staging implies a coordination of views, visualizations, and narrations that unfold in time, put in place for public or private effect and readiness for further movement and action.[22] Software, as Garcia and Sack stress, is obviously the key factor since there is not a great deal of variation in the hard infrastructure of such staging so far (such as screens of one form or another on your desk, at hand, or in your goggles).

Spatialization is shaped by organization and interaction. At the most obvious level, staging itself is an organization of presentation and narration. A less obvious instance is the organization of bodies of data and knowledge—and the relations between such bodies (see Ernst). There is also the organization of actions and practices within digital formations that have spatial implications (from downloading and file sharing to open source code distribution), as well as the organization of access that brings in or keeps out various actors and participants (see Sassen).

The latter bears on interaction and space understood as the realm of possible relations. Webs of exchange in tightly bounded, highly structured networks—as in Ernst's GFNs or Sassen's electronic markets—yield a spatiality that can take form as narrow channels of connectivity, where the options for sanctioned actions might be quite rich, but possibilities for disruptive interventions and actions are quite limited. On the other hand, the large-scale conversations analyzed by Sack or knowledge communities discussed by Bach and Stark produce a quite different space, which takes form more as a relatively open, loosely configured, discursive field susceptible to interventions that constitute serious breaks or ruptures, but which are more simple in nature compared to more highly structured and narrow spaces.[23]

[22] Cf. Laurel (1993) for the development of the theater-computer analogy. We do not, however, seek to go as far as making the connection to theater in toto, but only to the activity of staging. (Laurel is particularly focused on the dramatic aspects of "life on screen" such as MUDs.) As Sennett (1977: 34–42, 313) points out, the metaphor of "society as theater" is old. Some of the twentieth-century applications, Sennett argues—such as in the work of Erving Goffman (1959), where roles and social drama are emphasized—tend to take the social context and structure that produce the drama and roles for granted. As a result, the analysis, however insightful, tends to be conservative and narrow. Since this volume is focused on how and why digital formations come into place and with what political, social, and economic implications, we believe we avoid this pitfall.

[23] This is a classic trade-off between thick but highly bounded worlds and thin but open ones (Walzer 1994). Sack addresses the importance of breaks in meaning in shaping the course of conversations. See also Winnograd and Flores (1986), who in their analysis of artificial intelligence draw on Heidegger's (1962: 105–6) development of breaks in "referen-

It is important to emphasize that digital formations as such are not reducible to electronic networks or to social networks more generally. On one level, digital formations subsume both kinds of networks.[24] Electronic networks—which are sets of nodes, software, and technologies of transmission—are the part of the material manifestation of digital formations. IT-based social networks, composed of patterns and structures of social interaction, in turn represent one, albeit important, aspect of digital formations. On another level, a network, as a complex ensemble of not just interaction but space and organization, can represent one type of digital formation, as can a digitally based public sphere, community, or market. The network as type of digital formation appears, for example, in Latham's chapter—which focuses on the emergence of the Internet as the global computer-based communication system—where it takes shape in the many research networks that arose around the project of developing digital networking technology.[25] These research networks entailed more than just sets of electronic nodes and connections (although they entailed that as well). Computer networks such as the Arpanet constituted electronic spaces, modes of organization among institutions and resources (both material and knowledge), and webs (or networks) of interaction among researchers.[26]

In some cases an ensemble of space, organization, and interaction on the Internet constitutes not just a network but a community. Community, especially as thought about in electronic terms, is a complicated matter, but

---

tial contexts"—where things can literally break down—that open the way for transformation. There is also interesting resonance with the features of thin and thick networks specified by Granovetter (1983).

[24] We refer here only to social networks that are relevant to digital formations, and not to all social networks per se.

[25] Sociologists who do network analysis could accuse us of using the term network in a loose, metaphorical way. However, the employment of the term here is useful to distinguish a type of formation emerging out of configurations of direct and indirect connection among elements; a space that is shaped by those configurations (that is, by the channels of transmission and interaction); and logics of organization that arise in the ordering of relations and resources among elements. Besides Latham, see the chapters by Alker, Ernst, and Garcia. The point is to be able to contrast a network type of formation with other types such as electronic communities or markets. Overlap between types can be understood in two ways. One way is as a Venn diagram, where some networks, for example, shade over into a community form. The other way is as intertwinement because, as just pointed out, any digital formation involves electronic and social networks. Neither sort of overlap justifies reduction of all formations to the network form. We are trapped by the sediment of a sociological language that only helps us make distinctions that are ultimately clumsy.

[26] The use of the term electronic space is based on Sassen (1998: chap. 9). Although some people associate electronic space with media such as television, it is used here as it relates to digital formations rather than mass media.

we take it to mean that configurations of space, organization, and interaction sustain a common identity around shared goals and reciprocal relations among participants, and that such identity, goals, and reciprocity are an important and substantive aspect of each of participant's life, professional or personal.[27] While it might be the case that the experts involved in developing the Arpanet in the end constituted a professional community (not just a research network), the concept of community can clearly be applied to the open source movement, and Weber in his chapter adopts it explicitly. As Weber points out, the Internet was essential to the rise of the open source communities he analyzes as the communication medium of access, exchange, and interaction. Open source communities as digital formations take shape in the organizational logics of collective software production (analyzed in detail by Weber), the webs of interaction across wide geographical expanses, the constellation of sites and electronic postings that constitute the electronic space within which participants operate as code and ideas are exchanged.

Also using the category of community are Bach and Stark, who apply it to a type of digital formation they label "knowledge communities." They explore how such "knowledge communities" emerge around the activities of NGOs. In contrast to the production of software, NGO knowledge communities, composed mostly of activists, are organized around the pursuit and exchange of knowledge about various areas of human development and security, from economic development to minority rights. Bach and Stark consider how new social networks, organizational forms, and spaces are constituted through the practices of knowledge production and exchange, especially as tied to the activities of linking, searching, and interaction that are familiar ways of moving through the Internet. They argue that such otherwise simple practices can be associated with the rise of unprecedented connections among actors (webs of interaction), forms of deliberative associations (organization), and knowledge spaces that they contend are part of a transformation of global political life.

Sack also considers how a digital formation can emerge around the exchange of ideas. "Very large-scale conversations" (VLSCs) are quite literally conversations that unfold around a given topic involving a relatively large number of participants. Typically these conversations, which can be transnational in scale, manifest in forums, mail lists, and newsgroups. Sack shows that these innocent-looking forms actually involve a complex intersection of interpersonal networks, thematic organization, and ideational relationships that together yield an architecture of discursive space.

[27] This definition is consistent with those of analysts such as Wellman and Gulia (1999), Smith and Kollock (1998), and Calhoun (1998).

He thinks through the different ways that a VLSC can institutionalize linguistic meaning and "common sense" (a form of knowledge) and of course be shaped by linguistic institutions that form the context of discourse within a VLSC.

The production of meaning, histories, stories, themes, and knowledge is also central to the digital formation examined by Alker. He analyzes the design of digital information networks for the linked conflict early warning efforts of experts in various institutional settings from NGOs to intergovernmental organizations. These networks are meant to serve as expert information systems, the capacity of which to store and distribute case histories would allow for the rewriting of interpretations of conflicts and the conditions of conflict—as ideas evolve, new data is introduced, or new connections are established. These information networks are meant to constitute unique spaces of knowledge, organizations of data, and networks of interaction among practitioners that can exploit the collaborative power of contested and alternative views of deadly conflicts to produce better early warning practices.

A far more familiar application of digital information systems is detailed by Ernst in an exploration of a digital formation he calls "global flagship networks (GFNs)." These networks link and coordinate a set of far-flung firms and suppliers—around a global flagship firm—collaborating in R & D, production, distribution, and marketing through the exchange of knowledge about these economic activities. Database sharing, conferencing, e-mail, and control mechanisms are among the activities found on these networks. Across the electronic space of GFNs, whole new ways of organizing economic cooperation are emerging, along with new logics of interaction among a diversity of actors. By looking at GFNs, Ernst is able to move beyond the usual claims about flexible production and virtual corporations that have occupied reflections on economic globalization to uncover tensions among network actors, the generation of new hierarchies, and the limits of network strategies.

Another feature of economic globalization is the rise of massive electronic financial markets for credit, currency, equity, and commodity futures. Sassen seeks to specify the difference that digital networks and the digitizing of financial instruments make to transboundary financial markets that have been part of modern capitalism since its beginnings. What is new about the type of digital formations usually referred to as electronic markets is not only their much noted speed of operation and scale of connectivity. Perhaps more striking is the extent to which in such markets complex financial instruments have been developed to guide decision-making, based on powerful computer processing and algorithms. This in turn has opened the way for an explosion in financial innovations, most famously in the area of derivatives. Sassen contrasts the powerful and resource-rich world around global financial markets with the attempts—of chief concern in Bach and Stark—in the resource-poor world of activists, especially in the global South, to amplify their political effectiveness through global digital networks. Sassen's contrast underscores how similar tendencies toward interconnectivity and decentralized access can be associated with quite diverse types of formations because of differences regarding who and what is mobilized.

The disparity between centers and peripheries is what Garcia studies in her chapter. She explores the possibility that digital formations such as "virtual industrial districts" could be designed based on rural networks that would allow rural communities to agglomerate resources (knowledge and material) to overcome their historical disadvantages relative to cities. Electronic networking holds out considerable hope to rural areas that are sparsely populated and therefore do not enjoy the advantages of urban resource concentration, a sharp contrast with the flagship networks examined by Ernst. It will be necessary, Garcia argues, for these networks to be "decentrally" organized and inclusively interactive. Further, and crucially, these networks will have to be based on the imaginative construction of electronic spaces through innovative software development that, in effect, produces virtual cities.

The design of information technologies to integrate wide geographical regions is not limited to the economic realm. Cederman and Kraus concentrate on the effort of the European Union to construct a "communicative space" that would provide a democratic political realm, if not public sphere, for their Union. The hope is that within such a space information can be accessed and disseminated, conversational networks around policies initiated, and decisions influenced by such processes. The authors examine the assumptions underlying such an ambitious digital formation, drawing on an analysis of the rise of national politics. They force us to contemplate whether or not digital formations are relevant to such large-scale political projects, the vast stakes of which are defined upfront by designers. In contrast, the purposes of the very large-scale conversations examined by Sack seem to emerge organically.

Finally, Guthrie squarely confronts the relations between many of the digital formations mentioned above and the national polity—in his case China—not just as a model of formation, but as a field of transformation. Emerging networks of firms, knowledge communities among activists and educational institutions, and electronic social networks are among the formations touched on by Guthrie in his detailed analysis of effects on sovereignty, economic change, and the development of Chinese civil society.

Sociodigitization

There is nothing unique about digital formations being constituted by configurations of organization, interaction, and space. The same could be said about households, corporations, cities, states, nations, empires, or a dozen other social entities that populate modernity. What distinguishes digital formations besides their newness (as defined above) is their basis in digital technology. While a corporation, for example, can digitize its operations to a great extent, it is only when it becomes a "virtual corporation" that it can be said to owe its existence to the digital. In contrast, the global flagship networks portrayed by Ernst from the start are founded on digital technologies. We can imagine a GFN organized around nondigital information technologies, but it would no longer have the distinctive qualities that digitization entails, as we outlined above, and no longer represent a *digital* formation.

The fact that digital formations are grounded in information technologies raises the question of the relationship between the digital and nondigital. Central to this volume is the notion that it is not enough to focus on the digital. Crucial are the contexts and fields of social life, from finance to the environment of which digital formations are a part. Indeed, what is especially interesting about Ernst's chapter is not necessarily the workings of GFNs per se, but their relationship to the corporations and economies with which they are imbricated. Viewed in this way, the process of digital formation depends on the dynamics at play in the links between the digital and nondigital.

We believe the best way to view that process is through a concept we call "sociodigitization." This denotes the rendering of facets of social and political life in a digital form. These facets can vary from discourse about political events (Sack) and interpretations of conflicts (Alker) to regional economic practices (Garcia) and policy positions (Cederman and Kraus). "Digitization" as a concept has been around for some time as it is closely associated with the efforts of librarians, publishers, artists, and others to convert analog content to digital form.[28] The qualifier "socio" is added to distinguish from the process of content conversion, the broader process whereby activities and their histories in a social domain are drawn up into the digital codes, databases, images, and text that constitute the substance of a digital formation. As the various chapters below show, such drawing up can be a function of deliberate planning and reflexive ordering or of contingent and discrete interactions and activities. In this respect as well,

[28] There is a considerable literature on digitization linked to archiving and library science. See in particular Saxby (1990).

sociodigitization differs from digitization: what is rendered in digital form is not only information and artifacts but also logics of social organization, interaction, and space as discussed above. Ultimately, the character of digital formations depends on the social relationships, practices, institutions, and organizations that feed sociodigitization.

The drawing up of facets of social life into information systems is at least as old as writing itself and has been tied to processes of state formation as records, maps, and statistics produced potent forms of social knowledge.[29] Sociodigitization is on one level continuous with this long-standing development of social knowledge based on paper. But it strays from it because it allows actors other than states (and firms, since the early twentieth century) to generate, organize, and distribute substantial bodies of social knowledge. The most notable actors are the NGOs and social movements discussed in the chapters by Bach and Stark and Sassen. But the same can be said about the conflict experts in Alker's chapter; the researchers in Latham's; the software developers in Weber's; and the private citizens in Sack's.

What underlies the discontinuity of sociodigitization with past information media is the manipulative capacities engendered by digital technologies. Information and knowledge are subject to far greater levels of computation and organization. There are not only, as pointed out above, the complicated algorithms at play in the financial realm, but also the algorithms for producing semantic codes and structures explored by Sack. As increasingly sophisticated forms of manipulation and computation are put in the reach of nonstate actors through software, it is far from clear where disruptive practices and politics will go. Open-source development is so full of disruptive potential exactly because it can place control over augmentation into private and nonstate hands.

Another notable difference is the capacity to translocate information (of varying amounts) in digital form among various contexts.[30] This is a key to the mobility of knowledge described by Sassen. We see some of the implications of this mobility in the chapters by Ernst and by Bach and Stark. Ernst refers explicitly to the importance in GFNs of the modularization of knowledge, which allows for various units or nodes to work in a knowledgeable way on discrete portions of an economic process such as production.

It is impossible at this time to know what shape sociodigitization will

[29] Michel Foucault (1977) opened our eyes to this process. See also James Scott (1998) for a wide-ranging integrative perspective on relevant research and analysis in this area.

[30] The point is not to claim wholly new practices and capacities as these things were done prior to digitization. Innis (1951), for instance, differentiates the effects of writing on light media such as parchment from say stone based on the mobility the former affords. Differences regarding the digital are of degree, aggregating into differences of kind.

take in the future, and with what implications. The character of the information involved will likely be critical to developments and our understanding of them. On the one hand is the basic issue of the scope of information: what data will be drawn up into what formations. On the other is the question of the effects of that information, which depends to a large extent on how such information finds its way into evaluative statements that shape perceptions and actions.[31] As new algorithms are developed, they will open up new forms of information manipulation, aggregation, and distribution around which also new digital formations might emerge.

Analytic Operations

Three types of analytic operations allow us to factor in the intersection of digital technologies and social logics. These analytic operations should hold whether these technologies are derivative, transformative, or constitutive. They should hold for a broad range of specific instances of the intersection between society and technology. And they should hold for a variety of analytical frameworks. This would include framings in terms of independent-dependent variables, but also strategies that aim at capturing imbrications and mutual interaction. Again, these analytic operations can themselves conceivably assume multiple forms. We have opted for three such operations, sufficiently complex as to accommodate a broad range of outcomes. We specify these as a first approximation for constituting digital formations as an object of study. Constructed as objects of study, digital formations can then also function as analytic categories. Each chapter in this volume represents an elaboration of a particular type of digital formation and illustrates a particular research strategy and theoretico-empirical specification.

At the most general level we want to emphasize the importance of analytic categories and frames that allow us to capture the complex *imbrications* between the capacities of digital technologies—specifically computer-centered interactive technologies—and the contexts within which they are deployed or used. A second set of analytic operations concerns the *mediating practices and cultures* that organize the relation between these technologies and users. Until quite recently there was no critical elaboration of these mediations. The dominant assumption was that questions of access, competence, and interface design captured the full set of mediating experiences. A third set of analytic operations is aimed at

[31] This formulation integrates the discussion of information in the chapters by Latham and Sassen.

recognizing questions of *scaling,* an area where these particular technologies have evinced enormous transformative and constitutive capabilities. In the social sciences, scale has largely been conceived as a given or as context and has, in that regard, not been a critical category. The new technologies have brought scale to the fore precisely through their destabilizing of existing hierarchies of scale and notions of nested hierarchies. Thereby they have contributed to launch a whole new heuristic, which, interestingly, also resonates with developments in the natural sciences where questions of scaling have surfaced in novel ways. The next three subsections develop these issues very briefly.

Digital/Social Imbrications

As a first approximation we can identify three features of this process of imbrication. To illustrate we can use one of the key capabilities of these technologies, that of raising the mobility of capital and thereby changing the relationship between mobile firms and territorial nation-states. This is further accentuated by the sociodigitization of much economic activity. Digitization raises the mobility of what we have customarily thought of as barely mobile and renders mobile much of what we had considered immobile. Digitization can liquefy the nondigital. Once digitized, an entity can gain hypermobility—instantaneous circulation through digital networks with global span. Both mobility and digitization are usually seen as mere effects or at best functions of the new technologies. Such conceptions erase the fact that achieving this outcome requires multiple conditions, including such diverse ones as infrastructure and changes in the law.

The first feature is, then, that the production of capital mobility and the process of digitization requires capital fixity: state of the art built-environments, a professional workforce on the ground at least some of the time, legal systems, and conventional infrastructure—from highways to airports and railways. These are all partly place-bound conditions. Once we recognize that the hypermobility of the instrument had to be *produced,* we introduce nondigital variables in our analysis of the digital. Such an interpretation carries implications for theory and practice. For instance, it becomes quite evident that simply having access to these technologies is not enough: it will not necessarily alter the position of resource-poor countries or organizations in an international system with enormous inequality in resources.[32]

[32] Much of the work on global cities (Sassen 2001) has been an effort to conceptualize and document the fact that the global digital economy requires massive concentrations of physical and social resources in order to be what it is. Finance is an important intermediary

A second feature that needs to be recovered here is that the capital fixity needed for hypermobility and digitization is itself transformed in this process. The real estate sector illustrates some of these issues. Financial services firms have invented instruments that represent the value of real estate. This liquefies real estate, thereby facilitating investment and circulation of these instruments in markets other than the property market. While real estate remains very physical, this physicality has been transformed by its representation in highly liquid instruments that can circulate in global markets. It may look the same, it may involve the same bricks and mortar, it may be new or old, but it is a transformed entity.

These two properties signal that the hypermobility gained by an object through digitization is but one moment of a more complex condition. Representing such an object simply as hypermobile or as fixed is, then, a partial representation since it includes only some of the components of that object. The nature of the place-boundedness of this type of fixed capital differs from what it may have been one hundred years ago when it was far more likely to be a form of immobility. Today it is a place-boundedness that is, in turn, inflected or inscribed by the hypermobility of some of its components, products, and outcomes. Both capital fixity and mobility are located in a temporal frame where speed is ascendant and consequential. This type of capital fixity cannot be fully captured through a description confined to its material and locational features (Sassen 2001: chaps. 2 and 5).

A third feature in this process of imbrication can be captured through the notion of the social logics organizing the process. Many of the digital components of financial markets are inflected by the agendas that drive global finance, and these agendas are not technological per se. The same technical properties can produce outcomes that differ from those of electronic financial markets (see Sassen, this volume). Much of what we think of when it comes to electronic space would lack any meaning or referents if we were to exclude the nondigital world—cultures, material practices, systems of law, and imaginaries. It is necessary to distinguish between the technologies and the digital formations they help make possible.

In this regard, then, sociodigitization is multivalent. It brings with it an amplification of both mobile and fixed capacities. It inscribes the nondigital but is itself also inscribed by the nondigital. The specific content, implications, and consequences of each of these variants are empirical questions, and are objects for study in their own right. So are what is conditioning the outcome when digital technologies are at work and what is

---

in this regard: it represents a capability for liquefying various forms of nonliquid wealth and for raising the mobility (i.e., producing hypermobility) of that which is already liquid. But to do so, even finance needs significant concentrations of nondigital resources.

conditioned by the outcome. We have difficulty capturing this multivalence through our conventional categories, which tend to dualize and posit mutual exclusivity: an entity is either fixed or mobile. The example of real estate signals that the partial representation of real estate through liquid financial instruments produces a complex imbrication of the digital and nondigital moments of that which we continue to call real estate. And so does the partial endogeneity of physical infrastructure in electronic financial markets. Finally, capturing the imbrications of the digital with the nondigital allows us to capture this endogenizing of the social in the digital.

Mediating Practices and Cultures

One consequence of the above developed proposition about electronic space as embedded and not exclusively technological is that the articulations between electronic space and users—whether social, political, or economic actors—are constituted in terms of mediating cultures. Use is not simply a question of access and understanding how to use the hardware and the software. The mediating cultures through which use is constituted result partly from the values, projects, power systems, and institutional orders within which users are embedded.

There is a strong tendency in the literature to assume use to be an unmediated event, an unproblematized activity. There is in fact much more of a critical literature when it comes to questions of access than there is about cultures of use.[33] At best, recognition of a mediating culture has been confined to that of the "techie," one that has become naturalized rather than recognized as one particular type of mediating culture. Beyond this thick computer-centered use culture, there is a tendency to flatten the practices of users to questions of competence and utility. From the perspective of the social sciences, use of the technology should be problematized rather than simply seen as shaped by technical requirements and the necessary knowledge, even as this might be the perspective of the computer scientist and engineer.

Use—to be distinguished from access—is constructed or constituted in terms of specific cultures and practices through and within which users articulate the experience and/or utility of digital technology. Thus our concern here is not purely with the technical features of digital networks and what these might mean for users, nor is it simply with the impact of digital technology on users. The concern is, rather, with this in-between zone that constructs the articulations of users and digital technology.

[33] There are of course important exceptions, notably the work by Dale Eickelman and Jon Anderson (1999) on how these technologies get used by, for instance, scholars of the Koran.

The practices through which use is constituted partly derive their meanings from the aims, values, cultures, power systems, and institutional orders of the users and their settings. These mediating cultures also can produce a subject and a subjectivity that become part of the mediation. For instance, in open source networks (see Weber), much meaning is derived from the fact that these practitioners contest a dominant economic-legal system centered in private property protections; participants become active subjects in a process that extends beyond their individual work and produces a culture. The kinds of rural-user-oriented networks examined by Garcia partly result from an awareness of long-term historical and institutional disadvantages of rural areas compared to urban areas and an orientation toward overcoming this disadvantage. There are multiple ways of examining the mediating cultures organizing use. Among others, these can conceivably range from small-scale ethnographies to macrolevel surveys, from descriptive to highly theorized accounts, from a focus on ideational forms to one on structural conditions.

The Destabilizing of Older Hierarchies of Scale

Key technical properties of digital networks are contributing to destabilize current formalized hierarchies of scale. These hierarchies, mostly dating from the period that saw the consolidation of nation-states and the interstate system, continue to operate and remain prevalent. They are typically organized in terms of institutional scope and relative territorial size: from the international, down to the national, the regional, the urban, to the local, with the national scale as the main articulator of the other scales. Today's rescaling dynamics cut across institutional scope and across the institutional encasements of territory produced by the formation of national states (Taylor 2000; Brenner 1998; Ruggie 1993; Sassen 2004). This does not mean that the old hierarchies disappear, but rather that rescalings emerge alongside the old ones and that these can trump the latter. This is partly because the practices and objectives of key political and economic actors are beginning to operate at, and thereby contribute to constituting, subnational and global scales where before they might have been confined to the national domain. Further, new types of scalar actors and objectives have emerged.

Existing theory is not enough to map the multiplication of practices and actors that are constituting these rescalings. Included are a variety of nonstate actors and forms of cross-border cooperation and conflict—global business networks, the new cosmopolitanism, NGOs, diasporic networks, and transboundary public spheres. Several critical scholars have shown us how the disciplines concerned with transboundary or international processes tend to remain focused on the scale of the state at a time when we see a proliferation of nonstate actors, crossborder processes, and associated changes in the scope, exclusivity, and competence of state authority over its territory.[34]

With few exceptions, found most prominently in a growing scholarship in geography, the social sciences have lacked critical distance from the scale of the national. The consequence has been a tendency to scale as fixed, reifying it, and, more generally, to neutralize the question of scaling (or at best to reduce scaling to a hierarchy of size). Associated with this tendency is also the often uncritical assumption that these scales are mutually exclusive, most pertinently for the argument here, that the scale of the national is mutually exclusive with that of the global. A qualifying variant in the scholarship, though of a very limited sort, can be seen when scaling is conceived of as a nested hierarchy. The types of developments we focus on in this volume bring to the fore the historicity of scales and the limits of nested hierarchies.[35]

Digital networks strengthen the multiscalar character of many social processes, particularly processes that do not fit into nested hierarchies. An example of such a multiscalar system is the combination of the far-flung network of affiliates of a multinational firm and the strategic system-integration and management functions that tend to be concentrated in a very limited number of cities (e.g., Taylor et al. 2002).[36] This is a multiscalar system operating not only at a self-evident global scale, but also at a horizontal global scale (the network of affiliates). The latter is constituted as one step in a process of vertical integration, but it has its own scalar specificity, and it is useful to recognize its distinctiveness. It does not merely scale upward because of new communication capabilities that allow it to expand the scope of operations, going from local to global. Nor is it nested in a hierarchy of scales. Conceptualizing such systems entails distinguishing (1) the various scales that are constituted through global processes and practices,[37] and (2) the specific contents and institutional locations of this multiscalar globalization.[38]

Narrowing the discussion of scaling to the formation of transboundary domains, we can identify four types of scaling dynamics in the constitution of global digital formations. These four dynamics are not mutually

[34] Examples include Taylor (2000), Cerny (2000), Ferguson and Jones (2002), Hall and Bierstaker (2002), and Walker (1993).

[35] At the same time, it is important to recognize the risks of reification contained in exclusively scalar analytics in that it can lead to disregarding the thick and particularistic forces that are part of these dynamics (e.g., Amin 2002; Howitt 1993).

[36] See also the research network on globalization and world cities (GaWC) at http://www.lboro.ac.uk/gawc.

[37] See, for example, Taylor (2000), Swyngedouw (1997), and Amin and Thrift (1994).

[38] See, for example, Massey (1993), Hewitt (1993), Jonas (1994), and Brenner (1998).

exclusive, as becomes clear when we use the example of what is probably one of the most globalized and advanced instance of a digital formation, electronic financial markets. A first type of scaling dynamic is the formation of global domains that function at the self-evident global scale. Other instances might be some types of very large-scale conversations that are indeed global (Sack) and the knowledge spaces examined by Bach and Stark.

A second type of scaling can be identified in the fact that local practices and conditions become directly articulated with global dynamics, not having to move through the traditional hierarchy of jurisdictions. Electronic financial markets also can be used as an illustration here. The starting point is floor or screen-based trading in exchanges and firms that are part of a worldwide network of financial centers (e.g., Knorr-Cetina and Bruegger 2002). These localized transactions link up directly to a global electronic market. What begins as local gets rescaled at the global level. Similarly, we see this in the case of very large conversations (Sack), where the interaction of individual interventions leads to the formation of a space that can be global.

A third type of scaling dynamic results from the fact that interconnectivity and decentralized simultaneous access multiplies the cross-border connections among various localities. This produces a very particular type of global digital formation, one that is a kind of distributed outcome: it resides in the multiplication of lateral and horizontal transactions, or in the recurrence of a process in a network of local sites, without the aggregation that leads to an actual globally scaled digital formation as is the case with electronic markets. Instances are open source software communities (Weber), the early warning systems described by Alker, and the activist networks described by Sassen.

A fourth type of scaling dynamic results from the fact that global digital formations can actually be partly embedded in subnational sites and move between these differently scaled practices and organizational forms. For instance, the global electronic financial market is constituted both through electronic markets with global span and through locally embedded conditions, namely, financial centers and all they entail, from infrastructure to systems of trust (Zaloom 2005). So are the global communication flagships examined by Ernst.

The new digital technologies have not caused these developments, but they have in variable yet specific ways facilitated them and shaped them. The overall effect is to reposition the meaning of local and global (when internetworked) in that each of these will tend to be multiscalar. For example, much of what we might still experience as the "local" (an office building or a house or an institution right in our neighborhood or downtown) actually is a microenvironment with global span insofar as it is internetworked. Such a microenvironment is in many senses a localized entity, but it is also part of global digital networks that give it immediate far-flung span. To continue to think of this as simply local may not always be very useful. It is a multiscalar condition.

Design

Conjectures about the future are often part of analyses of contemporary developments around IT. The analytical operations discussed above and the chapters in this volume are no exception. However, we distinguish conjectures about the future overall shape of societies from conjectures about specific realms of human activity.[39] Conjectures of the latter sort can be understood through the lens of design. As Herbert Simon (1996: 114) so simply put it, design is about "devising artifacts to obtain goals." Design forces contemplation of the future. In thinking about digital formations, the authors confront design because what they are studying is *formative*. It might indeed be the case that digital formations are more variable than many other formations—especially those anchored in geo-corporeal space such as cities—because they are (as picto-textual forms) highly susceptible to (re)configuration. Design is thus always proximate. This places each chapter at the edge between—to use well-worn but problematic terminology—normative and positive analysis, with the former focused on aims and values in social life; the latter, on insights into the workings and history of social fabrics.[40] Even if an author did not start self-consciously thinking about design, understanding what is at stake in formations requires thinking through the possibilities and trajectories of their development, and what those trajectories impinge on.

Design does not sit easily within social science; the latter tends to force a division between normative and positive analysis. When the term design is used in social science it typically denotes strategies for the effective construction of social artifacts such as institutions.[41] While this is a meaningful use of the term, it problematizes the object of design rather than the category of design itself. One way the latter happens in the chapters that follow is through the analysis of the process of design. This is most visible in Weber's study of open source software design. Sack, in turn, explores the possibility of direct involvement in design by offering tech-

[39] An example of such a social vision is Negroponte (1994).

[40] The chapters vary regarding the degree to which they confront design. Guthrie does so the least, Alker and Sack the most.

[41] A recent example is a special issue of *International Organization* (2001).

nologies of social analysis that can become a part of the architecture of the very large-scale conversations he studies. Alker, in turn, thinks through what a design process bearing on the organization and application of knowledge can look like, emphasizing that the possibilities of *re*-designing narrative structures must be incorporated into a formation from the start.[42] Bach and Stark also highlight the importance of redesign as they consider ongoing processes of translation and negotiation in activist knowledge communities. Redesign is often critical for electronic activism in the global South if bandwidth-intensive formats for information from the global North are to be used (see Sassen).

Goals and values in design are generally articulated in this volume through the conceptual optic of the social purpose of digital formations. This comes out the strongest in Garcia's chapter, where she explores the terms upon which digital networks can serve the purposes of rural economic development. She forces us to think not only about who might control design processes—and thus shape digital formation—but also how such control might be enacted through international regimes, regional cooperatives, or some other governance form.[43]

Limits and Logics of Formation

Processes of design and sociodigitization do not unfold in a vacuum. They run up against an array of conditions and forces. For convenience sake, we can divide such forces and conditions into those that are endogenous to digital formations and the technologies they entail and those that are exogenous.[44] Endogenous conditions and forces are wide ranging. One set has to do with the character of technological change and sociotechnical systems. For example, not all moments in technological development are equally propitious for designs or susceptible to digitization. Garcia claims that today the rapid set of IT innovations associated with the 1990s' boom created an open moment for rethinking uses and applications of technology. The implication is that other moments might be less open and inopportune. Another related endogenous condition stems from

[42] This is consistent with Herbert Simon's (1996: 165) strategy for avoiding teleology: any given design is only ultimately a platform for further design.

[43] We do not mean to imply that digital formation can be controlled or that what actually forms is the result of conscious planning. Controlling or governing processes of design is only one factor in determining the process of digital formation. Cederman and Kraus underscore the limits of design in the case of the EU's pursuit of a communication space.

[44] This division is for heuristic purposes, recognizing that, in practice, any force or condition likely has both exogenous and endogenous aspects. We are pointing to tendencies and salience.

aspects of a sociotechnical system that may render it unresponsive to design ambitions. Latham argues that the Internet system leaves little room for "legislating" social purpose at the overall system level because—as a dumb network—the Internet offers few means of control at the global level.

More internal to the social configuration of digital formations are tensions that can emerge across the three key dimensions of interaction, organization, and space. Such tensions can arise as a function of change in one dimension that undermines or challenges structures and practices in another dimension. A new pattern of interaction, for instance, can be inconsistent with existing organizational strategies honed in an earlier period. Tensions between dimensions are found throughout this volume. A particularly clear illustration is in Ernst's chapter. He shows how new interfirm interactions can challenge previously organized relations among firms in a global flagship network.

Tensions can also emerge within the very logics of formation, as various configurations of interaction, organization, and space exhibit both distributed and concentrated tendencies. Ernst writes of "concentrated dispersion" within GFNs; Garcia, of strategies of rural concentration countering a history of deconcentration; Latham, of the concentration that can emerge out of distributed internetwork relations; and Weber, of the concentration of authority that attaches to leaders in open source communities. Sassen makes the double movement of concentration and distribution central to her chapter.[45] This double movement occurs on two levels. One is within the global financial realm; the other between the relatively concentrated world of global finance and the comparatively distributed world of transnational activism. The latter can, of course, also exhibit its own forms of concentration, as hinted at by Bach and Stark.

The double movement is important because processes of concentration force us to ask questions about who or what governs digital formations, and what is drawn up into them via sociodigitization and on what terms. It bears directly on issues of leadership, authority, and hierarchy that are crucial to thinking through these questions.

Endogenous conditions are important and interesting. But processes of design and sociodigitization are also shaped by exogenous forces. This is not only because digital formations are embedded in social contexts that determine their very social character, but also because sociodigitization is so dependent on the fields of human endeavor and activity that it draws upon. As we have defined it, a digital formation cannot subsume a given

[45] This parallels some of the dynamics posited by Sassen's analytic construct of the global city (2001), which gains its specification precisely because a massively distributed global economy requires points of concentration.

area of activity. There should always be aspects of human life "outside" its boundaries, whether such life is ready to enter through digitization or remain in the frontier zone of a formation. Even in such an IT-focused arena as the open source movement, Weber shows how crucial are socioemotional factors such as prestige, trust, leadership, and norms that draw on a host of realms of human interaction from family to work.

Especially important in social contexts are the deep institutional and historical trajectories that digital formations bump up against. Cederman and Kraus point to the trajectories of democratic state formation in Europe that are not easily transcended by new electronic communicative space; Sassen, to the institutionalized practices and rules of global finance and the technical constraints faced in the global South; Ernst, to the transformations in economic life around liberalization; Garcia, to the deep-seated histories of rural zones in national, regional, and global economies; Alker, to the habits of knowledge around conflict; Weber, to the tension-filled intersection of open source practices with longstanding institutions of property and logics of production; and Latham, to the institutional power of state telecom agencies. Guthrie's chapter makes the intersection of historico-institutional trajectories central to his analysis, as he argues for the importance of preexisting institutional change in shaping political and economic outcomes relating to IT in China.

Ultimately, we can understand that this line of argument is, in some regards, about the limits to IT and digital formations as forces of transformation (something argued quite explicitly in the chapters by Cederman and Kraus and by Guthrie). From our vantage point the identification of limits is a crucial step in understanding a phenomenon because it helps us see its boundaries and, with better accuracy, the way it is intertwined in social life.

We find that the concept of digital formation helps us think more productively about information technology as a social force. It tells us that IT itself is not a stable causal force but part of a process of social formation. Technologies are always in use or, as Latour (1987) says, "in action." The Internet, for instance, stands for a moving, mobile ensemble of uses, social entities, logics, tensions, and practices. But that does not mean IT is not a force shaping political, social, and economic life. The point is to recognize that IT does so in and through social entities such as the digital formations considered in this book, which are themselves part of broader social fabrics. It is this embedment that allows technologies to have effects across contexts and domains. In turn, we get to see more clearly how the structures and logics found in those social fabrics shape IT.

Digital formations as a category also helps us think about how specific configurations of organization, interaction, and space can emerge across national boundaries bearing on quite different issues, from economics to education. As research in this area moves forward, scholars should benefit from keeping in mind the tensions and limits that such emergence can encounter. And by making design more central to social science, scholars might open new ways to think about the social purpose of technologies. We believe that the chapters that follow are an important step toward such an analytical vision.

Conclusion

This volume is focused on digital information and communication structures that arise out of the intersection of technology and society. We use the construct "digital formation" to capture this outcome, one shaped both by endogenous technical properties and by endogenized social logics. There are multiple instantiations of this intersection, and these can be organized into several types of digital formations. Electronic Networks, communities, and markets are familiar types to social scientists, and they are central to the various chapters in the volume.

Constituting the object of examination as a digital formation requires us to go beyond the notion that to understand this intersection we can confine analysis to the *impacts* of these technologies *on* society. Impacts are only one of several forms of intersection. Others have to do with the constitution of new domains and with major transformations in old domains. Thus the locus of intersection can be variously conceived, ranging from conceptualizations in terms of independent and dependent variables to the specifying of new objects for study. Constructing digital formations as an object of study entails several tasks, some covered in this chapter and others in the rest of the volume. In this chapter we sought to construct an object of study—digital formations—and to specify its location in a conceptual field that allows us to capture both endogenous technical properties and endogenized social logics.

There are several analytic vocabularies that can be used or constructed to engage in this type of study. Identifying and constructing such vocabularies is part of the conceptual mapping of this field of inquiry and is part of the effort to generate research agendas on the subject. Each of the chapters contains a distinct vocabulary and is focused on a distinct puzzle or theme. We decided to go for a broad range of cases rather than one theme and multiple treatments, a decision that some might find problematic. Even if broad, the range of cases is clearly not exhaustive. It is impossible to cover the full range of pertinent themes. Ours is one possible selection. We look forward to the suggestions of our critics as to other options, not included here. We consider this volume one contribution to an emergent field of inquiry.

References

Abbott, Andrew. 1995. Things of Boundaries. *Social Research* 62 (4) (Winter): 857–82.

Amin, Ash. 2002. Spatialities of Globalisation. *Environment and Planning A* 34(3): 385–99.

Amin, Ash, and Nigel Thrift. 1994. *Globalization, Institutions and Regional Development in Europe*. Oxford: Oxford University Press.

Barfield, Woodrow, and Furness III, Thomas, eds. 1995. *Virtual Environments and Advanced Interface Design*. New York: Oxford University Press.

Bell, Daniel. 1973. *The Coming of Post-Industrial Society*. New York: Basic Books.

Benedikt, Michael, ed. 1991. *Cyberspace: First Steps*. Cambridge: MIT Press.

Beniger, James R. 1986. *The Control Revolution: Technological and Economic Origins of the Information Society*. Cambridge: Harvard University Press.

Brenner, Neil. 1998. Global Cities, Glocal States: Global City Formation and State Territorial Restructuring in Contemporary Europe. *Review of International Political Economy* 5:1–37.

Brown, John Seely, and Paul Duguid. 2002. *The Social Life of Information*. Cambridge: Harvard University Press.

Calhoun, Craig. 1998. Community without Propinquity Revisited: Communications Technology and the Transformation of the Urban Public Sphere. *Sociological Inquiry* 68(3): 373–97.

Castells, Manuel. 1996. *The Information Age: Economy, Society and Culture*. Vol. 1: *The Rise of the Network Society*. Oxford: Blackwell.

Cederman, Lars-Eric. 2002. Computational Models of Social Forms: Advancing Generative Macro Theory. Conference paper prepared for Social Agents: Ecology, Exchange, and Evolution, Chicago.

Cerny, P. G. 2000. Structuring the Political Arena: Public Goods, States and Governance in a Globalizing World. In *Global Political Economy: Contemporary Theories*, edited by Ronen Palan, 21–35. London: Routledge.

Chandler, Jr., Alfred D., and James W. Cortada, eds. 2000. *A Nation Transformed by Information*. New York: Oxford University Press.

Choucri, Nazli. 2000. Introduction: Cyberpolitics in International Relations. *International Political Science Review* 21(3): 243–63.

Deibert, Ronald. J. 2000. International Plug 'n Play? Citizen Activism, the Internet, and Global Public Policy. *International Studies Perspectives* 1(3): 255–72.

Der Derian, James. 2001. *Virtuous War: Mapping the Military-Industrial-Media-Entertainment Network*. Boulder: Westview Press.

Deutsch, Karl. 1953. *Nationalism and Social Communication*. Cambridge: MIT Press.

Deutsch, Karl, et al. 1957 *Political Community and the North Atlantic Area*. Princeton: Princeton University Press.

Eickelman, Dale F., and Jon W. Anderson, eds. *New Media in the Muslim World: The Emerging Public Sphere*. Bloomington: Indiana University Press.

Emirbayer, Mustapha. 1997. Manifesto for a Relational Sociology. *American Journal of Sociology* 103(2) (September): 281–317.

Ferguson, Yale H., and R. J. Barry Jones, eds. 2002. *Political Space: Frontiers of Change and Governance in a Globalizing World.* Albany: SUNY Press.
Foucault, Michel. 1977. *Discipline and Punish: The Birth of the Prison.* New York: Vintage.
Goffman, Erving. 1959. *The Presentation of Self in Everyday Life.* New York: Anchor.
Granovetter, Mark. 1983. The Strength of Weak Ties: A Network Theory Revisited. *Sociological Theory* 1:201–33.
Hall, Rodney Bruce, and Thomas J. Biersteker, eds. 2002. *The Emergence of Private Authority in Global Governance.* Cambridge: Cambridge University Press.
Hargittai, E. 1998. Holes in the Net: Internet and International Stratification. Paper presented at INET '98 Conference: The Internet Summit, Geneva, Switzerland, July 21–24.
Heidegger, Martin. 1962. *Being and Time.* New York: Harper & Row.
Howitt, Richard. 1993. A World in a Grain of Sand: Towards a Reconceptualisation of Geographical Scale. *Australian Geographer* 24:33–44.
Innis, Harold. 1951. *The Bias of Communication.* Toronto: University of Toronto Press.
International Organization. 2001. 55(4).
Jervis, Robert. 1976. *Misperception in International Politics.* Princeton: Princeton University Press.
Johnston, R. J., Peter J. Taylor, and Michael Watts. 2002. *Geographies of Global Change: Remapping the World.* Oxford: Blackwell Publishers
Jonas, Andrew. 1994. The Scale Politics of Spatiality. *Environment and Planning D: Society and Space* 12(3): 257–64.
Knorr-Cetina, Karin, and Urs Bruegger. 2002. Global Microstructures: The Virtual Societies of Financial Markets. *American Journal of Sociology* 107(4): 905–50.
Laguerre, Michel. 2000. *The Global Ethnopolis: Chinatown, Japantown and Manilatown in American Society.* New York: Macmillan Press.
Latham, Robert. 2002. Information Technology and Social Transformation. *International Studies Review* 4(1): 101–15.
———. 2003. *Bombs and Bandwidth: The Emerging Relationship between Information Technology and Security.* New York: The New Press.
Latour, Bruno. 1987. *Science in Action.* Cambridge: Harvard University Press.
———. 1991. Technology Is Society Made Durable. In *A Sociology of Monsters,* edited by J. Law. London: Routledge.
Laurel, Brenda. 1993. *Computers as Theater.* Reading, MA: Addison-Wesley.
Lefebvre, Henri. 1974. *The Production of Space.* Translated by Donald Nicholson-Smith. Oxford: Blackwell.
Loader, Barry, ed. 1998. *Cyberspace Divide: Equality, Agency and Policy in the Information Age.* London: Routledge.
Mackenzie, Donald 1999. Technological Determinism. In *Society on the Line: Information Politics in the Digital Age,* edited by W. H. Dutton. Oxford: Oxford University Press.
MacKenzie, Donald A., and Judy Wajcman eds. 1999. *The Social Shaping of Technology.* Buckingham, UK: Open University Press.

Manovich, Lev. 2001. *The Language of New Media.* Cambridge: MIT Press.
Massey, Doreen. 1993. Politics and Space/Time. In *Place and the Politics of Identity,* edited by M. Keith and S. Pile, 141–61. London: Routledge.
May, Christopher. 2002. *The Information Society: A Skeptical View.* Cambridge: Polity.
Millennium: Journal of International Studies (1999). Special issue on Territorialities, Identities and Movement in International Relations. 28(3).
Munker, Stefan, and Alexander Roesler, eds. 1997. *Mythos Internet.* Frankfurt: Suhrkamp.
Nardi, Bonnie A., and Vicki L. O'Day. 1999. *Information Ecologies: Using Technology with Heart.* Cambridge: MIT Press.
Negroponte, Nicholas. 1994. *Being Digital.* New York: Knopf.
Nettime. 1997. *Net Critique.* Compiled by Geert Lovink and Pit Schultz. Berlin: Edition ID-Archive.
Nye, Joseph, and Robert Keohane. 1971. Transnational Relations and World Politics: An Introduction. *International Organization* 25(3): 329–49.
Poster, Mark. 1997. Cyberdemocracy: Internet and the Public Sphere. In *Internet Culture,* edited by D. Porter, 93–116. London: Routledge.
Ruggie, John. 1993. *Multilateralism Matters: The Theory and Praxis of an Institutional Form.* New York: Columbia University Press.
Sassen, Saskia. 2001. *The Global City: New York, London, Tokyo.* 2d edition. Princeton: Princeton University Press.
———. 1998. *Globalization and Its Discontents: Essays on the New Mobility of People and Money.* New York: The New Press.
———. 2005. *Denationalization: Territory, Authority, and Rights in a Global Digital Age.* Princeton: Princeton University Press.
Saxby, Stephen. 1990. *The Age of Information: The Past Development and Future Significance of Computing and Communications.* London: Macmillan.
Scott, James C. 1998. *Seeing Like a State: How Certain Schemes to Improve the Human Condition Have Failed.* New Haven: Yale University Press.
Sennett, Richard. 1977. *The Fall of Public Man.* New York: Knopf.
Simon, Herbert A. 1996. *The Sciences of the Artificial.* 3d ed. Cambridge: MIT Press.
Smith, Marc, and Kollock, Peter, eds. 1998. *Communities in Cyberspace.* London: Routledge.
Swyngedouw, Erik. 1997. Neither Global nor Local: "Glocalization" and the Politics of Scale. In *Globalization: Reasserting the Power of the Local,* edited by K. R. Cox, 137–66. New York: Guilford.
Taylor, Peter. 2000. World Cities and Territorial States under Conditions of Contemporary Globalisation. *Political Geography* 19:5–32.
Taylor, Peter J., D.R.F. Walker, and J. V. Beaverstock. 2002. Firms and Their Global Service Networks. In *Global Networks, Linked Cities,* edited by Saskia Sassen. New York: Routledge.
Tilly, Charles. 1995. To Explain Political Processes. *American Journal of Sociology* 100(6) (May): 1594–1610.
Turkle, Sherry. 1995. *Life on the Screen: Identity in the Age of the Internet.* New York: Touchstone.

Wajcman, Judy. 2002. Addressing Technological Change: The Challenge to Social Theory. *Current Sociology* 50(3): 347–63.
Walker, R.B.J. 1993. *Inside/Outside: International Relations as Political Theory.* Cambridge: Cambridge University Press.
Walzer, Michael. 1994. *Thick and Thin: Moral Argument at Home and Abroad.* Notre Dame: University of Notre Dame Press.
Webster, Frank. 1995. *Theories of the Information Society.* London: Routledge.
Wellman, Barry, and Gulia, Milena. 1999. Net Surfers Don't Ride Alone. In *Networks in the Global Village,* edited by Barry Wellman, 331–66. Boulder: Westview Press.
Wilson, Ernest J., III. 2004. *The Information Revolution and Developing Countries.* Cambridge: MIT Press.
Winnograd, Terry, and Flores, Fernando. 1986. *Understanding Computers and Cognition: A New Foundation for Design.* Menlo Park, CA: Addison-Wesley.
Zaloom, Caitlin. 2005. Trapped in the Pits: Open-outcry Technology and the Chicago Board of Trade. In *Frontiers of Capital: Ethnographic Reflection on the "New Economy,"* edited by Melissa Fisher and Greg Downey. Durham, NC: Duke University Press.

SPACES OF KNOWLEDGE

Recombinant Technology and New Geographies of Association

JONATHAN BACH AND DAVID STARK

Introduction

Forms of social organization trade on the illusion of permanence while constantly renegotiating their relationships; their stability rests in part on their ability for transformation. The global state system is a famously reified form of social organization, its defining doctrinal characteristic of state sovereignty based on an increasingly anachronistic single-point perspective (Ruggie 1993: 159). Today the social ordering functions of state sovereignty are under duress, global issues exist beyond the control of any one state, and the global political system is undergoing a significant transformation. Global political space is increasingly defined by networks that operate fluidly; enhance flows of money, people, commodities, ideas, and weapons; and accelerate trends. At the core of this oft-noted phenomenon of spatio-temporal compression is the co-evolution of organizational forms with interactive technologies (IT), a process that rearranges the ways firms produce, states fight wars, and people structure their lives. Changes to the organization of global political space are symbiotically linked to the emergence of new organizational forms of our epoch.

These forms, whether benign or malevolent, reflect a shift from the hierarchical, bureaucratic concept of "mass" (mass production, mass media) to distributed, networked forms of production and communication. It is within this shift that the bones of the sovereign state system creak while trying to regulate transborder flows with institutions evolved to regulate life within territorial borders And it is within this shift that nonstate organizations (of all kinds) emerge to reterritorialize transborder flows in various ways (Sassen 1998; Strange 1996). Nongovernmental organizations (NGOs) are one of the most complex nonstate actors to emerge in this process, engaged directly or at the margins in the transformation of national, international, and transnational political space. NGOs are boundary objects, drawing upon their ambiguous status along the public-private spectrum to operate as informal shapers of international norms in both oppositional and partnership modes. NGOs can be seen historically as an effect of the state system, drawing their legitimacy from claims to represent civil society and addressing issues that require state intervention (Toulmin 1994). As the organization of global political space changes to contend with global governance issues, however, NGOs have become new actors seeking to confront the diversification and reproduction of the decentralized, distributed power that was once considered the domain of the sovereign state (Sassen 1999).

NGOs' expanded role has been enhanced through their use of interactive technology, and many NGOs have rushed to embrace and encourage the use of interactive technology, with mixed results. Conventionally, the role of interactive technology is thought of as a tool to improve existing functions. In this chapter we take up the relation of interactive technology to NGOs from a different angle: what is often called information technology is less a tool to be correctly applied than a logic of interaction that contains within it a new relationship to organizational innovation. Our approach is part of a growing body of social science research that seeks to overcome the artificial divide between "society" and "technology" by viewing the social as consisting of humans and nonhumans (objects, things, artifacts).[1] Accordingly, new technologies do not simply allow organizations to communicate faster or to perform existing functions more effectively, they also present opportunities to communicate in entirely new ways and to perform radically new functions. Especially because these technologies are interactive, their adoption becomes an occasion for innovation that restructures interdependencies, reshapes interfaces, and transforms relations.

The first section below addresses the relation of interactive technology to NGOs and argues that the commonly employed information broker model is insufficient to understand how the multiplicative properties of the Internet are changing the form and function of NGOs. The second section argues that the dynamics of collaboration afforded by interactive technology are resulting in new associative relations as NGOs move from pseudo-autarky to collaboration, a change that enables their structural role in globalization to become increasingly prominent.

[1] This approach draws on the work of French sociologists Michel Callon (1998) and Bruno Latour (1991) and social scientists in the United States who have been working with similar concepts. Hutchins (1995), for example, argues that cognition is distributed across a network of persons and instruments. Suchman's (1987) pathbreaking work on human-machine interaction similarly resonates with the work of Callon and Latour and provides the basis for further studies on distributed design.

Elective Affinities?

At first glance NGOs possess a superficial isomorphism with the perceived properties of interactive technology, since for many NGOs the concept of network is closely intertwined with their operational logic. When viewed mainly as a tool for processing information, interactive technology increases NGOs' communication and facilitates networking by enhancing the core tasks of getting information to constituents, channeling and interpreting information from varied sources, aggregating information and demands, transmitting them to diverse audiences, and mobilizing individuals and groups.[2] Interactive technology thus seemed ideal for lowering transaction costs, increasing participation and impact, and streamlining operations. The democratic rhetoric that accompanied the early years of the Internet was also a strong plus for NGOs—social and organizational change could be seen as complementing each other.

It would be an error, however, to see NGOs as having an elective affinity with interactive technology, and then to use this a priori affinity to claim that NGOs plus IT equals new organizational forms capable of transforming global space if only the forces of friction are sufficiently overcome. This, however, is the undertone that pervades much popular discussion about NGOs. Technology is often appended to a constellation of factors that are used to explain the recent growth and prominence of NGOs, such as the retrenchment of the welfare state, the end of the cold war, and a rise in private donations (Lindenberg and Bryant 2001: 8–12). In nearly all of these scenarios, interactive technology appears in a diffusionist fashion as either speeding up the process, presenting obstacles, or both. In these representations NGOs' use of interactive technology is discussed within the confines of an information broker model.

The information broker model is a reasonable and conditioned reaction from the age of mass communication and mass production. Modern society is organized along lines of access to quantifiable information brokered between those who have information and those who want or need it. It has an hourglass structure, with information passing through the broker in the middle on the way from A to B, similar to Burt's (1992) bridges across structural holes or Latour's (1987) obligatory passage points. This can take the ruthless form of a monopolistic corporation or the benevolent form of an NGO seeking to spread formerly guarded information. Structurally, however, brokers work in the same way by exploiting gaps and, accordingly, gaining rents. They have a vested interest

[2] Increased communication, however, is in itself not a good. Not everything works better with e-mail (O'Mahoney and Barley 1999).

in maintaining the gap between information producers and consumers. The affordances of interactive technology can be used to maximize this brokering role, along with the power (and perils) that comes with it.

NGOs gain power through an enhanced brokering role, even while they do not mimic those who "hold" power in principle, such as states or rulers. NGOs' power can be understood in Latour's (1986: 273) sense, where power accrues to "those who practically define or redefine what 'holds' everyone together." Engaging in this practical redefinition enhances NGOs' power. Transnational NGOs are particularly important in this respect. To the extent that NGOs become obligatory passage points, power can be exerted through the discursive production of the subjects they claim to represent, be they aid recipients, organizations to be included in a civil society database, or the creation of a regional identity.[3] As Paige West (2001: 29) documents in her study of environmental NGOs in Papua New Guinea, NGOs use their structural and rhetorical power 'to discursively produce 'local peoples,' 'indigenous peoples,' 'peasants' . . . and have their productions taken very seriously."[4]

But since translation is always also misunderstanding, NGOs do not only produce identities but renegotiate them. And since interactive technology affords the ability to shift from information as a discrete property to "knowledge" that requires a knowing subject, there is more out there than the brokerage model. Much of the literature, however, views technology as an external actant and therefore misses the way in which intelligence is distributed across actors and artifacts (Hutchins 1995).

Unlike information brokering, where the emphasis is on possession of information and rent-seeking, what we call knowledge facilitation emphasizes not information per se but communication and distributed intelligence. Knowledge, unlike "information," cannot exist independently of a subject and cannot be conceived of independent of the communication network in which it is both produced and consumed (thus blurring the notion itself of producer and consumer). This does not displace or solve the practical and epistemological problems occasioned by "information" (e.g., how to process large amounts of data, how to insure data protection, how to ascribe meaning to data), but raises different questions of an ontological nature. These question the very a priori (diffusionist) assumptions of the institutional and organizational forms that order our world. As Neff and Stark (2003) show for what they call "permanently beta" organizations, information technology can enable users and producers alike to reshape technology and organizations, blurring the lines

[3] This bears similarities to how nonprofits in the United States helped construct the categories and stigma of welfare recipients (Cruikshank 1999).

[4] See also our discussion of meta-NGOs in Bach and Stark (2002).

between user and producer (or agency and clients) while constituting new organizational structures.

NGOs themselves transform when shifting their emphasis from brokering information to facilitating knowledge. This could make a difference for their potential to be genuinely transformative of social structure. Facilitating knowledge is powerful for forming associations that are not just linked communities, but what we can call knowledge communities—communities that use a recombinant and multiplicative logic of link, search, interact to sustain themselves and grow.

We refer to this as the logic of link, search, interact to express concisely what it is about interactive technology—particularly its most widespread instantiation in the Internet—that makes it resonate deeply in the NGO community and in so many registers across the globe. This is certainly not the first technology to enable each of these functions: using a telephone, you can search by dialing the operator to get "information" and can then use the same phone to link with a party with whom you interact. But consider the popular search engine Google: when it suggests sites to match your query, it is also performing a search and establishing a link. To prioritize your answer, it considers all the other sites that have linked to the potentially relevant sites that match your query and ranks them based on patterns of links (i.e., the site with the highest number of links to them is considered more relevant). In other words, it searches based on the pattern of links. For the telephone the process of link, search, and interact is merely additive.[5] For Google it is multiplicative and recombinatory: each of these processes forms the basis for the other.

This recombinant technology allows searches not only on the pattern of links, but also on the pattern of interactions. If you are even a casual user of Amazon.com, the web site will suggest titles to you based on a book or CD you are looking at. This is done not by matching terms in the title or abstract of the book, which would entail a high degree of potentially humorous error, but by tracking patterns of purchase and preferences and then using an algorithm to determine that "people who bought this book also bought. . . ."[6] The output of Google or Amazon, of course, is web sites or books, while the output of the telephone is interaction with a person. What if you could harness the properties of the web's recombinatory logic to suggest interaction with people?

This would be desirable even at a merely practical level; the glut of information available on the web is such that even if you know what you are looking for, you need a way to find the most relevant information ex-

[5] Which is not to downplay linking by itself—after all, we do have a very real use for the one-to-one technology of the telephone.

[6] This form of search is known as collaborative filtering (Gladwell 1999).

peditiously. Since the creators of all this content are *people,* not machines, it stands to reason that asking the right person might be the best way to find the information you are looking for. Researchers have developed such "word of mouth" software (one is appropriately named "gab," as in talk, but also for Group Asynchronous Browsing) (Wittenburg 1998). But there is an even more compelling reason to prefer a recombinatory over an additive approach—when you *don't* know what you are looking for but would recognize it when you find it (e.g., what happens every night at a singles bar). Unlike finding a phone number from "information," this way you find things you did not know and come into contact with people whom you do not know. Most people would probably balk at interacting directly with other customers of Amazon, but there are communities where this would be quite an asset—for example, a doctor who wants to know who else is treating patients for similar rare diseases or a member of an NGO community that wants to share best practices. "During the Gujarat earthquake," recounts Paul Mylea, the editor of an NGO website called Alternet.org that facilitates collaboration among humanitarian aid agencies, "a member was based very close to the center—and they were experienced in drought relief rather than earthquake relief. A member from our advisory board contacted the member on the ground because he had experience of earthquake relief and was able to offer advice and guidance on how to deal with the crisis. They went off site and spoke on the phone" (Lewis 2001).

Using the patterns of search or interact, one can link social structures (who knows whom) and knowledge networks (who knows what). Amazon.com's collaborative filtering software is a commercial variant of similar programs such as the aptly named Yenta, Beehive, or the browser Alexa.[7] For members of an NGO or nonprofit community, this could help develop and promote their respective knowledge networks. Working with a group of 285 such organizations in the Midwest, researchers at the University of Illinois developed a software program that could help the organizations identify those in the community who shared common or complementary interests and show how they may be directly or indirectly connected.[8] This software, based on a tool called IKNOW, is distinctive because the users can find out not only "who knows whom" and "who knows what," but also "*who* knows who knows whom," and "*who* knows who knows what" (Contractor et al. 1998).[9] This works by cap-

[7] See, respectively, http://foner.www.media.mit.edu/people/foner/Yenta/; http://info.alexa.com/; ftp://parcftp.xerox.com/pub/dynamics/beehive.html.

[8] PrairieNet communityware can be seen at http://www.tec.spcomm.uiuc.edu/nosh/prairienet.

[9] IKNOW stands for Inquiring Knowledge Networks on the Web. The IKNOW web site is http://www.tec.spcomm.uiuc.edu/nosh/IKNOW.

turing network data of both knowledge networks (based on links between actors' web sites, on common links from their web sites to third party sites, on similarity in content between different web sites, and on an inventory of skills and expertise provided by the actors) and communication networks (based on an inventory of existing task and project links between them).

From social structures and knowledge networks we thus get at cognitive social structures and cognitive knowledge networks (*who* knows whom or what). The cognitive perceptions of the members of a knowledge community taken individually may be incomplete or inaccurate, but together they form a transactive memory system that shares domains of knowledge (Contractor et al. 1998; Contractor 2000). This hints at a larger significance for what at first might seem like just a good way to sell books: communities of knowledge can be not only identified, but also created. IKNOW does not just enable dyadic relationships in the manner of personal ads, but also facilitates communities of knowledge.

In a similar vein, a group of researchers is working on Augmented Social Networks, or ASN. Unlike IKNOW, ASN is not software, and unlike Alternet.org, it is not a web site. Rather, ASN seeks to establish a model for a "persistent online identity" for individuals moving between different Internet communities. This identity can be the centerpiece for enhancing "the power of social networks by using interactive digital media to exploit the transitive nature of trust through the principle of six degrees of connection. As a result, people will be able to inform themselves and self-organize more effectively—in non-hierarchical, rhizomatic social formations—leading to more opportunities for engaged citizenship" (Jordan et al. 2003: 2). The idea for ASN builds on the work of Robert Metcalfe, whose Metcalfe's Law holds that "The total value of a network where each node can reach every other node grows with the square of the number of nodes," and on research on Group Forming Networks by David Reed, who studied the exponential growth in new, and previously unknown, types of value created by the online interconnection of social networks. ASN seeks specifically to support civil society and citizen participation in governance structures through its model and is developing software, protocols, open standards, and principles of implementation (Jordan et al. 2003).

The Geography of Association

Whether idealistic, as with ASN, or practical, as with Alternet, the rise of knowledge communities opens up a space dissimilar to the established means of communication because it integrates discursive and nondiscursive elements and in doing so creates a new basis for association. What we can call an associative space is as much a space *within* which something happens as it is a space *for* something to happen (Johnson 1997). As a space within which something happens, we can trace empirically the circulation and creation of knowledge communities. As a space for something to happen, we can speculate that new forms of social organization, including new social bonds (Levy 1997: 10–13), will develop on the basis of a relation to knowledge (for example, by the relocating of ties in social structures such as the family or the workplace, the valorization of programming skills and the mobility of electronic labor, and so forth). Such a transformation does not imply that knowledge is a function of interactive technology, any more than exchange is a function of capitalism. But just as exchange acquired specific characteristics under capitalism that became the basis for a complex system, so does knowledge acquire new characteristics in our age.

Three of these characteristics are of particular importance in understanding how NGOs are embedded in a changing geography where knowledge is increasing as a resource for creating enduring associations (i.e., as a source of power). The first is related to the organization of global political space, specifically the shift among states and intergovernmental organizations from a concern about the sanctity of sovereignty to a concern about the enforcement of universal norms. This can be viewed cynically or hopefully, through the lens of empire or enlightenment. Certainly not all governments embrace such a shift (ironically, the United States is foremost among the obstructionists while also one of the greatest proselytizers of universal principles), but an agenda that prioritizes humanitarian, environmental, and even economic justice issues has established itself as a global discourse. NGOs were in the forefront in the shift from sovereign sanctity to universal norms, particularly in the realms of the environment and human rights. The stunning successes of Doctors Without Borders and the Campaign to Ban Landmines, both of which won the Nobel Peace Prize, gave NGOs publicity and legitimacy that far surpassed previous efforts. From a different angle, the anti-WTO protests in Seattle and similar "antiglobalization" protests from Ottawa to Prague criticized the distributed modes of production and called attention to the new forms of connectedness under globalization. In an intriguingly isomorphic fashion the protesters, especially the more radical of them, also used a distributed logic to achieve their seeming chaotic but well-orchestrated effect: the weird coalitions of the antiglobalization movement, as Katharine Viner (2000) notes, are also wired coalitions.

It is not only protesters, however, that use distributed logic, which can be seen in the networks formed in support of a variety of causes, such as humanitarian relief efforts for earthquake and war victims, preserving the

Arctic wildlife reservation from oil drilling, or pressing for minority rights. This is the second shift: from decentralized to distributed structures. Decentralized governing structures emerged to (over)compensate for the inability of centralized forms of government and market to efficiently provide the resources or results deemed necessary for the good life, resulting in privatization or political structures such as subsidiarity and devolution. Decentralized production enabled capital to increase its mobility. But decentralization is an effect. Distribution, on the other hand, is the capacity for a collective actor to act strategically based on an emergent effect of the patterns of association and not on the basis of a single person alone, or even a network of humans (Hutchins 1995; Suchman 1987; Law and Hassard 1999; Girard and Stark 2002). Adopting a distributed structure does not mean that competition between, or hierarchy within, NGOs has disappeared. But the isolation of NGOs diminishes as networks become increasingly standard operating procedure, especially when linked by the Internet, as most of them are. This allows the leveraging of knowledge across multiple logics and ordering principles, creating new opportunities and conundrums, including the thorny problem of how to make *networks* accountable.[10]

This leveraging of knowledge through distributed cognition allows NGOs to engage in translation as one of their major functions.[11] However, since (as Latour reminds us) a site of translation is always also a misunderstanding, it is where negotiations of meaning take place. NGOs occupy a particularly strategic position in this regard: they work upward with governments and corporations (e.g., through lobbying, media campaigns, protest, and participation in policy processes) and downward with local and marginalized populations (e.g., through in-country projects, training, regranting and consciousness raising). They thus are in a position to embody *the tension* between diffusion and translation. This corresponds to a third shift, this time in the analytical methodology that informs (social) scientific development from what Latour identified as a diffusion model to a model of translation (Latour 1986: 266–69). The diffusion model is a model of inertia and friction, where changes are explained by theorizing about what retards or accelerates an order or object's trajectory—for example, the idea of the nation-state as a stable, given combination of traits and territory whose trajectory can be explained by a mixture of hard times that slow down its progress (perhaps

[10] Because authority is distributed, accountability becomes highly problematic, especially when thought of in the juridical sense of locating responsibility in a figure or specific institution of authority. See Stark and Bruszt (1998).

[11] Compare the concept of translation with Fox and Brown's (1998) "bridging individuals."

covetous neighbors who invade their territory) or good times that speed it up (such as economic boom, or the nation-state's own military conquests).[12] The nation is merely transmitted from one generation to the next with a rich history of (and potential for future) friction. A translation model dispenses with inertia and sees an object or order as being continuously transformed by the actors themselves who engage in continuous reinterpretation.[13] In more fashionable terms, a translation model could be seen as a process akin to social construction, where, since translation is also always a misunderstanding, the translation site is also the site of interpretation, contention, and renegotiation.

These shifts are harbingers of a new geography of association that involves negotiations across ordering principles and multiple logics (Stark and Bruzst 1998: 109–36). As Charles Sabel (1992) points out in his study of economic developmental associations, no state can possibly have knowledge superior to that of economic actors or coordinate restructuring better than regional developmental associations—it is the associations, not the states, that do the developing. Likewise, as NGOs become deliberative associations, they can play a greater role in both develop*ment* (in the traditional sense) and develop*ing* global, regional, and national structures and institutions. This is because deliberative associations lead to new associations, both in the literal sense of new networks and the figurative sense of a mental connection between ideas.[14]

An example of how NGOs engendered deliberative associations that changed them from information brokers to knowledge facilitators is the story of development NGOs in India. As Bishwapriya Sanyal (1994: 37)

[12] See here Appadurai's (2000) notion of process geographies and trait geographies, and Stephen Toulmin's (1990) notion of a Newtonian image of power exerted with a central force through sovereign agencies.

[13] Latour (1986: 266–67) uses the example of rugby players and a rugby ball: "The initial force of the first in the chain is no more important than that of the second, or the fortieth, or of the four hundredth person. Consequently, it is clear that the energy cannot be hoarded or capitalized; if you want the token to move on you have to find fresh sources of energy all the time; you can never rest on what you did before, no more than rugby players can rest for the whole game after the *first* player has given the ball its *first kick*." Latour's preference for a translation model is that it allows power to be seen as a consequence and not a cause of collective action, a point we will return to later.

[14] The "antiglobalization" movement that emerged from the protests in Seattle is an example of how a rally of disparate agendas morphed into a community of deliberative associations where the lines between environment, economic development, and human rights increasingly blurred. A much smaller-scale example of an associative solution is a Roma Rights organization in Hungary, which began solely by trying to link disparate organizations and individuals to each other. As a result of the subsequent interaction, the one-time clients moved from being serviced by the organization to claiming the organization as their own, eventually becoming involved in its governance. From its origins as an information broker, the organization transformed into a knowledge community (Bach and Stark 2002).

explains, NGOs in India were privileged in the 1970s as "the most appropriate catalytic agent for fostering development from below because their organizational priorities and procedures are diametrically opposed to those of the institutions at 'the top.'" To fulfill this avant garde role, NGOs valorized a form of pseudo-autarky for two negative reasons and one positive reason: Collaboration with the state was ruled out because it was seen as leading to control or cooptation, while collaboration with the market would poison community solidarity bondings. In both cases legitimacy and effectiveness were thought to suffer. These were negative reasons for maintaining independence. A positive reason was that the principles of self-sufficiency, self-reliance, and social innovation would become the motor for self-reproduction. The basic analytic unit was the isolated NGO engaged in a form of autopoesis. There was indeed a self-generating quality to this approach, but what it generated was isolation and contradictions. NGOs competed fiercely with each other for money and avoided forming institutional linkages with government, the commercial sector, or even other NGOs. The lack of institutional support doomed all but the smallest projects and precluded replication or expansion. When they began to fall apart as a result of these incapacities, it only intensified competitiveness and isolation and made a mockery of the attempt to create a broad base "from below" (Sanyal 1994).

The relative success and high growth of NGOs in the latter part of the 1980s and especially the 1990s can be attributed not only, or even primarily, to increased externalities, but to the NGOs' shift from self-imposed isolation to collaboration. NGOs moved to collaboration as they began to recognize that success, when it happened, came because they were already engaging in semiconscious forms of collaboration that went unacknowledged. For example, NGOs' own leaders were drawn from an elite with informal linkages to all the types of institutions—banks, bureaucracies, and parties—that form the "top." Sanyal (1994: 45) gives the example of the founders of the Grameen Bank, Drs. Yunus and Latifee, who are mythologized as visionaries whose efforts resulted in this paradigmatic development from below. They doubtless possessed great vision, but, as Sanyal points out, they also had an institutional association with the top university that provided both salary and legitimacy, and Yunus's efforts to convince the bank to make loans were made not on the strength of his grassroots organizing ability but because of his family's longstanding relationship as a major depositor. As the project expanded and became the famous Grameen Bank, it was on the firm basis of a tripartite alliance among NGOs, government, and market institutions.[15]

[15] See also Sanyal's (1994) accounts of the Bangladeshi NGO Proshika, and the Indian NGO SEWA (Self-Employed Women's Association).

The need to be self-sustaining caused conflicts within NGOs because of the siren call of alliances with the market as a source of generating independent income, especially as foundations began to require better accountability and plans for sustainability. Over the last fifteen years, in the search for self-sustainability some NGOs have indeed turned to income-generation alternatives that mimic commercial enterprises. For example the "dot-corg" dual enterprise model combines social and business ventures, separating revenue generation from the NGOs' social mission and evaluating it according to business metrics. There is also a minority of NGOs who, from early on, set their long-term goal as evolution into a socially oriented, for-profit venture, such as many Internet Service Providers in Eastern Europe who began as nonprofits and grew into viable businesses (Peizer 2000). When you consider the early resistance of NGOs to allying themselves too closely with the market, it is striking (or even shocking) to watch partnerships emerge such as CARE-Starbucks (Lindenberg and Bryant 2001: 164–65; Austin 2000; and, for a critique, Gereffi et. al 2001) or the "Libraries Online Partnership" between Microsoft Corporation and the nonprofit American Library Association (Sagawa and Segal 2000).[16]

Alliances with the market certainly do open new forms of sustainability and even synergy and cannot be dismissed out of hand. If NGOs reject cooperation with state and market forces too completely, they risk slipping into an exclusively oppositional role with diminished opportunities for agenda-setting (though some may relish precisely this oppositional role). Yet the benefits of collaboration do not mean that old problems of cooptation have disappeared—on the contrary, they may even be exacerbated by the new hybrid forms. The values of the market and of the nonprofit world remain antagonistic. As NGOs spread their accountability unevenly among constituents, board members, donors, and the public, they find themselves faced with a proliferation of performance criteria that catches them between the value systems of business (efficiency, solvency) and social mission (adherence to principles, ideological agenda) (Edwards and Hulme 1996b). In the best case they may exploit these contradictions, but the danger is real that actors who are accountable according to many principles become accountable to none (Stark 2001).[17]

Most importantly, success for NGOs came less from developing innovative ideas than from basing their efforts "on relatively old ideas which

[16] Of course Microsoft and Starbucks were themselves once upon a time anti-establishment upstarts. On the phenomenon of voluntary-commercial cooperation and its attendant challenges, see Edwards and Hulme (1996a) and Bendell (2000).

[17] Because the state and market themselves are not static but are undergoing fundamental changes, an even bigger problem may be distinguishing cooperation from co-optation in certain cases (Bach and Stark 2002).

may have been tried, even by the government, in another context. . . . Successful NGOs did not pursue only a decentralized approach . . . their success was due to a skilful blending of centralization and decentralization of decisions, cooperation and competitiveness" (Sanyal 1994: 43). In other words, successful NGOs used logics that are distributed and recombinatory.

Conclusion

When we employ analytical concepts that bridge the society/technology divide, NGOs appear as a molecular technology, a large, self-organizing community of deliberative associations (Latour 1991; Levy 1997: 41). They translate (i.e., misunderstand, interpret, and renegotiate) between multiple logics, such as indigenous peoples and government bureaucrats. They also translate between an older spatio-temporal order (the cold war, the sovereign state system, Fordism, etc.) and what we have provisionally marked as an associative space. This space is transforming what Agnew calls the field-of-forces model of political power based on a spatiality of political power frozen into state territorial units (Agnew 2002). The geography of association rests on a different epistemological premise than the dyadic conception of power that hypostasizes the sovereign nation-state—associations are based on recombinant principles derived from social network theory rather than billiard ball models of classic international relations theory.

As we described above, knowledge communities assume a central place in the geographies of association as circuits of social (re)production at the local and global level. NGOs can be significant in this regard because their liminal role between local, national, and global situates them strategically within the technospatial: the technologically mediated social and material orders that are defined by new boundaries of place and technosocial practices. For instance, the mix of face-to-face and virtual interaction now standard within and between many NGOs blurs the line between serendipitous and intentional contact, turning what were once primarily disjunctive interactions into contiguous experiences. This is happening in a context where interactive (especially mobile) technologies are changing our notions of personal and social space.[18] The recombinatory aspects of

[18] See Ito and Okabe (2003) for an empirical study of the way mobile technology alters the sense of place, including a discussion of serendipitous and intentional contact and co-presence among Tokyo mobile phone users. Ito and Okabe refer to "technosocial situations" to describe the situations where boundary-spanning technologies restructure social identity and practice.

the term "association" that we highlighted in this chapter can thus acquire at least three additional salient meanings: NGOs as part of a shifting landscape of reputation ("by association"), as part of a network of more or less formal societies (associations in the German sense of *gesellschaften*), and embedded in a field of technosocial practices that privilege nonlinearity (as in "associative thinking").

It would be a mistake, however, to assume that associative spaces predetermine any a priori normative outcome for NGOs—as mentioned earlier, the problems of accountability alone present substantial challenges to future development. How are alliances, much less networks, to be held accountable? As NGOs move from confronting businesses to partnering with them, how will this affect their justificatory claims to representing civil society? Could not NGOs operate nefariously as the moral instruments of a new global society of control precisely *because* they are networked, molecular structures, functioning as "the capillary ends of the contemporary networks of power" (Hardt and Negri 2000: 313)? In the growing literature NGOs appear alternatively as an incipient global civil society, functional equivalents of democracy, as tools of the ruling class, or as the vanguard for globalization from below (Warkentin 2001; Rosenau 1998; Falk 1999; Appadurai 2000). NGOs are diverse enough to incorporate all these contradictory interpretations, yet too often the discussion proceeds as if NGOs' form were given and only their effect remains to be worked out. In the new geography of association, NGOs' most striking function is their renegotiation of the justificatory regimes upon which the global spatio-temporal order is based. In this uncertain process they will continue to assume an increasingly central and controversial role as co-constituents of the organization of global political space.

References

Agnew, John A. 2002. *Making Political Geography.* London: Arnold; New York: Oxford University Press.

Appadurai, Arjun. 2000. Grassroots Globalization and the Research Imaginaton. *Public Culture* 12(1): 1–21.

Austin, J. 2000. *The Collaboration Challenge.* San Fransisco: Jossey-Bass.

Bach, Jonathan, and David Stark. 2002. Innovative Ambiguities: NGOs Use of Interactive Technology in Eastern Europe. *Studies in International and Comparative Development* 37(2): 3–23.

Bendell, J. 2000. *Terms for Endearment: Business, NGOs and Sustainable Development.* Sheffield: Greenleaf.

Beunza, Daniel, and David Stark. 2003. The Organization of Responsiveness: Innovation and Recovery in the Trading Rooms of Lower Manhattan. *Socio-Economic Review* 1(2): 135–64.

Burt, Ronald. 1992. *Structural Holes.* Cambridge: Harvard University Press.

Callon, Michel. 1998. *The Laws of the Markets.* Oxford: Blackwell Publishers.

Castells, Manuel. 1999. *The Network Society.* London: Blackwell.

Contractor, Noshir. 2000. Social Network Formulations of Knowledge and Distributed Intelligence: Using Computational Models to Extend and Integrate Theories of Transactive Memory and Public Goods. Paper presented at workshop on Heterarchies: Distributed Intelligence and the Organization of Diversity. Santa Fe, NM.

Contractor, Noshir, et al. 1998. IKNOW: A Tool to Assist and Study the Creation, Maintenance, and Dissolution of Knowledge Networks. In *Community Computing and Support Systems, Lecture Notes in Computer Science 1519,* edited by T. Ishida, 210–17. Berlin: Springer Verlag.

Cruikshank, Barbara. 1999. *The Will to Empower.* Ithaca: Cornell University Press.

Deleuze, Gilles, and Felix Guattari. 1987. *A Thousand Plateaus.* Minneapolis: University of Minnesota Press.

Edwards, Michael, and David Hulme, eds. 1996a. *Too Close for Comfort: NGOs, States and Donors.* London: Earthscan Press.

———. 1996b. *Beyond the Magic Bullet: NGO Performance and Accountability in the Post Cold War World.* West Hartford, CT: Kumarian Press.

Falk, Richard. 1999. *Predatory Globalization.* Oxford: Polity Press.

Fox, Jonathan A., and L. David Brown. 1998. *The Struggle for Accountability: The World Bank, NGOs, and Grassroots Movements.* Cambridge: MIT Press.

Gereffi, Gary, Ronie Garcia-Johnson, and Erika Sasser. 2001. "The NGO-Industrial Complex." *Foreign Policy* 125:56–65.

Girard, Monique, and David Stark. 2002. Distributing Intelligence and Organizing Diversity in New Media Projects. *Environment and Planning A* 34(11): 1927–49.

Gladwell, Malcolm. 1999. The Science of the Sleeper. *The New Yorker.* October 4, 1999. Available at http://www.gladwell.com/1999/1999_10_04_a_sleeper.htm.

Greider, William. 2000. Global Agenda: After the WTO Protest in Seattle, It's Time to Go on the Offensive. Here's How. *The Nation.* January 31.

Hardt, Michael, and Antonio Negri. 2000. *Empire.* Cambridge: Harvard University Press.

Hutchins, Edwin. 1995. *Cognition in the Wild.* Cambridge: MIT Press.

Ito, Mizuko, and Daisuke Okabe. 2003. Technosocial Situations: Emergent Structurings of Mobile Email Use. Available at http://www.itofisher.com/PEOPLE/mito/mobileemail.pdf.

Johnson, Stephen. 1997. *Interface Culture: How New Technology Transforms the Way We Create and Communicate.* San Francisco: HarperCollins.

Jordan, Ken, et al. 2003. The Augmented Social Network: Building Identity and Trust into the Next-Generation Internet. New York: The Link Tank.

Latour, Bruno. 1986. Powers of Association. In *Power, Action, and Belief: A New Sociology of Knowledge,* edited by J. Law, 264–80. New York: Routledge.

———. 1987. *Science in Action. How to Follow Scientists and Engineers through Society.* Milton Keynes: Open University Press.

———. 1991. Society Is Technology Made Durable. In *A Sociology of Monsters: Essays on Power, Technology, and Domination,* edited by J. Law, 103–31. New York: Routledge.

Law, John, and John Hassard, eds. 1999. *Actor Network Theory and After.* Oxford: Blackwell.

Levy, Pierre. 1997. *Collective Intelligence.* New York: Plenum Trade.

Lewis, Ellen. 2001. Red Alert. London: *The Guardian.* November 26.

Lindenberg, Marc, and Coralie Bryant. 2001. *Going Global.* Bloomfield, CT: Kumarian Press.

Neff, Gina, and David Stark. 2003. Permanently Beta: Responsive Organization in the Internet Era. In *Society Online: The Internet in Context,* edited by Phillip E. N. Howard and S. Jones. Thousand Oaks, CA: Sage.

O'Mahoney, Siobhan, and Stephen R. Barley. 1999. Do Digital Telecommunications Affect Work and Organization? *Research in Organizational Behavior* 21:125–61.

Peizer, Jonathan. 2000. Sustainable Development in the Digital Age. New York: Media Channel. Available at http://www.mediachannel.org/views/oped/values.shtml.

Rosenau, James. 1998. Governance and Diplomacy in a Globalizing World. In *Reimagining Political Community,* edited by D. Archibugi, D. Held, and M. Köhler. Stanford: Stanford University Press.

Ruggie, John. 1993. Territoriality and Beyond: Problematizing Modernity in International Relations. *International Organization* 47(1): 139–74.

Sabel, Charles. 1992. Studied Trust: Building New Forms of Cooperation in a Volatile Economy. In *Industrial Districts and Local Economic Regeneration,* edited by F. Pyke and W. Sengenberger, 215–49. Geneva: International Labor Organization.

Sagawa, Shirley, and Eli Segal. 2000. *Common Interest Common Good: Creating Value through Business and Social Sector Partnerships.* Cambridge: Harvard Business School Press.

Sanyal, Bishwapriya. 1994. *Cooperative Autonomy: The Dialectic of State-NGOs Relationship in Developing Countries.* Geneva: International Labor Organization.

Sassen, Saskia. 1998. *Globalization and Its Discontents: Essays on the New Mobility of People and Money.* New York: New Press.

———. 1999. *Losing Control? Sovereignty in an Age of Globalization.* New York: Columbia University Press.

Stark, David. 2001. Ambiguous Assets for Uncertain Environments: Heterarchy in Postsocialist Firms. In *The Twenty-First-Century Firm: Changing Economic Organization in International Perspective,* edited by P. DiMaggio, 69–104. Princeton: Princeton University Press.

Stark, David, and Laszlo Bruzst. 1998. *Postsocialist Pathways: Transforming Politics and Property in East Central Europe.* Cambridge: Cambridge University Press.

Strange, Susan. 1996. *The Retreat of the State: The Diffusion of Power in the World Economy.* Cambridge: Cambridge University Press.

Suchman, Lucy. 1987. *Plans and Situated Actions: The Problem of Human-Machine Communication.* Cambridge: Cambridge University Press.

Toulmin, Stephen. 1990. *Cosmopolis: The Hidden Agenda of Modernity.* Chicago: University of Chicago Press.

———. 1994. The Role of Transnational NGOs in Global Affairs. Paper presented to the conference on The UN and Japan in an Age of Globalization, Peace Research Institute, International Christian University, Tokyo.

Viner, Katharine. 2000. "Luddites" We Should Not Ignore: Instead of Vilifying the Prague Protesters, We Could Learn from Them. London: *The Guardian.* September 29.

Warkentin, Craig. 2001. *Reshaping World Politics: NGOs, the Internet and Global Civil Society.* New York: Rowman and Littlefield.

West, Paige. 2001. Environmental NGOs and the Nature of Ethnograpic Inquiry. *Social Analysis* 45(2): 55–77.

Wittenburg, Kent, et al. 1998. Group Asynchronous Browsing on the World Wide Web. Available at http://www.w3.org/Conferences/WWW4/Papers/98/.

Electronic Markets and Activist Networks: The Weight of Social Logics in Digital Formations

SASKIA SASSEN

INTERACTIONS BETWEEN digital technology and social logics can produce a third condition that is a mix of both. When this mixed domain gets structured in electronic space, we call it a digital formation (see Latham and Sassen, this volume). This chapter focuses on two such formations: the global market for capital and global electronic activist networks. In both cases my organizing question concerns the operation of social logics and how they shape and are in turn shaped by these technologies. The focus is, then, on both the transformative capacities of these new computer-centered technologies and their conditioning by social logics. The two very different types of cases examined in this chapter make legible the variable ways in which this sociotechnical interaction produces outcomes.

Both cases are part of global dynamics, and both have been significantly shaped by the three properties of digital networks—decentralized access/distributed outcomes, simultaneity, and interconnectivity. But, I will argue, these technical properties have produced strikingly different outcomes in each case. In one case, these properties contribute to distributive outcomes—greater participation of local organizations in global networks—and thereby help in constituting elementary forms of transboundary public spheres or forms of globality centered in multiple localized types of struggles and agency. In the second case, these same properties have contributed to higher levels of control and concentration in the global capital market even though the growth in the levels of capital drawn up into these financial electronic networks rests on a kind of distributed power, that is, millions of investors and their millions of decisions.

This difference points to the possibility that networked forms of power are not inherently distributive, as is often theorized when the focus is exclusively on technical properties. Intervening mechanisms that may have little to do with the technology per se can reshape what is, technically, a primary outcome of these networks. These two cases show us that the trajectory followed by what begins in each as the distributed power we associate with computer-centered networks can take on many forms. In the case of the global capital market, it winds up as concentrated power. This indicates that technology alone does not explain outcomes: each case constitutes a distinct domain through specific imbrications of technical and social logics. We can expect these imbrications to range from simple to complex, depending on the type of case. One way of describing this interaction is to posit that the new technologies are partly embedded in institutional environments that have the power to inscribe technology. As a result, the outcome does not reflect exclusively the features of the particular technology at work.

To capture the interactions between the technical and social logics at work in producing the distinct outcomes of each case, we need to identify appropriate indicators. One type of indicator is the counterfactual, in this case, to a purely technology-driven outcome. In the case of this chapter, it would be that which disproves the technological logic. For the global capital market, one such counterfactual would be to posit a lumpy rather than seamless electronic space: The social logics operating in this electronic, transjurisdictional, globally interconnected market can alter the outcomes we might deduct from the technical capacities at work in these electronic networks. The effort then becomes one of laying bare the ways in which this electronic market is embedded and conditioned. The new technologies have had a deeply transformative effect, but they do not dislodge the fact of substantive agendas organizing market actors. The argument I develop below is that today's global capital market is a complex formation markedly different from earlier global financial markets because of its extensive digitization, but that this does not necessarily mean that it is disembedded. In the case of electronic activist networks of local organizations, the indicator would function in precisely the opposite direction: the local can constitute the nonlocal, specifically in this case global networks and global agendas. The effort here is to understand how highly specific local environments and agendas can constitute global scalings.

Both cases make legible how digitization can destabilize nested, formalized hierarchies of scale: a global electronic space is shown to be multiscalar, and situated local struggles are shown to be drawn up into a global electronic space. In the first case, the multiscalar nature of the electronic global capital market comes about through its embeddedness in a network of financial centers located in highly institutionalized national environments. In the second case, the multiscalar nature of the local comes about through its constituting global networks, which in turn maximize connectivity and interaction among localities. Localized entities become microenvironments with global span. Local organizations confined to localized actions gain cognition of the recurrence of these types of actions in locality after locality, thereby contributing to the reshaping of these global networks for communication into global zones for interactivity.

The global capital market is a particularly helpful case for examining these dynamics of transformation and embeddedness. It represents an enormously complex series of imbrications of digital and nondigital factors that can actually be traced given a high level of institutionalization and a considerable amount of evidence. In contrast, the global network of local organizations represents rather simple types of imbrications, at least at this point, and is far more difficult to trace given low, if any, institutionalization; as a field for research, it has also suffered from a northern perspective that has misinterpreted and/or overlooked key aspects of global south electronic activist networks. However, the case of electronic activist networks helps us understand the fact of different trajectories and thereby illuminates the variability and specificity of the transformative capacities of these technologies (see introduction, this volume).

The Locational and Institutional Embeddedness of Electronic Financial Markets

In seeking to understand the role of the new technologies in shaping today's market for capital, it is important to recognize that there has long been a global market for capital and that there clearly would have continued to be one even if these technologies never had come about.[1] The question then becomes one of understanding the specific ways in which computer-centered technologies have reshaped financial markets, and distinguishing between merely derivative changes and genuinely transformative ones.

There are, in my reading, two major sets of differences that distinguish today's global market for capital from that of earlier periods.[2] One has to do with the level of formalization and institutionalization of the global market for capital today, partly an outcome of the interaction with national regulatory systems that themselves gradually have become far more elaborate over the last hundred years. I will not focus especially on this aspect here. The second set of differences concerns the transformative role of digital networks and the possibility of digitizing financial instruments

[1] A strong line of interpretation in the literature is that today's market for capital is nothing new and represents a return to an earlier global era at the turn of the century (Hirst and Thompson 1996; Wade 2004). I argue that this holds only at a high level of generality, but when we factor in the specifics of today's capital market, especially digitization, some significant differences emerge with those past phases (Sassen 2001: chaps. 4, 5, and 7). There is an emerging literature focused on electronic financial markets (e.g., Knorr-Cetina and Bruegger 2002; Barrett and Scott 2004; Callon 1998; MacKenzie and Millo 2003; Zaloom 2005).

[2] Neither of these has been addressed by those who argue that the current global market for capital represents a return to an older form and that hence it is nothing new.

(both henceforth called digitization). In combination with the various dynamics and policies we usually refer to as globalization, they have constituted the capital market as a distinct institutional order, one different from other major markets and circulation systems such as global trade and foreign direct investment.

One of the key and most significant outcomes of digitization in finance has been the jump in orders of magnitude and the extent of worldwide interconnectedness. I argue that there are basically three ways in which digitization has contributed to this outcome (for a greater elaboration of this argument, see Sassen 2001: chaps. 5 and 7; Sassen 2005: chaps. 5 and 7). One is the use of sophisticated software, a key feature of the global financial markets today and a condition that in turn has made possible an enormous amount of innovation. It has raised the level of liquidity as well as increased the possibilities of liquefying forms of wealth hitherto considered nonliquid. This can require enormously complex instruments; the possibility of using computers not only facilitated the development of these instruments but also enabled the widespread use of these instruments insofar as much of the complexity can be contained in the software. It enables users who might not fully grasp either the mathematics or the software design issues of financial instruments. Development of these instruments is further enhanced by the fact that their softwaring facilitates proprietary rights.

Second, the distinctive features of digital networks can maximize the implications of global market integration by producing the possibility of simultaneous interconnected flows and transactions, as well as decentralized access by investors and by financial exchanges in a growing number of countries. The key background factor here is that, since the late 1980s, the trend has been for more and more countries to de- and reregulate their economies according to a particular set of criteria that has ensured cross-border convergence and the global integration of their financial centers. This nondigital condition amplified the new capabilities introduced by the digitization of markets and instruments.

Third, because finance is particularly about transactions rather than simply flows of money, the technical properties of digital networks assume added meaning. Interconnectivity, simultaneity, decentralized access, and softwared instruments all contribute to multiply the number of transactions, the length of transaction chains (i.e., distance between instrument and underlying asset), and thereby the number of participants. The overall outcome is a complex architecture of transactions.[3]

[3] Elsewhere (Sassen 2005: chap. 7) I have developed this thesis of finance today as being increasingly about transactions rather than about money per se. In my reading financial centers become even more important today because they contain the capabilities for managing this transactivity precisely at a time when the latter assumes whole new features, given digitization.

These three features of today's global market for capital are inextricably related to the new technologies. The difference they have made can be seen in two consequences. One is the multiplication of specialized financial markets. It is a question not only of global markets for equities, bonds, futures, currencies, but also of the proliferation of enormously specialized global submarkets for each of these. This proliferation is a function of increased complexity in the instruments and simultaneous market integration, in turn made possible by digitization of, respectively, instruments and markets.

The second consequence is that the combination of these conditions has contributed to the distinctive position of the global capital market in relation to several other components of economic globalization. We can specify two major traits, one concerning orders of magnitude and the second the spatial organization of finance. In terms of the first, indicators are the actual monetary values involved and, though more difficult to measure, the growing weight of financial criteria in economic transactions, sometimes referred to as the financializing of the economy. Since 1980 the total stock of financial assets has increased three times faster than the aggregate GDP of the twenty-three highly developed countries that formed the OECD for much of this period. The volume of trading in currencies, bonds, and equities has increased about five times faster and now surpasses it by far. This aggregate GDP stood at about U.S. $30 trillion in 2000 while the worldwide value of internationally traded derivatives reached over $65 trillion in the late 1990s, a figure that rose to $168 trillion in 2001 and $192 trillion in 2002. To put this in perspective, we can make a comparison with the value of other major high-growth components of the global economy, such as the value of cross-border trade (ca. $8 trillion in 2000) and global foreign direct investment stock ($6 trillion in 2000) (IMF 2001; BIS 2002). Foreign exchange transactions were ten times as large as world trade in 1983, but seventy times larger in 1999, even though world trade also grew sharply over this period.[4]

As for the second major trait, the spatial organization of finance, it has been deeply shaped by regulation. In theory, regulation has operated as one of the key locational constraints keeping the industry, its firms, and markets from spreading to every corner of the world.[5] The wave of dereg-

[4] The foreign exchange market was the first one to globalize, in the mid-1970s. Today it is the biggest and in many ways the only truly global market. It has gone from a daily turnover rate of about U.S. $15 billion in the 1970s, to $60 billion in the early 1980s, and an estimated $1.3 trillion today. In contrast, the total foreign currency reserves of the rich industrial countries amounted to about $1 trillion in 2000.

[5] Wholesale finance has historically had strong tendencies toward cross-border circulation, whatever the nature of the borders might have been. Venice-based Jewish bankers had multiple connections with those in Frankfurt, and those in Paris with those in London; the Hawala system in the Arab world was akin to the Lombard system in western Europe. For a detailed discussion see Arrighi (1994).

ulations that began in the mid-1980s has lifted many of these formal constraints on the geographic spread of the industry. Further, being a highly digitized industry today, financial outputs can circulate instantaneously worldwide, financial transactions can be executed digitally, and both circulation and transactions can cut across conventional borders. In principle this generates locational options that are quite specific to finance and diverge from those of most other globalized economic sectors (see, e.g., Budd 1995). The large-scale deregulation of the industry in a growing number of countries since the mid-1980s has indeed brought with it a sharp increase in access to what were still largely national financial centers and has enabled innovations that, in turn, facilitated the industry's expansion both geographically and institutionally. This possibility of locational and institutional spread also brings with it a heightened level and diversification of risk, a marking feature of the current phase of the market for capital. Yet, as I will discuss below, the geography of its spread is lumpy rather than seamless because of the substantive agendas guiding the sector and its dependence on a network of at least partly nondigital financial centers. Financial centers are "agents" through which specific utility logics are drawn into the global electronic market.

The Distinctiveness of Today's Capital Market

Though there is little agreement on the subject, in my reading these current conditions make for important differences between today's global capital market and the period of the gold standard before World War I. In some ways the international financial market from the late 1800s to the interwar period was as massive as today's. This appears to be the case if we measure the volume of long-term flows as a share of national economies. The international capital market in that earlier period was large and dynamic, highly internationalized, and backed by a healthy dose of Pax Britannica to keep order. The extent of its internationalization can be seen in the fact that in 1920, for example, Moody's rated bonds were issued by about fifty governments to raise money in the American capital markets (Sinclair 1994). The depression brought on a radical decline in the extent of this internationalization, and it was not until very recently that Moody's was once again rating the bonds of about fifty governments. Indeed, as late as 1985, only fifteen foreign governments were borrowing in the U.S. capital markets. Not until after 1985 did the international financial markets reemerge as a major factor.[6]

But there are significant differences. One is the volume of short-term financial flows that has grown sharply and outstrips long-term flows. Fur-

[6] Switzerland's international banking was, of course, the exception. But this was a very specific type of banking and does not represent a global capital market, particularly given the fact of basically closed national financial systems at the time.

ther, this has brought with it the rise of types of financial institutions almost exclusively involved in such flows and hence highly speculative. More generally, there has been a growing concentration of market power in institutions, including more conservative ones such as pension funds and insurance companies.

Institutional investors are not new. What is different beginning in the 1980s is the diversity of types of funds, the rapid escalation of the value of their assets, and the sharp rise of extremely speculative institutions. There are two phases in this short history, one going into the early 1990s and the second one taking off in the later 1990s. Focusing briefly on the first phase and considering pension funds, for instance, their assets more than doubled in the United States from $1.5 trillion in 1985 to $3.3 trillion in 1992. Pension funds grew threefold in Britain and fourfold in Japan over that same period, and they more than doubled in Germany and Switzerland. In the United States, institutional investors as a group came to manage two-fifths of U.S. households' financial assets by the early 1990s, up from one-fifth in 1980. Another feature is that today the global capital market is increasingly a necessary component of a growing range of transactions, such as the diversity of government debts that now get financed through the global market: increasingly kinds of debt that were thought to be basically local, such as municipal debt, are entering this market. The overall growth in the value of financial instruments and assets also is evident with U.S. institutional investors whose assets had risen from 59 percent of GDP in 1980 to 126 percent by 1993.

TABLE 1
Financial Assets of Institutional Investors, 1990 to 2001, selected countries, (bn USD)

Country	*1993*	*1999*	*2001*
Canada	435.9	757.3	794.3
France	906.3	1691.1	1701.3
Germany	729.8	1529.0	1478.4
Japan	3610.7	4928.2	3644.8
Netherlands	465.2	799.3	722.3(2)
United Kingdom	1543.6	3321.3	2743.3
United States	9051.7	19274.0	19257.7
Percentage of OECD(1)	90.6%	87.2%	86.3%

Source: Based on OECD, Institutional Investor Statistical Yearbook, 2003, Table S.1, pp. 20

(1) Percentages based on author's calculation. Percentage indicates the proportion these seven countries represent.

(2) Netherlands figure for 2001 excludes non-life insurance.

As for the phase that began in the late 1990s, besides the growth of older types of institutional investors there is a proliferation of institutional investors with extremely speculative investment strategies. Hedge funds are among the most speculative of these institutions; they sidestep certain disclosure and leverage regulations by having a small private clientele and, frequently, by operating offshore. While they are not new, their size and their capacity to affect the functioning of markets certainly grew enormously in the 1990s, and they emerged as a major force by the late 1990s. According to some estimates, they numbered 1,200 with assets of over $150 billion by mid-1998 (BIS 2000), which was more than the $122 billion in assets of the total of almost 1,500 equity funds as of October 1997 (UNCTAD 1998). To put these figures in perspective, both of these types of funds need to be distinguished from asset management funds, of which the top ten were estimated to have $10 trillion under management in 2000.[7]

It is particularly in the world of short-term flows and speculative investors that digitization has had transformative consequences. Two sets of properties need to be emphasized here. One set—instantaneous transmission, interconnectivity, and speed—has transformed the character of financial transactions. A major consequence has been the sharp jump in the volume and the overall value of transactions. The other set of properties has to do with computerization, specifically, the possibility of computerizing mathematics. This has enabled the development of enormously complex financial instruments and, very importantly, their widespread use in that they could be packaged into reasonably simple-to-use software. One major consequence has been the increase in the industry's capacities to liquefy assets.

These two sets of properties have contributed to a third major difference, the explosion in and demand for financial innovations. Innovations are not new to finance, nor is the fact that an effect of innovations is to raise the supply of financial instruments that are tradeable—sold on the open market. The crucial difference between earlier phases and the contemporary phase is one of thresholds and the extent to which a change in thresholds can be interpreted as a qualitative transformation. The increased digitization of both transactions and instruments discussed above has enabled the work of producing innovations and the workability of a variety of new but also older innovations. While it is true that much of this innovation centers on derivatives and that the concept of the deriva-

[7] The level of concentration is enormous among these funds, partly as a consequence of mergers and acquisitions (M&As) driven by the need for firms to reach what are de facto the competitive thresholds in the global market today. (For more details, see Sassen 2001: chap. 7.)

tive is an old one, today we have seen a multiplication of types of derivatives and a sharp increase in the complexity of many of these types of derivatives.[8] This in turn has led to what we might describe as the growing importance of academic economics in financial instrument development (for a critical account, see MacKenzie and Millo 2003; Callon 1998). Digitization of transactions and instruments has been central to this multiplication of types of derivatives and their increased complexity. The overall result has been a massive increase in the extent to which the financial industry has been able to securitize various forms of what were previously considered untradeable assets or were simply not considered as assets, such as many forms of debt.[9] Mediated through these specifics of contemporary finance and financial markets, digitization can be seen as having contributed to a vast increase in the number of transactions, which in turn has translated into increased volumes and values.

At a macroinstitutional level, the proliferation of innovative derivatives has furthered the linking of national markets by producing specific types of incentives. For instance, various kinds of derivatives make it easier to exploit, or arbitrage, price differences among diverse financial instruments. One indicator is the growing importance of cross-border transactions measured in terms of their value as a percentage of GDP in the leading developed economies (table 2). For instance, the value of such transactions in the United States represented 4 percent of GDP in 1975, 35 percent in 1985 when the new financial era is in full swing, a quadrupling by 1995, and 230 percent in 1998. Other countries show even sharper increases. In Germany this share grew from 5 percent in 1975 to 334 percent in 1998; in France, it went from 5 percent in 1980 to 415 percent in 1998. In part, this entails escalating levels of risk and innovation driving the industry; indeed, it is only over the last decade and a half that we see this acceleration.

The drive to produce innovations is one of the marking features of the financial era that begins in the 1980s. The history of finance is in many ways a long history of innovations. But what is perhaps different today is the intensity of the current phase and the multiplication of instruments that lengthen the distance between the financial instrument and the actual underlying asset. This is reflected, for instance, in the fact that stock mar-

[8] While currency and interest-rate derivatives did not exist until the early 1980s and represent two of the major innovations of the current period, derivatives on commodities, so-called futures, have existed in some version in earlier periods. Famously, Amsterdam's stock exchange in the seventeenth century—when it was the financial capital of the world—was based almost entirely on trading in commodity futures.

[9] There are significant differences by country in the extent to which these innovations have been implemented. For instance and in general terms, securitization is well advanced in the United States, but just beginning in most of Europe.

TABLE 2
Cross-border Transactions in Bonds and Equities*, as % of GDP, 1975 to 2002 (selected years)

	as a percentage of GDP						
	1975	*1980*	*1985*	*1990*	*1995*	*2000*	*2002*
United States	4	9	35	89	135	229	292
Japan	2	8	62	119	65	96	106
Germany	5	7	33	57	172	447	464
France	N.A.	5	21	54	187	398	430
Italy	1	1	4	27	253	782	821**
Canada	3	9	27	65	187	241	311

Source: Bank for International Settlements, Annual Report 1999, 1 April 1998-31 March 1999, Table VI.5; IMF 2004, Table A.3

Note: * denotes gross purchases and sales of securities between residents and non-residents.

**Year for Italy is 2002

ket capitalization and securitized debt, before the financial crisis of 1997–98, in North America, the European Union, and Japan amounted to $46.6 trillion in 1997, while their aggregate GDP was $21.4 trillion and global GDP was $29 trillion. Further, the value of outstanding derivatives that same year in these same sets of countries stood at $68 trillion, which was about 146 percent of the size of their underlying capital markets. (For a full description of assumptions and measures, see IMF 1999: 47).

In The Digital Era: More Concentration than Dispersal?

A second major set of issues about the transformative capacities of digitization has to do with the limits of technologically driven change, or, in other words, with the point at which this global electronic market for capital runs into the walls of its embeddedness in nondigital conditions. There are two distinct issues here. One is the extent to which the global market for capital, even though global and digital, is actually embedded in multiple environments, some indeed global in scale but others subnational, that is, the actual financial centers within which the exchanges are located. A second issue is the extent to which it remains concentrated in a limited number of the most powerful financial centers, notwithstanding its character as a global electronic market.

In theory, the intensification of deregulation and the instituting of policies in various countries aimed at creating a supportive cross-border environment for financial transactions could have dramatically changed the locational logic of the industry. This is especially the case because it is a digitized and globalized industry that produces highly mobile outputs. It could be argued that the one major feature that could keep this industry from having locational constraints would be regulation. With deregulation, that constraint should be disappearing. Other factors, such as the premium paid for location in major cities, should be a deterrent to locate there, and with the new developments of telecommunications there should be no need for such central locations. Further, even accepting the notion that this market needs financial centers, given the costs of operating in major centers we might expect a shift of operations to lower-order financial centers given their lower prices compared to the major centers; thus, we would expect a shift from the leading to lesser centers.

Today, then, we might expect the actual spatial organization of the industry to be a much better indicator of its market-driven locational dynamics than was the case in earlier phases with more regulation and less digitization. We have seen considerable deregulation in the industry, the incorporation of a growing number of national financial centers into a global market, and the sharp increase in digitization of transactions and instruments. This would hold especially for the international level given the earlier prevalence of highly regulated and closed national markets.

There has, indeed, been geographic decentralization of certain types of financial activities, aimed at securing business in the growing number of countries becoming integrated into the global economy. Many of the leading investment banks have operations in more countries than they had twenty years ago. The same can be said for the leading sister industries, such as accounting, legal, and other specialized corporate services that now need to deliver a global service to their corporate clients; a good indicator of this is the explosive growth in these firms' networks of overseas affiliates (Taylor et al. 2002, see generally GAWC). And it can be said for some markets: for example, in the 1980s all basic wholesale foreign exchange operations were in London. Today these are distributed between London and several other centers (even though the number of centers is far smaller than the number of countries whose currency is being traded).

But empirically what stands out in the evidence about the global financial markets after a decade and a half of deregulation, worldwide integration, and major advances in electronic trading is the extent of locational concentration and that firms are willing to pay a premium to be in major financial centers. Large shares of many financial markets are disproportionately concentrated in a few financial centers. This trend toward consolidation in a few centers also is evident within countries. Further, this pattern toward the consolidation of one leading financial center per country is a function of rapid growth in the sector, not of decay in the losing cities.

The sharp concentration in leading financial markets can be illustrated with a few facts.[10] London, New York, Tokyo (notwithstanding a national economic recession), Paris, Frankfurt, and a few other cities regularly appear at the top and represent a large share of global transactions. This holds even after the September 11 attacks in New York that destroyed the World Trade Center (albeit that it was not largely a financial complex) and damaged over fifty surrounding buildings home to much financial activity. Many saw the level of damage as a wake-up call about the vulnerabilities of sharp spatial centralization in a limited number of sites.[11] The fact that the global capital market is a global digital market does not seem to reduce the need for being present in the actual centers where the exchanges are located. London, Tokyo, New York, Paris (now consolidated with Amsterdam and Brussels as EuroNext), Hong Kong, and Frankfurt account for a major share of worldwide stock market capitalization. London, Frankfurt, and New York account for an enormous world share in the export of financial services. London, New York, and Tokyo account for over one-third of global institutional equity holdings, this as of the end of 1997 after a 32 percent decline in Tokyo's value over 1996. London, New York, and Tokyo account for 58 percent of the foreign exchange market, one of the few truly global markets; together with Singapore, Hong Kong, Zurich, Geneva, Frankfurt, and Paris, they ac-

[10] Among the main sources of data for the figures cited in this section are the International Bank for Settlements (Basle); IMF national accounts data; specialized trade publications such as *Wall Street Journal*'s *WorldScope*; *MorganStanley Capital International*; *The Banker;* data listings in the *Financial Times* and *The Economist;* and, especially for a focus on cities, the data produced by Technimetrics, Inc. (now part of Thomson Financials). Additional names of standard, continuously updated sources are listed in Sassen (2001).

[11] The case of New York after September 2001 requires clarification. The destruction of a considerable amount of the office space of several financial firms in addition to the destruction of communications infrastructure forced many firms to either fully or partly move out of lower Manhattan. Some of these firms will not return; some have already returned to either lower or mid-Manhattan. Most are likely to keep their strategic operations centered in Manhattan. But there is now a broader geography to the Manhattan financial sector than was the case before September 2001: it includes growing concentrations of at least partial components of firms in specific areas of New Jersey and Connecticut. In my interpretation there are two issues to factor in. One of these is that the destruction of office space can be seen as a brutal elimination of inertia in the financial sector, where many of these firms have grown enormously and have kept huge workforces—of ten thousand employees in several cases—when only a fraction of these need to be located in a major financial center. The second issue is that, given digitization, "spatial centrality" can be constituted through diverse actual geographies (Sassen 2001: chap. 5). The geography of the downtown business center is but one of these. A second type of geography of "centrality" is that of the larger metropolitan area where a variety of dense business nodes are connected via state-of-the-art conventional infrastructure and digital networks. A third is the network of global cities, constituted through the multiple digital and other transactions among various firms in these cities. In the case of the Manhattan financial industry, all three geographies of centrality were present throughout the 1980s and 1990s; September 11 strengthened the second type.

TABLE 3
The Twelve Biggest Stock Markets in the World by Market Capitalization, 2003 and 2000

Stock Market	*2003 Market Capitalization*	*2003 Percentage of Members Capitalization*	*2000 Market Capitalization*	*2000 Percentage of Members Capitalization*
NYSE	11,329.0	36.3%	11,534.5	37.1%
Nasdaq	2,844.2	9.1%	3,597.1	11.6%
Tokyo	2,953.1	9.5%	3,193.9	10.3%
London	2,460.1	7.9%	2,612.2	8.4%
Euronext	2,076.4	6.7%	2,271.7	7.3%
Osaka	1,951.5	6.3%	(1)	(2)
Deutsche Borse	1,079.0	3.5%	1,270.2	4.1%
Toronto	888.7	2.8%	766.2	2.5%
Spanish Exchanges	726.2	2.3%	(1)	(2)
Swiss Exchange	727.1	2.3%	792.3	2.5%
Hong Kong	714.6	2.3%	623.4	2.0%
Italy	614.8	2.0%	768.4	2.5%
Percentage of Total Capitalization for Top 12		90.9%		90.8% (2)

Compiled from World Federation of Exchanges Annual Statistics for 2001 (pp. 92) and 2003 (pp. 82), year end figures with calculations of percentages added

(1) The top 12 for 2000 did not include Osaka or the Spanish Exchanges (BME). Instead the Spain, consisting of Madrid, was 11th with market capitalization of $504.2 billion and Australia was 12th with market capitalization of $372.8 billion.

(2) This figure indicates the percentage represented by the top 12 exchanges in terms of market capitalization for 2000 (including the exchanges in Spain, 1.6%, and Australia, 1.2%)

count for 85 percent in this, the most global of markets. These high levels of concentration do not preclude considerable activity in a large number of other markets, even though the latter may account for a small global share.

This trend toward consolidation in a few centers, even as the network of integrated financial centers expands globally, also is evident within countries. In the United States, for instance, the leading investment banks are concentrated in New York, with only one other major international financial center, Chicago, in this enormous country. Sydney and Toronto have gained power in continent-sized countries and have taken over functions and market share from what were once the major commercial centers—Melbourne and Montreal, respectively. So have Sao Paulo and Bombay, which have gained share and functions from, respectively, Rio

TABLE 4
Foreign Listings in Major Stock Exchanges, 2003 and 2000

Exchange	*2003 Number of Foreign Listings*	*2003 Percentage of Foreign Listings*	*2000 Number of Foreign Listings*	*2000 Percentage of Foreign Listings*
Nasdaq	343	10.4%	488	10.3%
NYSE	466	20.2%	433	17.5%
London	381	14.2%	448	18.9%
Deutsche Borse	182	21.0%	241	24.5%
Euronext*	346	24.9%	—	—
Swiss Exchange	130	31.0%	164	39.4%
Tokyo	32	1.5%	41	2.0%

Compiled from WFE Annual Statistics 2001, p. 86; 2003 p. 83 (Calculations of percentages added).
*Euronext includes Brussels, Amsterdam, and Paris
All year end figures

de Janeiro in Brazil and New Delhi and Calcutta in India. One might have thought that such huge countries could sustain multiple major financial centers, but even though many of the secondary centers may be thriving, the point is that the leading centers have gained national share. This pattern is evident in many countries, including the leading economies of the world.[12] Again, consolidation of one leading financial center in each country is an integral part of the growth dynamics in the sector rather than only the result of losses in the losing cities.

There is both consolidation in fewer major centers across and within countries and a sharp growth in the number of centers that become part of the global network as countries deregulate their economies and the global economy expands accordingly. Bombay, for instance became incorporated in the global financial network in the early 1990s after India (partly) deregulated its financial system. This mode of incorporation into the global network is often at the cost of losing functions that these cities may have had when they were largely national centers. Today the leading, typically foreign, financial, accounting, and legal services firms enter their markets to handle many of the new cross-border operations. Incor-

[12] In France, Paris today concentrates larger shares of most financial sectors than it did ten years ago, and once important stock markets like Lyon have become "provincial," even though Lyon is today the hub of a thriving economic region. Milan privatized its exchange in September 1997 and electronically merged Italy's ten regional markets. Frankfurt now concentrates a larger share of the financial market in Germany than it did in the early 1980s, and so does Zurich in Switzerland, which once had Basel and Geneva as significant competitors.

poration in the global market typically happens without these cities showing a gain in their global share of the particular segments of the market they are in, even as capitalization in their stock market may increase, often sharply, and even though they add to the total volume in the global market.

Why is it that at a time of rapid growth in the network of financial centers, in overall volumes, and in electronic networks, there is such high concentration of market shares in the leading global and, within countries, in the leading national centers. It should not be this way since both globalization and electronic trading are about expansion and dispersal beyond what had been the confined realm of national economies and floor trading. Indeed, one might well ask why financial centers matter at all.

The Continuing Utility of Spatial Agglomeration

The continuing weight of major centers is, in a way, countersensical, as is, for that matter, the existence of an expanding network of financial centers. The rapid development of electronic exchanges, the growing digitization of much financial activity, the fact that finance has become one of the most deregulated sectors in a growing number of countries, and that it produces a digital, hypermobile product all suggest that location should not matter. In fact, geographic dispersal would seem to be a good option given the high cost of operating in major financial centers. Further, the last ten years have seen an increased geographic mobility of financial experts and financial services firms.

There are, in my view, at least three reasons that explain the trend toward consolidation in a limited number of centers rather than massive dispersal—that is, a lumpy geography that reflects the social logics of finance. (For a more detailed examination of the technical reasons, including risk-management see Sassen 2005: chap. 7.)

THE IMPORTANCE OF SOCIAL CONNECTIVITY AND CENTRAL FUNCTIONS

First, while the new communication technologies do indeed facilitate geographic dispersal of economic activities without losing system integration, they have also strengthened the importance of central coordination and control functions for firms and even markets.[13] Indeed, for firms in any sector, operating a widely dispersed network of branches and affiliates and doing business in multiple markets has made central functions far more complicated. Their execution requires access to top talent, to innovative milieux—in technology, accounting, legal services, economic

[13] This is one of the seven organizing hypotheses through which I specified my global city model. For a full explanation see Sassen (2001), especially the preface to the new edition.

forecasting, and all sorts of other specialized, often new, corporate services. Major centers have massive concentrations of these and other state-of-the-art resources that allow firms to maximize the benefits of the new communication technologies and collectively to manage the new conditions for operating globally. Even electronic markets such as NASDAQ and E*Trade rely on traders and banks that are located somewhere, with at least some in a major financial center. The question of risk and how it is perceived and handled is yet another factor that has an impact on how the industry organizes itself, where it locates operations, what markets become integrated into the global capital market, and so on. The outcome has been a lumpy geography that combines seamless electronic networks and thick financial centers.

It is increasingly evident that to maximize the benefits of the new information technologies, firms need not only infrastructure but also a complex mix of other resources. In my analysis, organizational complexity is a key variable allowing firms to maximize the utility and benefits they can derive from using digital technology (Sassen 2001: 115–16). In the case of financial markets, we could make a parallel argument. Most of the value added that these technologies can produce for advanced service firms runs through so-called externalities, that is, material and human resources—state-of-the-art office buildings, top talent, and the social networking infrastructure that maximizes connectivity. Fiber optic cables are not enough (Garcia 2002).

A second fact that is emerging with greater clarity concerns the meaning of "information." There are two types of information (Sassen 2001: chap. 5). One is the datum, which may be complex yet is standard knowledge: the level at which a stock market closes, a privatization of a public utility, the bankruptcy of a bank. But there is a far more difficult type of "information," akin to an interpretation, evaluation, or judgment. It entails negotiating a series of standardized datums and a series of interpretations of a mix of datums in the hope of producing a higher-order datum. Access to the first kind of information is now global and immediate from just about any place in the highly developed world, thanks to the digital revolution. But it is the second type of information, which requires a complicated mixture of elements—the social infrastructure for global connectivity—that gives major financial centers a leading edge.

It is possible, in principle, to reproduce the technical infrastructure anywhere. Singapore, for example, has technical connectivity matching Hong Kong's. But does it have Hong Kong's social connectivity? At a higher level of global social connectivity, we could probably say the same for Frankfurt and London. When the more complex forms of information needed to execute major international deals cannot be retrieved from existing databases, no matter what one can pay, then one needs the social information loop and the associated de facto interpretations and inferences that come with interaction among talented, informed, and experienced people. It is the weight of this input that has given a whole new importance to credit rating agencies, for instance. Part of the rating has to do with interpreting and inferring. When this interpreting becomes "authoritative," it becomes "information" available to all. The process of making inferences and interpretations into "information" takes quite a mix of talents and resources.

In brief, financial centers provide the social connectivity that allows a firm or market to maximize the benefits of its technical connectivity.

ALLIANCES AMONG CENTERS AS PART OF THE ORGANIZATIONAL INFRASTRUCTURE OF ELECTRONIC MARKETS

Besides the familiar mergers and acquisitions of firms,[14] I would argue that an important trend in the global capital market is the "merger" of electronic exchanges that connect select groups of centers. A number of networks connecting markets have been set up in the last few years. NASDAQ, the second largest U.S. stock market after the New York Stock Exchange (NYSE), set up NASDAQ Japan in 1999 and NASDAQ Canada in 2000. This gives investors in Japan and Canada direct access to the market in the United States. Europe's more than thirty stock exchanges have been seeking to shape various alliances. Euronext (NEXT) is Europe's largest stock exchange merger, an alliance among the Paris, Amsterdam, and Brussels bourses. The Toronto Stock Exchange allied with the NYSE to create a separate global trading platform. The NYSE is a founding member of a global trading alliance, Global Equity Market (GEM), which includes ten exchanges, among them Tokyo and NEXT. Small exchanges are also merging: in March 2001 the Tallinn Stock Exchange in Estonia and its Helsinki counterpart created an alliance. A novel pattern is hostile takeovers, not of firms, but of exchanges, such as the attempt by the owners of the Stockholm Stock Exchange to buy the London Stock Exchange (for a price of U.S. $3.7 billion).

These developments may well ensure the consolidation of a stratum of select financial centers at the top of the worldwide network of thirty to forty global cities through which the global financial industry operates.[15]

[14] Global firms and markets in the financial industry need enormous resources, a trend that is leading to rapid mergers and acquisitions of firms and strategic alliances among markets in different countries. These are happening on a scale and in combinations few would have foreseen as recently as the early 1990s. There are growing numbers of mergers among financial services firms, accounting firms, law firms, and insurance brokers—in brief, firms that need to provide a global service. A similar evolution is also possible for the global telecommunications industry, which will have to consolidate in order to offer a state-of-the-art, globe-spanning service to its global clients, among which are the financial firms.

[15] We now also know that a major financial center needs to have a significant share of

An indicator such as equities under management shows a similar pattern of spread and simultaneous concentration at the top of the hierarchy. The worldwide distribution of equities under institutional management is spread among a large number of cities that have become integrated in the global equity market along with deregulation of their economies and the whole notion of "emerging markets" as an attractive investment destination. Thomson Financials (1999) has estimated that at the end of 1999 (latest available data), twenty-five cities accounted for about 80 percent of the US$14 trillion controlled by institutional money managers around the world. These twenty-five cities also accounted for roughly 48 percent of the total stock market capitalization of the world, which stood at $24 trillion at the end of 1999. On the other hand, this global market is characterized by a disproportionate concentration in the top six or seven cities. London, New York, and Tokyo together accounted for a third of the world's total equities under institutional management in 1999.

These developments make clear a second important trend that in many ways characterizes the current global era. These various centers do not just compete with each other: there is collaboration and division of labor. In the international system of the postwar decades, each country's financial center, in principle, covered the universe of necessary functions to service its national companies and markets. The world of finance was, of course, much simpler than it is today. In the initial stages of deregulation in the 1980s, there was a strong tendency to see the relation among the major centers as one of straight competition when it came to international transactions. New York, London, and Tokyo, then the major centers in the system, were seen as competing. But in my research in the late 1980s on these three top centers, I found clear evidence that a division of labor already existed then. They remain the major centers in the system today, with the addition of Frankfurt and Paris in the 1990s, and there is a fairly specialized division of functions and advantages among them. What we are seeing now is an additional pattern whereby the cooperation or division of functions is somewhat institutionalized: strategic alliances exist not only between firms across borders but also among exchanges. There is competition, strategic collaboration, and hierarchy. Together these trends indicate the emergence of global formations where before there were interactions among national centers, but global formations partly embedded in networks of financial centers.

---

global operations to be such. If Tokyo does not succeed in getting more of such operations, it is going to lose standing in the global hierarchy, notwithstanding its importance as a capital exporter. It is this same capacity for global operations that will keep New York at the top levels of the hierarchy even though it is largely fed by the resources and the demand of domestic (though state-of-the-art) investors.

TOWARD DENATIONALIZED FINANCIAL CENTERS

It is important to recognize that national financial centers have themselves been transformed by these developments. National attachments and identities are becoming weaker for global firms and their customers. This is particularly strong in the West but may develop in Asia as well. Deregulation and privatization have reduced the need for *national* centers. The nationality question does not disappear (e.g., Salzinger 2003; Corbridge, Thrift, and Martin 1994), but it plays differently in these sectors from the way it did even a decade ago. Global financial products are accessible in national markets, and national investors can operate in global markets. For instance, some of the major Brazilian firms now list on the New York Stock Exchange and bypass the Sao Paulo exchange, a new practice that has caused somewhat of an uproar in specialized circles in Brazil. While it is as yet inconceivable in the Asian case, this may well change given the growing number of foreign acquisitions of major firms in several countries discussed earlier (see note 16). Another indicator of this trend is the fact that the major U.S. and European investment banks have set up specialized offices in London to handle various aspects of their global business making London probably the most denationalized of the major financial centers. Even French banks have set up some of their global specialized operations in London, inconceivable a decade ago and still not avowed in national rhetoric.

One way of describing this process is as an incipient and highly specialized denationalization of particular institutional arenas (Sassen 1996: chap. 1; Sassen 2005). It can be argued that such denationalization is a necessary condition for economic globalization as we know it today. The sophistication of this system lies in the fact that it needs to involve only strategic institutional areas—most national systems can be left basically unaltered. China is a good example. It adopted international accounting rules in 1993, necessary to engage in international transactions. To do so it did not have to change much of its domestic economy. Japanese firms operating overseas adopted such standards long before Japan's government considered requiring them. In this regard, the "wholesale" side of globalization is quite different from the global consumer markets, in which success necessitates altering national tastes at a mass level. This process of denationalization has been strengthened by state policy enabling privatization and foreign acquisition. The Asian financial crisis has functioned as a mechanism to denationalize, at least partly, control over key sectors of economies that, while allowing the massive entry of foreign investment over the last two decades, never relinquished that control.[16]

[16] For instance, Lehman Brothers bought Thai residential mortgages worth half a billion dollars for a 53 percent discount. This was the first auction conducted by the Thai govern-

Major international business centers produce what we could think of as a new subculture, a move from the "national" version of international activities to the "denationalized" version. The longstanding resistance in Europe to M&As, especially hostile takeovers, and to foreign ownership and control in East Asia signals national business cultures that are somewhat incompatible with the new global economic system. I would posit that major cities, and the variety of so-called global business meetings (such as those of the World Economic Forum in Davos), contribute to the denationalizing of corporate elites. Whether this is good or bad is a separate issue, but it is, I would argue, one of the conditions for setting in place the systems and subcultures necessary for a global economic system, especially in global finance. It is, then, a denationalized lumpy geography.

Politics of Places on Global Circuits: The Local as Multiscalar

The issue I want to highlight here concerns the ways in which particular instantiations of the local can actually be constituted at multiple scales and thereby construct global formations that tend toward lateralized and horizontal networks. I examine this through a focus on various political practices and the technologies used. Of particular interest is the possibility that local, often resource-poor organizations and individuals can become part of, and constitute global networks and struggles. These practices are constituting a specific type of global politics, one that runs through localities and is not predicated on the existence of global institutions. The engagement can be with global institutions, such as the International Monetary Fund or World Trade Organization, or with local institutions, such as a particular government or local police force charged with human rights abuses. Theoretically these types of global politics illuminate the distinction between a global network and the actual transactions that constitute it: the global character of a network does not necessarily imply that its transactions are equally global, or that it all has to happen at the global level. It shows the local to be multiscalar in a parallel to the preceding section, which showed the global to be multiscalar—

ment's Financial Restructuring Authority, which is managing the sale of $21b of financial companies' assets. Lehman Brothers also acquired the Thai operations of Peregrine, the Hong Kong investment bank that failed. The fall in prices and in the value of the yen has made Japanese firms and real estate attractive targets for foreign investors. Merrill Lynch bought thirty branches of Yamaichi Securities; Societé Generale Group is buying 80 percent of Yamaichi International Capital Management; Travelers Group is now the biggest shareholder of Nikko, the third largest brokerage; and Toho Mutual Insurance Co. announced a joint venture with GE Capital. These are but some of the best-known examples. Much valuable property in the Ginza—Tokyo's high-priced shopping and business district—is now being considered for acquisition by foreign investors, in a twist on Mitsubishi's acquisition of New York's Rockefeller Center a decade earlier.

that is, partly embedded in a network of localities, specifically, financial centers.

Computer-centered technologies have also made all the difference here; in this case, the particular form of these technologies is mostly the public-access Internet.[17] The latter matters not only because of low-cost connectivity and the possibility of effective use (via e-mail) even with low bandwidth availability, but also, and most importantly, because of some of its key features. Simultaneous decentralized access can help local actors have a sense of participation in struggles that are not necessarily global but are, rather, globally distributed in that they recur in multiple localities. In so doing, these technologies can also help in the formation of cross-border public spheres for these types of actors, and can do so without the necessity of running through global institutions,[18] and without forms of recognition that depend on much direct interaction and joint action on the ground. Among the implications of these options are the feasibility of forming global networks that bypass central authority and, especially significant for resource-poor organizations, the possibility that those who may never be able to travel can nonetheless be part of global struggles and global publics.

Such forms of recognition are not new. Yet there are two specific matters that signal the need for empirical and theoretical work on their ICT-enabled form. One is that much of the conceptualization of the local in the social sciences has assumed physical/geographic proximity and thereby a sharply defined territorial boundedness, with the associated implication of closure. The other, partly a consequence of the first, is a strong tendency to conceive of the local as part of a hierarchy of nested scales, especially once there are national states. To a very large extent, these conceptualizations continue to hold for most of the instantiations of the local

[17] While the Internet is a crucial medium in these political practices, it is important to emphasize that beginning in the 1990s, and particularly since the mid-1990s, we have entered a new phase in the history of digital networks, one in which powerful corporate actors and high-performance networks are strengthening the role of private digital space and altering the structure of public-access digital space (Sassen 2002). Digital space has emerged not simply as a means for communicating, but as a major new theater for capital accumulation and the operations of global capital. Yet civil society—in all its various incarnations—is also an increasingly energetic presence in electronic space. (For a variety of angles, see, e.g., Rimmer and Morris-Suzuki 1999; Poster 1997; Frederick 1993; Miller and Slater 2000; Laguerre 2005). The greater the diversity of cultures and groups, the better for this larger political and civic potential of the Internet, and the more effective the resistance to the risk that the corporate world might set the standards. (For cases of ICT use by different types of groups, see, e.g., APCWNSP 2000; Allison 2002; WomenAction 2000; Yang 2003; Camacho 2001; Esterhuysen 2000).

[18] For instance, in centuries past organized religions had extensive, often global, networks of missionaries and clerics. But these partly depended on the existence of a central authority.

today—more specifically, for most of the actual practices and formations likely to constitute the local in most of the world. But there are also conditions today that contribute to destabilizing these practices and formations and hence invite a reconceptualization of the local that can accommodate a set of instances that diverge from dominant patterns. Key among these current conditions are globalization and/or globality as constitutive not only of cross-border institutional spaces but also of powerful imaginaries enabling aspirations to transboundary political practice even when the actors involved are basically localized.

Computer-centered interactive technologies have played an important role, precisely in the context of globalization, including global imaginaries. These technologies facilitate multiscalar transactions and simultaneous interconnectivity among those largely confined to a locality. They can be used to further develop old strategies (e.g., Tsaliki 2002; Lannon 2002) and to develop new ways of organizing, notably electronic activism (Denning 1999; Smith 2001; Yang 2003). Internet media are the main type of ICT used. E-mail is perhaps the most widely used, partly because organizations in the global South often have little bandwidth and slow connections, making visual and audio intensive (e.g., the WWW) software a far less usable and effective option. To achieve the forms of globality that concern me in this chapter, it is important that there be a recognition of these constraints among major transnational organizations dealing with the global South: for instance, this means making text-only databases, with no visuals or HTML, no spreadsheets, and none of the other facilities that demand considerable bandwidth and fast connections (e.g., Pace and Panganiban 2002: 113).[19]

As has been widely recognized by now, new ICTs do not simply replace existing media techniques. The evidence is far from systematic, and the object of study is continuously undergoing change. But we can basically identify two patterns. On the one hand, it might mean no genuine need for these particular technologies given the nature of the organizing, or it

[19] There are several organizations that have taken on the work of adjusting to these constraints or providing adequate software and other facilities to disadvantaged NGOs. For instance, Bellanet (2002), a nonprofit set up in 1995, aims to help such NGOs gain access to online information and to disseminate information to the South. To that end it has set up web-to-e-mail servers that can deliver web pages by e-mail to users confined to low bandwidth. It has developed multiple service lines. For example, Bellanet's Open Development service line seeks to enable collaboration among NGOs through the use of open source software, open content, and open standards, so it customized the Open Source PhP-Nuke software to set up an online collaborative space for the Medicinal Plants Network. Bellanet has adopted open content for all forms of contents on its web site, freely available to the public, and supports the development of an open standard for project information (International Development Markup Language or IDML). The value of such open standards is that they enable information sharing.

might come down to underutilization. (For studies of particular organizations, see, e.g., Tsaliki 2002; Lannon 2002).[20] For instance, a survey of local and grass-roots human rights NGOs in several regions of the world found that the Internet makes exchange of information easier and is helpful in developing other kinds of collaboration but does not help launch joint projects (Lannon 2002: 33). On the other hand, there is evidence of highly creative ways of using the new ICTs along with older media, recognizing the needs of particular communities. A good example is using the Internet to send audio files that can then be broadcast over loudspeakers to groups who lack access to the Internet or are illiterate. The M. S. Swamintham Research Foundation in southern India has supported this type of strategy by setting up Village Knowledge Centers catering to populations that, although mostly illiterate, know exactly what types of information they need or want. When we consider mixed uses, it becomes clear that the Internet can often fulfill highly creative functions by being used with other technologies, whether old or new. Thus Amnesty International's International Secretariat has set up an infrastructure to collect electronic news feeds via satellite, which it then processes and redistributes to its staff workstations.

But there is also evidence that use of these technologies has led to the formation of new types of organizations and activism. For instance, Yang (2003) found that what were originally exclusively online discussions among groups and individuals in China concerned with the environment evolved into active NGOs. Further, one result of this genesis is that their membership is national, distributed among different parts of the country. The variety of online hacktivisms examined by Denning (1999) involve largely new types of activisms. To mention what is perhaps one of the most widely known cases of how the Internet made a strategic difference, the Zapatista movement became two organizational efforts, one a local rebellion in Mexico, the other a transnational civil society movement. The latter saw the participation of multiple NGOs concerned with peace, trade, human rights, and other social justice struggles. It functioned through both the Internet and conventional media (Cleaver 1998; Arquilla and Ronfeldt 2001), putting pressure on the Mexican government. Importantly, it shaped a new concept for civil organizing: multiple rhizomatically connected autonomous groups (Cleaver 1998).

But what is far less known is that the local rebellion of the Zapatistas operated basically without e-mail infrastructure (Cleaver 1998). Comandante Marcos was not on e-mail, let alone able to join collaborative workspaces on the web. Messages had to be hand-carried, crossing military

[20] In a study of the web sites of international and national environmental NGOs in Finland, Britain, Netherlands, Spain, and Greece, Tsaliki (2002: 15) concludes that the Internet is mainly useful for intra- and interorganizational collaboration and networking, mostly complementing already existing media techniques for issue promotion and awareness raising.

lines in order to bring them to others for uploading to the Internet; further, the solidarity networks themselves did not all have e-mail, and local communities sympathetic to the struggle often had problems with access (Mills 2002: 83). Yet Internet-based media did contribute enormously, in good part because of preexisting social networks (see also Garcia 2002). Among the electronic networks involved, LaNeta played a crucial role in globalizing the struggle. LaNeta is a civil society network established with support of a San Francisco–based NGO, the Institute for Global Communication (IGC). In 1993 LaNeta became a member of APC and began to function as a key connection between civil society organizations in and outside Mexico. In this regard, it is interesting to note that a local movement made LaNeta into a transnational information hub.

There is little doubt that the gathering, storage, and dissemination of information are crucial functions for these kinds of organizations (Meyer 1997; Tuijl and Jordan 1999). Human rights, large development, and environmental organizations are at this point the leaders in the effort to build online databases and archives (see, for example, Human Rights Internet at www.hri.ca; Greenpeace's web site; and Oxfam's web site). Oxfam has also set up knowledge centers on its web site—specialized collections around particular issues, such as the Land Rights in Africa site and its related resource bank (Warkentin 2001: 136). Specialized campaigns, such as those against the WTO, for the banning of landmines, or for canceling the debt of hyperindebted countries (the Jubilee 2000 campaign), have also been effective at this type of work since it is crucial for their efforts. Special software can be designed to address the specific needs of organizations or campaigns. For example, the HR Information and Documentation Systems International (HURIDOCS), a transnational network of human rights organizations, aims at improving access to, dissemination of, and use of human rights information. It runs a program to develop tools, standards, and techniques for documenting violations.

The evidence on NGO use of Internet media also shows the importance of institutional mechanisms and the use of appropriate software. Amnesty International has set up an institutional mechanism to help victims of human rights abuses use the Internet to contact transnational organizations for help: its Urgent Action Alert is a world wide e-mail alerting system with seventy-five networks of letter-writing members who respond to urgent cases by immediate mailings to key and pertinent entities.[21]

[21] Another, very different case is Oxfam America's effort to help its staff in the global South manage and electronically publish information quickly and effectively, no easy aims in countries with unreliable, slow connections and other obstacles to working online. To that end Oxfam adopted a server-side Content Management System and a client-side ArticleBuilder, called Publ-X, that allows end users to create or edit local XML articles while offline and submit them to the server when work has been completed. An editor on the server side is then promptly notified, ensuring that the information immediately becomes public.

All of this facilitates a new type of cross-border politics, one centered in multiple localities yet intensely connected digitally. Adams (1996), among others, shows us how telecommunications create new linkages across space that underline the importance of networks of relations and partly bypass older hierarchies of scale. Activists can develop networks for circulating place-based information (about local environmental, housing, political conditions) that can become part of political work and strategies addressing a global condition—the environment, growing poverty and unemployment worldwide, lack of accountability among multinationals, and so forth. The issue here is not so much the possibility of such political practices: they have long existed in other mediums and with other velocities. The issue is rather one of orders of magnitude, scope, and simultaneity: the technologies, institutions, and imaginaries that mark the current global digital context inscribe local political practice with new global meanings and new potentialities.[22]

There are many examples that illustrate the new possibilities and potentials for action. Besides some of the cases discussed above, there is the vastly expanded repertory of actions that can be taken when electronic activism is also an option. The "New Tactics in Human Rights Project" of the Center for Victims of Torture has compiled a workbook with 120 antitorture tactics, including exclusively online forms of action (www.cvt.org/new_tactic/tools/index.html). The web site of the New York–based Electronic Disturbance Theater, a group of cyberactivists and artists, contains detailed information about electronic repertories for action (www.thing.net/-rdom/ecd/EDTECD.html). The International Campaign to Ban Landmines, officially launched in 1992 by six NGOs from the United States, France, Britain, and Germany, evolved into a coalition of over one thousand NGOs in 60 countries. It succeeded when 130 countries signed the Landmines Ban Treaty in 1997 (Williams and Goose 1998). The campaign used both traditional techniques and ICTs. Internet-based media provided mass distribution better and cheaper than telephone and fax (Scott 2001; Rutherford 2002). Jubilee 2000 used the In-

[22] Elsewhere (2002) I have posited that we can conceptualize these "alternative" networks as countergeographies of globalization because they are deeply implicated with some of the major dynamics and capabilities constitutive of, especially, economic globalization yet are not part of the formal apparatus or the objectives of this apparatus, such as the formation of *global* markets. The existence of a global economic system and its associated institutional supports for cross-border flows of money, information, and people has enabled the intensification of transnational and translocal networks and the development of communication technologies that can escape conventional surveillance practices. (For one of the most critical and knowledgeable accounts, see, e.g., World Information Order 2002; Nettime 1997). These countergeographies are dynamic and changing in their locational features. And they include a broad range of activities, including a proliferation of criminal activities.

ternet to great effect. Its web site brought together all the information on debt and campaign work considered necessary for the effort, and information was distributed via majordomo listserve, database, and e-mail address books.[23] Generally speaking preexisting online communication networks are important for these types of actions and for e-mail alerts aiming at quick mobilization. Distributed access is crucial: once an alert enters the network from no matter what point of access, it spreads very fast through the whole network. Amnesty's Urgent Action Alert described above is such a system. However, anonymous web sites are definitely part of such communication networks: this was the case with S.11.org, a website that can be used for worldwide mobilizations insofar as it is part of multiple online communication networks. The Melbourne mobilization against the regional Asian meeting of the World Economic Forum (WEF) (Sept. 11–13, 2000) brought activist groups from around Australia together on this site to coordinate their actions, succeeding in paralyzing a good part of the gathering, a first in the history of the WEF meetings (Redden 2001). There are by now several much studied mobilizations that were organized online, such as those against the WTO in Seattle in 1999 and against Nike.[24]

An important feature of this type of multiscalar politics of the local is that it is not confined to moving through a set of nested scales from the local to the national to the international but can directly access other such local actors, whether in the same country or across borders. One Internet-based technology that reflects this possibility of escaping nested

[23] But, it must be noted, even in this campaign, centered as it was on the global South and determined as it was to communicate with global South organizations, the latter were often unable to access the sites (Kuntze, Rottmann, and Symons 2002).

[24] There are many other, somewhat less-known campaigns. For instance, when Intel announced that it would include a unique personal serial number in its new PentiumIII processing chips, privacy advocacy groups objected to this invasion of privacy. Three groups in different locations set up a joint web site called Big Brother Inside to provide an organizational space for advocacy groups operating in two different countries, thereby also enabling them to use the place-specific resources of the different localities (Leizerov 2000). The Washington, DC, based group Public Citizen put an early draft of the MAI agreement (a confidential document being negotiated by the OECD behind closed doors) on its web site in 1997, launching a global campaign that brought these negotiations to a halt about eight months later. And these campaigns do not always directly engage questions of power. For instance, Reclaim the Streets started in London as a way to contest the Criminal Justice Act in England that granted the police broad powers to seize sound equipment and otherwise discipline ravers. One tactic was to hold street parties in cities across the world: through Internet media, participants could exchange notes, tactics on how to deal with the police, and create a virtual space for coming together. Finally, perhaps one of the most significant developments is Indymedia, a broad global network of ICT-based alternative media groups located around the world. Other such alternative media groups are MediaChannel.org, Zmag.org, Protest.net, and McSpotlight.org.

hierarchies of scale is the online workspace, often used for Internet-based collaboration. Such a space can constitute a community of practice (Sharp 1997) or knowledge network (Creech and Willard 2001). An example of an online workspace is the Sustainable Development Communications Network, also described as a knowledge space (Kuntze, Rottmann, and Symons 2002), set up by a group of civil society organizations in 1998; it is a virtual, open, and collaborative organization aiming at joint communications activities to inform broader audiences about sustainable development and build members' capacities to use ICT effectively. It has a trilingual Sustainable Development Gateway to integrate and showcase members' communication efforts. The network contains links to thousands of member-contributed documents, a job bank, and mailing lists on sustainable development. It is one of several NGOs whose aim is to promote civil society collaboration through ICTs; others are the Association for Progressive Communications (APC), One World International, and Bellanet.

At the same time, this possibility of exiting or avoiding hierarchies of scale does not preclude the fact that powerful actors can use the existence of different jurisdictional scales to their advantage (Morrill 1999) and the fact that local resistance is constrained by how the state deploys scaling through jurisdictional, administrative, and regulatory orders (Judd 1998). On the contrary, it might well be that the conditions analyzed by Morrill and Judd, among others, force the issue, so to speak. Why work through the power relations shaped into state-centered hierarchies of scale? Why not jump ship if this is an option? This combination of conditions and options is well illustrated by research showing how the power of the national government can subvert the legal claims of first nation-people (Howitt 1998; Silvern 1999), which has in turn led the latter increasingly to seek direct representation in international fora, bypassing the national state (Sassen 1996: chap. 3).[25] In this sense, then, my effort here is to recover a particular type of multiscalar context, one characterized by direct local-global transactions or by a multiplication of local transactions as part of global networks. Neither type is marked by nested scalings.

There are many examples of such types of cross-border political work. We can distinguish two forms of it, each capturing a specific type of scalar interaction. In one the scale of struggle remains the locality, and the object is to engage local actors—for example, a local housing or environmental agency—but with the knowledge and explicit or tacit invocation of multiple other localities around the world engaged in similar localized struggles with similar local actors. It is this combination of multiplication

[25] Though with other objectives in mind, a similar mix of conditions can also partly explain the growth of transnational economic and political support networks among immigrants (e.g., Smith 1994; Smith 1997; Cordero-Guzmán et al. 2001).

and self-reflexivity that contributes to constitute a global condition out of these localized practices and rhetorics. It means, in a sense, taking Cox's notion of scaled "spaces of engagement" constitutive of local politics and situating it in a specific type of context, not necessarily the one Cox himself might have had in mind. Beyond the fact of relations between scales as crucial to local politics, it is perhaps the social and political construction itself of scale as social action (Howitt 1993; Swyngedouw 1997; Brenner 1998) that needs emphasizing.[26] Finally, and crucial to my analysis, is the actual thick and particularized content of the struggle or dynamic that gets instantiated.

These features can be illustrated with the case of the Society for the Promotion of Area Resources (SPARC). This organization began as an effort to organize slum dwellers in Bombay to get housing. Its purpose is to organize urban and rural poor, especially women, so as to develop their capabilities to organize around issues of concern. The focus is local, and so are the participants and those whom they seek to reach, usually local governments. But they have established multiple networks with other, similar organizations and efforts in other Asian countries, and now also some cities in Latin America and Africa. The various organizations making up the broader network do not necessarily gain power or material resources from this global networking, but they gain strength for themselves and vis-à-vis the agencies to which they make their demands.

The second form of multiscalar interaction is one where localized struggles are aiming at engaging global actors, such as the WTO, IMF, or multinational firms, either on a global scale or in multiple localities. Local initiatives can become part of a global network of activism without losing the focus on specific local struggles (e.g., Cleaver 1998; Espinoza 1999; Ronfeldt et al. 1998; Mele 1999).[27] This is one of the key forms of critical politics that the Internet can make possible: a politics of the local with a big difference—these are localities that are connected with each other across a region, a country, or the world. From struggles around human rights and the environment to workers' strikes and AIDS campaigns against the large pharmaceutical firms, the Internet has emerged as a pow-

[26] Some of these issues are well developed in Adams' (1996) study of the Tiananmen Square uprisings of 1989, the popular movement for democracy in the Philippines in the mid-1980s, and the U.S. civil rights movement in the 1950s. Protest, resistance, autonomy, and consent can be constructed at scales that can escape the confines of territorially bounded jurisdictions.

[27] One might distinguish a third type of political practice along these lines, one that turns a single event into a global media event, which in turn serves to mobilize individuals and organizations around the world in support of that initial action or around similar such occurrences elsewhere. Among the most powerful of these actions, and now emblematic of this type of politics, are those by the Zapatistas. The possibility of a single human rights abuse case becoming a global media event has been a powerful tool for human rights activists.

erful medium for nonelites to communicate, support each other's struggles, and create the equivalent of insider groups at scales going from the local to the global.[28] The possibility of doing so transnationally at a time when a growing set of issues are seen as escaping the bounds of nation-states makes this even more significant.

Yet another key scalar element here is that political activists can use digital networks for global transactions but also for strengthening local communications and transactions inside a city. The architecture of digital networks, primed to span the world, can actually serve to intensify transactions among residents of a city or region. It can serve to make them aware of neighboring communities and to gain an understanding of local issues that resonate positively or negatively with communities that are right in the same city, rather than with those that are at the other end of the world (Lagverre 2005, Riemens and Lovink 2002). Recovering how the new digital technology can serve to support local initiatives and alliances inside a locality is conceptually important given the almost exclusive emphasis on their global scope and deployment in the representation of these technologies.[29]

Coming back to Howitt's (1993) point about the constructing of the geographical scales at which social action can occur, let me suggest that electronic space is, perhaps ironically, a far more concrete space for social struggles than that of the national political system. It becomes a place where nonformal political actors can be part of the political scene in a way that is much more difficult in national institutional channels. Nationally, politics needs to run through existing formal systems, whether the electoral political system or the judiciary (taking state agencies to court). Nonformal political actors are rendered invisible in the space of national politics. Electronic space can accommodate a broad range of social struggles and facilitate the emergence of new types of political subjects that do not have to go through the formal political system.[30] Indi-

[28] The Internet may continue to be a space for democratic practices, but it will be so partly as a form of resistance against overarching powers of the economy and of hierarchical power (e.g., Calabrese and Burgelman 1999; see also Warf and Grimes 1997), rather than the space of unlimited freedom that is part of its romantic representation. The images we must bring into this representation increasingly need to deal with contestation and resistance to commercial and military interests, rather than simply freedom and interconnectivity (Sassen 2002).

[29] One instance of the need to bring in the local is the issue of what data bases are available to locals. Thus the World Bank's Knowledge Bank, a development gateway aimed at spurring ICT use and applications to build knowledge, is too large according to some (Wilks 2001). A good example of a type and size of database is Kubatana.net, an NGO in Zimbabwe that provides web site content and ICT services to national NGOs. It focuses on national information in Zimbabwe rather than going global.

[30] I have made a parallel argument for the city, especially the global city, being a more concrete space for politics. In many ways, the claim-making politics evident today in elec-

viduals and groups that have historically been excluded from formal political systems and whose struggles can be partly enacted outside those systems can find in electronic space an enabling environment both for their emergence as nonformal political actors and for their struggles.

The types of political practice discussed here are not the cosmopolitan route to the global.[31] They are global through the knowing multiplication of local practices. These are types of sociability and struggle deeply embedded in people's actions and activities. They are also forms of institution-building work with global scope that can come from localities and networks of localities with limited resources and from informal social actors. We see here the potential transformation of actors "confined" to domestic roles into actors in global networks without having to leave their work and roles in their communities. From being experienced as purely domestic and local, these "domestic" settings are transformed into microenvironments located on global circuits. They do not have to become cosmopolitan in this process; they may well remain domestic and particularistic in their orientation and remain engaged with their households and local community struggles, and yet they are participating in emergent global politics. A community of practice can emerge that creates multiple lateral, horizontal communications, collaborations, solidarities, supports. I interpret these as microinstances of partial and incipient denationalization.

Conclusion

The two cases focused on in this chapter reveal two parallel developments associated with particular technical properties of the new ICTs that have become crucial for both financial markets and electronic activism. And they reveal a third, radically divergent outcome, one I interpret as signaling the weight of the specific social logics at work in each case.

First, perhaps the most significant feature in both cases is the possibility of expanded decentralization and simultaneous integration. The fact that local political initiatives can become part of a global network parallels the articulation of the capital market with a network of financial centers. The fact that the former rely on public access networks and the latter on private dedicated networks does not alter this technical outcome.

---

tronic space resonates with many of the activisms proliferating in large cities: struggles against police brutality and gentrification, for the rights of the homeless and immigrants, for the rights of gays, lesbians, and queers.

[31] This has become an issue in my current work: the possibility of forms of globality that are not cosmopolitan. It stems partly from my critique of the largely unexamined assumption that forms of politics, thinking, and consciousness that are global are ipso facto defined as cosmopolitan (see Sassen 2005).

Among the technical properties that produce the specific utility in each case is the possibility of being global without losing the focus on specific local conditions and resources. As with the global capital market, there is little doubt that digital networks have had a sharp impact on resource-poor organizations and groups engaged in cross-border work.

Second, once established, this condition of expanded decentralization and simultaneous integration enabled by global digital networks produces threshold effects. Today's global electronic capital market can be distinguished from earlier forms of international financial markets due to some of the technical properties of the new ICTs, notably the orders of magnitude that can be achieved through decentralized simultaneous access and interconnectivity, and through the softwaring of increasingly complex instruments. In the second case, the threshold effect is the possibility of constituting transboundary publics and imaginaries rather than being confined to communication. Insofar as the new network technologies strengthen and create new types of cross-border activities among nonstate actors, they enable the constitution of a distinct and only partly digital condition variously referred to as global civil society, global publics, and commons.

Third, the significant difference lies in the substantive rationalities, values, objectives, and conditionings to which each of these two types of cases is subject. Once we introduce these issues, we can see a tendency toward cumulative causation in each case leading to a growing differentiation in outcomes. The constitutive capabilities of the new ICTs actually lie in a combination of digital and nondigital variables. It is not clear that the technology by itself could have produced the outcome. The nondigital variables differ sharply between these two cases, even as digitization is crucial to constituting the specificity of each case. The divergence is evident in the fact that the same technical properties produced greater concentration of power in the case of the capital market and greater distribution of power in the case of global digital networks.

References

Adams, Paul C. 1996. Protest and the Scale Politics of Telecommunications. *Political Geography* 15:419–41.

Allison, Julianne Emmons, ed. 2002. *Technology, Development, and Democracy: International Conflict and Cooperation in the Information Age.* Albany: SUNY.

APCWNSP (Association for Progressive Communications Women's Networking Support Programme). 2000. Women in Sync: Toolkit for Electronic Networking. *Acting Locally, Connecting Globally: Stories from the Regions.* Vol. 3. Available at http://www.apcwomen.org/netsupport/sync/sync3.html.

Arquilla, John, and David F. Ronfeldt. 2001. *Networks and Netwars: The Future of Terror, Crime, and Militancy.* Santa Monica, CA: Rand.

Arrighi, Giovanni. 1994. *The Long Twentieth Century: Money, Power, and the Origins of Our Times.* London: Verso.

Bank for International Settlements (BIS). 2000. *BIS Quarterly Review: International Banking and Financial Market Developments.* Basel: BIS Monetary and Economic Development.

———. 2002. *Annual Report.* Basel: BIS.

Barrett, Michael, and Susan Scott. 2004. Electronic Trading and the Process of Globalization in Traditional Futures Exchanges: A Temporal Perspective. *European Journal of Information Systems* 13:65–79.

Bellanet. 2002. *Report on Activities 2001–2002.* Available at http://home.bellanet.org.

Brenner, Neil. 1998. Global Cities, Glocal States: Global City Formation and State Territorial Restructuring in Contemporary Europe. *Review of International Political Economy* 5:1–37.

Budd, Leslie. 1995. Globalisation, Territory, and Strategic Alliances in Different Financial Centres *Urban Studies* 32:345–60.

Calabrese, Andrew, and Jean-Claude Burgelman. 1999. *Communication, Citizenship, and Social Policy: Rethinking the Limits of the Welfare State.* Lanham, MD: Rowman & Littlefield Publishers.

Callon, Michel. 1998. *The Laws of the Markets.* Oxford: Blackwell Publishers.

Camacho, Kemly. 2001. The Internet, A Great Challenge for Civil Society Organizations in Central America. *Fundacion Acceso.* Available at http://www.acceso.or.cr/publica/challenges.shtml.

Cleaver, Harry. 1998. The Zapatista Effect: The Internet and the Rise of an Alternative Political Fabric. *Journal of International Affairs* 51:621–40.

Corbridge, Stuart, Nigel Thrift, and Ron Martin, eds. 1994. *Money Power and Space.* Oxford: Blackwell.

Cordero-Guzmán, Hector R., Robert C. Smith, and Ramón Grosfoguel. 2001. *Migration, Transnationalization, and Race in a Changing New York.* Philadelphia: Temple University Press.

Creech, Heather, and Terry Willard. 2001. *Strategic Intentions: Managing Knowledge Networks for Sustainable Development.* Winnipeg: International Institute for Sustainable Development.

Denning, Dorothy. 1999. *Information Warfare and Security.* New York: Addison-Wesley.

Espinoza, V. 1999. Social Networks among the Poor: Inequality and Integration in a Latin American City. In *Networks in the Global Village: Life in Contemporary Communities,* edited by Barry Wellman, 147–84. Boulder: Westview Press.

Esterhuysen, Anriette. 2000. Networking for a Purpose: African NGOs Using ICT. *Rowing Upstream: Snapshots of Pioneers of the Information Age in Africa.* Available at http://www.sn.apc.org/Rowing_Upstream/chapter1/ch.1.html.

Frederick, Howard. 1993. Computer Networks and the Emergence of Global Civil Society. In *Global Networks: Computers and International Communications,* edited by Linda M. Harasim, 283–95. Cambridge: MIT Press.

Garcia, Linda. 2002. Architecture of Global Networking Technologies. In *Global Networks, Linked Cities,* edited by Saskia Sassen, 39–70. London: Routledge.
Globalization and World Cities (GaWC). Available at http://www.lboro.ac.uk/gawc/.
Hirst, Paul Q., and Grahame Thompson. 1996. *Globalization in Question: The International Economy and the Possibilities of Governance.* Cambridge: Polity Press, Blackwell Publishers.
Howitt, Richard. 1993. "A World in a Grain of Sand": Towards a Reconceptualisation of Geographical Scale. *The Australian Geographer* 24:33–44.
———. 1998. Recognition, Respect and Reconciliation: Steps Towards Decolonisation? *Australian Aboriginal Studies* 1:28–34.
International Monetary Fund (IMF). 1999. *International Capital Markets: Developments, Prospects, and Key Policy Issues.* Washington, DC: IMF.
———. 2001. *Balance of Payments Statistics Yearbook.* Washington, DC: IMF.
———. 2004. Toward a Framework for Safeguarding Financial Stability. Prepared by Aerdt Houben, Jan Kakes, and Garry Schinasi. IMF Working Paper, WP/04/101. Washington, D.C.: IMF.
Judd, Dennis R. 1998. The Case of the Missing Scales: A Commentary on Cox. *Political Geography* 17:29–34.
Knorr-Cetina, Karin, and Urs Bruegger. 2002. Global Microstructures: The Virtual Societies of Financial Markets. *American Journal of Sociology* 107:905–50.
Kuntze, Marco, Sigrun Rottmann, and Jessica Symons. 2002. *Communications Strategies for World Bank and IMF-Watchers: New Tools for Networking and Collaboration.* London: Bretton Woods Project and Ethical Media. Available at http://www.brettonwoodsproject.org/strategy/Commsrpt.pdf.
Laguerre, Michel S. 2005. *The Digital City: The American Metropolis and Information Technology.* Basingstoke: Palgrave Macmillan Press.
Lannon, John. 2002. *Technology and Ties That Bind: The Impact of the Internet on Non-Governmental Organizations Working to Combat Torture.* Unpublished MA thesis completed at the University of Limerick.
Leizerov, Sagi. 2000. Privacy Advocacy Groups versus Intel: A Case Study of How Social Movements Are Tactically Using the Internet to Fight Corporations. *Social Science Computer Review* 18:461–83.
MacKenzie, Donald, and Yuval Millo. 2003. Constructing a Market, Performing Theory: The Historical Sociology of a Financial Derivatives Exchange. *American Journal of Sociology* 109:107–145.
Mele, Christopher. 1999. Cyberspace and Disadvantaged Communities: The Internet as a Tool for Collective Action. In *Communities in Cyberspace,* edited by Marc A. Smith and Peter Kollock, 290–310. London: Routledge.
Meyer, Carrie A. 1997. The Political Economy of NGOs and Information Sharing. *World Development* 25:1127–40.
Miller, Daniel, and Don Slater. 2000. *The Internet: An Ethnographic Approach.* Oxford: Berg.
Mills, Kurt. 2002. Cybernations: Identity, Self-Determination, Democracy, and the "Internet Effect" in the Emerging Information Order. *Global Society* 16:69–87.
Morrill, Richard. 1999. Inequalities of Power, Costs and Benefits across Geo-

graphic Scales: The Future Uses of the Hanford Reservation. *Political Geography* 18:1–23.

Nettime. 1997. *Net Critique*. Compiled by Geert Lovink and Pit Schultz. Berlin: Edition ID-Archive.

Notzke, Claudia. 1995. A New Perspective in Aboriginal Natural Resource Management: Co-Management. *Geoforum* 26:187–209.

Pace, William R., and Rik Panganiban. 2002. The Power of Global Activist Networks: The Campaign for an International Criminal Court. In *Civil Society in the Information Age,* edited by Peter I. Hajnal. Aldershot, 109–25. UK: Ashgate.

Poster, Mark. 1997. Cyberdemocracy: Internet and the Public Sphere. In *Internet Culture,* edited by D. Porter, 201–18. London: Routledge.

Redden, Guy. 2001. Networking Dissent: The Internet and the Anti-Globalization Movement. *MotsPluriels* 18. Available at http://www.arts.uwa.edu.au/MotsPluriels/MP1801gr.html.

Riemens, Patrice, and Geert Lovink. 2002. Local Networks: Digital City Amsterdam. In *Global Networks, Linked Cities,* edited by Saskia Sassen, 327–45. London: Routledge.

Rimmer, P. J., and T. Morris-Suzuki. 1999. The Japanese Internet: Visionaries and Virtual Democracy. *Environment and Planning A* 31:1189–1206.

Ronfeldt, David, et al. 1998. *The Zapatista "Social Netwar" in Mexico*. Santa Monica, CA: RAND, MR-994-A.

Rutherford, Kenneth R. 2002. Essential Partners: Landmines-Related NGOs and Information Technologies. In *Civil Society in the Information Age,* edited by Peter I. Hajnal. Aldershot, 95–107. UK: Ashgate.

Salzinger, Leslie. 2003. Market Subjects: Traders at Work in the Dollar/Peso Market. Presented at American Sociological Association Conference, 16–19 August 2003.

Sassen, Saskia. 1996. *Losing Control?: Sovereignty in an Age of Globalization*. New York: Columbia University Press.

———. 2001. *The Global City: New York, London, Tokyo*. 2d ed. Princeton: Princeton University Press.

———. 2002. Towards a Sociology of Information Technology. *Current Sociology* 50:365–88.

———. 2005. *Denationalization: Territory, Authority, and Rights in a Global Digital Age*. Princeton: Princeton University Press.

Scott, Matthew J. O. 2001. Danger-Landmines! NGO-Government Collaboration in the Ottawa Process. In *Global Citizen Action,* edited by Michael Edwards and John Gaventa, 121–33. Boulder: Lynne Rienner Publishers.

Sharp, John. 1997. Communities of Practice: A Review of the Literature. Available at http://www.tfriend.com/cop-lit.htm.

Silvern, Steven E. 1999. Scales of Justice: Law, American Indian Treaty Rights and Political Construction of Scale. *Political Geography* 18:639–68.

Sinclair, Timothy, J. 1994. Passing Judgment: Credit Rating Processes as Regulatory Mechanisms of Governance in the Emerging World Order. *Review of International Political Economy* 1:133–59.

Smith, Michael Peter. 1994. Can You Imagine? Transnational Migration and the Globalisation of Grassroots Politics. *Social Text* 39:15–33.

Smith, Peter J. 2001. The Impact of Globalization on Citizenship: Decline or Renaissance. *Journal of Canadian Studies* 36:116–40.

Smith, Robert C. 1997. Transnational Migration, Assimilation and Political Community. In *The City and the World: New York's Global Future,* edited by Margaret Crahan and Alberto Vourvoulias-Bush, 110–32. New York: Council of Foreign Relations.

Swyngedouw, Erik. 1997. Neither Global nor Local: "Glocalization" and the Politics of Scale. In *Spaces of Globalization: Reasserting the Power of the Local,* edited by Kevin R. Cox, 137–66. New York: Guilford.

Taylor, Peter J., D.R.F. Walker, and J. V. Beaverstock. 2002. Firms and Their Global Service Networks. In *Global Networks, Linked Cities,* edited by Saskia Sassen, 93–115. New York: Routledge.

Thomson Financials. 1999. *1999 International Target Cities Report.* New York: Thomson Financial Investor Relations.

Tsaliki, Liza. 2002. Online Forums and the Enlargement of the Public Space: Research Findings from a European Project. *The Public* 9:95–112.

Tuijl, Peter van, and Lisa Jordan. 2000. *Political Responsibility in Transnational NGO Advocacy.* Washington, DC: Bank Information Center. *World Development* 28:2029–198.

United Nations Conference on Trade and Development (UNCTAD). 1998. World Investment Report: Trends and Determinants. New York: UNCTAD.

Wade, Robert. 2004. *Governing the Market: Economic Theory and the Role of Government in East Asian Industrialization.* 2d ed. Princeton: Princeton University Press.

Warf, B., and J. Grimes. 1997. Counterhegemonic Discourses and the Internet. *The Geographical Review* 87:259–74.

Warkentin, Craig. 2001. *Reshaping World Politics: NGOs, the Internet, and Global Civil Society.* Lanham, MD: Rowman & Littlefield Publishers.

Wilks, Alex. 2001. A Tower of Babel on the Internet? The World Bank's Development Gateway. *Bretton Woods Project.* Available at http://www.brettonwoodsproject.org/topic/knowledgebank/k22gatewaybrief.pdf.

Williams, Jody, and Stephen Goose. 1998. The International Campaign to Ban Landmines. In *To Walk without Fear: The Global Movement to Ban Landmines,* edited by M. A. Cameron, R. J. Lawson, and Brian W. Tomlin. Ontario, CA: Oxford University Press.

WomenAction. 2000. Alternative Assessment of Women and Media Based on NGO Reviews of Section J, Beijing Platform for Action. Available at http://www.womenaction.org/csw44/altrepeng.htm.

World Information Order. 2002. *World-Information Files. The Politics of the Info Sphere.* Vienna: Institute for New Culture Technologies, and Berlin: Center for Civic Education.

Yang, Guobin. 2003. Weaving a Green Web: The Internet and Environmental Activism in China. *China Environment Series,* no 6. Washington, DC: Woodrow Wilson International Centers for Scholars.

Zaloom, Caitlin. 2005. Trapped in the Pits: Open-outcry Technology and the Chicago Board of Trade. In *Frontiers of Capital: Ethnographic Reflection on the "New Economy,"* edited by Melissa Fisher and Greg Downey. Durham: Duke University Press.

The New Mobility of Knowledge: Digital Information Systems and Global Flagship Networks

DIETER ERNST

DIGITAL INFORMATION systems (DIS) are electronic systems that integrate software and hardware to enable communication and collaborative work (Chandler and Cortada 2000). These systems are not developed in a vacuum. They are a response to transformations in economic institutions and structures that determine industrial dynamics. "Globalization" is a widely used shorthand for those transformations.

How does globalization interact with DIS? To answer that question, we need to open the black box of "globalization." I define globalization as the integration, across borders, of markets for capital, goods, services, knowledge, and labor. Barriers to integration continue to exist, of course, in each of these different markets (especially for low-wage labor), so integration is far from perfect. But there is no doubt that a massive integration has taken place across borders that, only a short while ago, seemed to be impenetrable.

This raises the question: Who are the "integrators"? States obviously play an important role in reshaping institutions and regulations. Equally important are private actors, especially large global corporations. Both sets of actors increasingly interact through complex digital formations, as outlined elsewhere in this book. The study of these formations allows us to identify what is "new" about the global economy.

This chapter focuses on digital formations centered in the corporate sector. It explores the link between transformations in international business organization and industry dynamics. The approach that I have chosen focuses on international knowledge diffusion through an extension of firm organization across national boundaries. A central argument is that two interrelated transformations in the organization of international business may *gradually* reduce constraints on international knowledge diffusion: the evolution of cross-border forms of corporate networking practices, especially global flagship networks (GFNs), and the increasing use of digital information systems to manage these networks. GFNs expand interfirm linkages across national boundaries, increasing the need for knowledge diffusion, while DIS not only enhance information exchange, but also provide new opportunities for the sharing and joint utilization and creation of knowledge.

This argument runs counter to a widespread belief, formalized by agglomeration and innovation economists and network sociologists, that knowledge is stickier in space (i.e., less mobile) than markets, finance, or production facilities (e.g., Markusen 1996; Archibugi and Michie 1995; Breschi and Malerba 2001). This is said to be true in particular for higher-level, mostly tacit forms of "organizational knowledge" required for learning and innovation. This chapter demonstrates that, in the emerging global network economy, we need to reconsider and amend the "stickiness-of-knowledge" proposition.

I first introduce two conceptual building blocks: a framework that links GFNs, DIS, and knowledge diffusion, and a stylized model of forces that drive the development of GFNs. Next I look at the economic structure and peculiar characteristics of the flagship network model that foster the new mobility of knowledge. I explore how two distinctive characteristics of GFN, which are enhanced by DIS, shape the scope for international knowledge diffusion: a rapid yet concentrated dispersion of value chain activities, and, simultaneously, their integration into hierarchical networks.

Finally, I explore some inherent contradictions of GFNs that reflect the increasingly complex nature of digital formations in the corporate sector. I argue that the combined forces of DIS and GFNs are gradually reducing constraints to international knowledge diffusion. This might actually make it easier for less advanced countries to access and use state-of-the-art knowledge. It may also provide new opportunities for "late innovation" strategies in these countries that attempt to redress the imbalance between excellence in manufacturing and a weak basis for knowledge creation (Ernst 2004a, 2004b, Ernst and Lundvall, 2004). The crucial issue is how this will affect the geographic distribution of "innovative capabilities," defined as the skills, knowledge, and management approaches needed to create, change, or improve products, services, equipment, and processes. As we will see, much of the new mobility of knowledge is focused on the redeployment of "blue-collar" forms of knowledge production to locations with lower costs of knowledge workers (Ernst 2004b): knowledge diffusion has created new "cost-and-time-reduction centers" in lower-income regions that thrive on the timely provision of knowledge support services like supply chain management, design services, and detailed engineering. Yet the sources of knowledge creation remain concentrated in a few global "centers of excellence" that combine unique capabilities in research, global branding, standard definition, and system integration (e.g, Pavitt 2003; Ernst 2004a).

Conceptual Framework

A GFN integrates a flagship's dispersed production, customer, and knowledge bases. Covering both intrafirm and interfirm transactions and forms of coordination, the network links together the flagship's own subsidiaries, affiliates, and joint ventures with its subcontractors, suppliers, service providers, as well as partners in strategic alliances. While equity ownership is not essential, network governance is distinctively asymmetric. The new mobility of knowledge is an unintended consequence of the evolution of these corporate networks. Global corporations (the "network flagships") construct these networks to gain quick access to skills and capabilities at lower-cost overseas locations that complement the flagships' core competencies. Furthermore, flagships need to transfer techni-

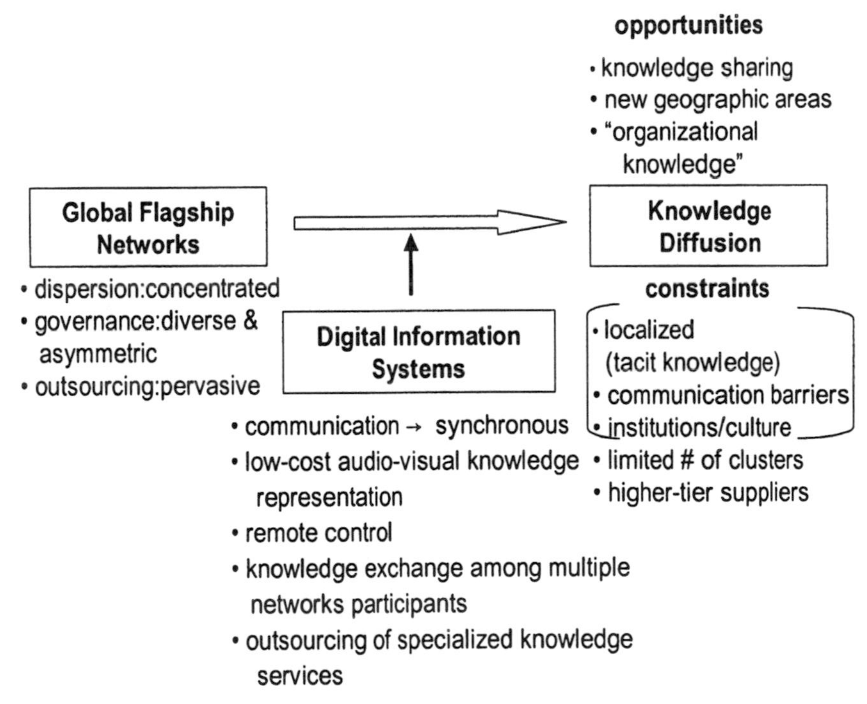

Figure 1. GFNs, DIS, and knowledge diffusion. Describes a simple framework to explore the links between GFNs, DIS, and knowledge diffusion.

cal and managerial knowledge to local suppliers. This is necessary to upgrade the suppliers' technical and managerial skills, so that they can meet the technical specifications of the flagships. Originally this involved primarily operational skills and procedures required for routine manufacturing and services. Over time, knowledge sharing also incorporates higher-level, mostly tacit forms of "organizational knowledge" required for learning and innovation (Ernst and Kim 2002a). The more dispersed and complex these networks, the more demanding their coordination requirements. Knowledge sharing is the necessary glue that keeps these networks growing (Ernst 2002a). In short, knowledge exchange penetrates new geographic areas, and the contents of knowledge become more complex.

The use of DIS as a management tool can enhance the scope for knowledge sharing among multiple network participants at distant locations. But these changes will occur only gradually, as a long-term, iterative learning process, based on search and experimentation. The digitization of knowledge implies that it can be delivered as a service and built around open standards. This has fostered the specialization of knowledge creation, giving rise to a process of modularization, very much like earlier modularization processes in hardware manufacturing. As a result, one of the most important recent developments that affect international knowledge diffusion is the rapidly growing trade in intellectual property rights (IPR) (Yau and Das 2001).

Under the heading of "e-business," a new generation of networking software provides a greater variety of tools for representing knowledge, including low-cost audiovisual representations (Foray and Steinmueller 2001). Those programs also provide flexible information systems that support not only information exchange among dispersed network nodes, but also the creation, utilization, and sharing of knowledge among multiple network participants at remote locations (Jørgensen and Krogstie 2000). New forms of remote control are emerging for manufacturing processes, quality, supply chains, and customer relations. Equally important are new opportunities for the joint production across distant locations of knowledge support services (e.g., software engineering and development, business process outsourcing, maintenance and support of information systems, as well as skill transfer and training).

While much of this is still at an early stage of trial and error, international business now faces a huge potential for extending knowledge exchange across organizational and national boundaries. But, as Sassen outlines in her chapter, the uncertainties and complexities of operating in global markets mean that there are agglomeration economies to be derived from dense spatial concentrations of specialized network suppliers.

Forces Driving Global Flagship Networks

A defining characteristic of digital formations in the corporate sector is the transition from vertically integrated multinational corporations (MNCs), with their focus on stand-alone, equity-controlled overseas investment projects, to global flagship networks that integrate their geographically dispersed supply, knowledge, and customer bases (Ernst 2002b). This contrasts with centuries of economic history where MNCs were the main drivers of international production (e.g., Braudel 1992; Wilkins 1970). Typically, the focus of MNCs has been on the penetration of protected markets through tariff-hopping investments, and on the use of assets developed at home to exploit international factor cost differentials, primarily for labor (e.g., Dunning 1981). This has given rise to a peculiar pattern of international production: stand-alone offshore production sites in low-cost locations are linked through triangular trade with the major markets in North America and Europe (e.g., Dicken 1992).

What forces have driven the shift in industrial organization from MNCs to GFNs? To answer this question, we highlight three interrelated explanatory variables: institutional change through liberalization; changes in competition and industrial organization; and information and communications technologies that gave rise to DIS.

Institutional Change: Liberalization

Liberalization dates back to the early 1970s: it thrived in response to the breakdown of fixed exchange rate regimes and the failure of Keynesianism to cope with pervasive stagflation. To a large degree, it has been initiated by government policies. But there are also other actors that have played an important role: financial institutions, rating agencies, supranational institutions like bilateral or multilateral investment treaties, and regional integration schemes, like the European Union or North American Free Trade Agreement. In some countries with decentralized devolution of political power, regional governments can also play an important role.

Liberalization imposes far-reaching changes on the economic institutions, that is, the rules of the game that structure economic interactions. These institutions shape the allocation of resources, the rules of competition, and firm behavior.[1] Liberalization covers four main policy areas:

[1] Liberalization affects all aspects of institutions, but at different speed. North (1996: 12) distinguishes formal rules (statute law, common law, regulations), informal constraints (conventions, norms of behavior, self-imposed codes of conduct), and the enforcement characteristics of both. While liberalization will first affect formal rules, informal constraints and enforcement mechanisms are more difficult to change. This implies that there is no homogeneous model of liberalization, but many different and often hybrid forms.

trade, capital flows, foreign direct investment (FDI), and privatization. While each of these has generated separate debates in the literature, they hang together. Earlier success in trade liberalization has sparked an expansion of trade and FDI, increasing the demand for cross-border capital flows. This has increased the pressure for a liberalization of capital markets, forcing more and more countries to open their capital accounts. In turn, this has led to a liberalization of FDI policies, and to privatization tournaments.

The overall effect of liberalization has been a considerable reduction in the cost and risks of international transactions and a massive increase in international liquidity. Global corporations (the network flagships) have been the primary beneficiaries: liberalization provides them with a greater range of choices for market entry among trade, licensing, subcontracting, franchising, and so forth (*locational specialization*) than otherwise; it provides better access to external resources and capabilities that a flagship needs to complement its core competencies (*vertical specialization*); and it has reduced the constraints for a geographic dispersion of the value chain (*spatial mobility*).

Competition and Industrial Organization

As liberalization has been adopted as an almost universal policy doctrine, this has drastically changed the dynamics of competition. Again, we reduce the complexity of these changes and concentrate on two impacts: a broader geographic scope of competition, and a growing complexity of competitive requirements. Competition now cuts across national borders—a firm's position in one country is no longer independent of its position in other countries (e.g., Porter 1990). This has two implications. The firm must be present in all major growth markets (*dispersion*). It must also integrate its activities on a worldwide scale, in order to exploit and coordinate linkages between these different locations (*integration*). Competition also cuts across sector boundaries and market segments: mutual raiding of established market segment fiefdoms has become the norm, making it more difficult for firms to identify market niches and to grow with them.

This has forced firms to engage in complex strategic games to preempt a competitor's move. This is especially the case for knowledge-intensive industries like electronics (Ernst 2002b). Intense price competition needs to be combined with product differentiation, in a situation where continuous price wars erode profit margins. Of critical importance, however, is speed-to-market: getting the right product to the largest-volume segment of the market right on time can provide huge profits. Being late can be a disaster and may even drive a firm out of business. The result has been an increasing uncertainty and volatility, and a destabilization of established market leadership positions (Richardson 1996; Ernst 1998).

This growing complexity of competition has changed the determinants of location, as well as industrial and firm organization. Take first location decisions. While both market access and cost reductions remain important, it became clear that they have to be reconciled with a number of equally important requirements that encompass the exploitation of uncertainty through improved operational flexibility (e.g., Kogut 1985; Kogut and Kulatilaka 1994); a compression of speed-to-market through reduced product development and product life cycles (e.g., Flaherty 1986); learning and the acquisition of specialized external capabilities (e.g., Antonelli 1992; Kogut and Zander 1993; Zander and Kogut 1995; Zanfei 2000; Dunning 2000); and a shift of market penetration strategies from established to new and unknown markets (e.g., Christensen 1997).

Equally important are changes in industrial organization. No firm, not even a dominant market leader, can generate all the different capabilities internally that are necessary to cope with the requirements of global competition. Competitive success thus critically depends on vertical specialization: a capacity to selectively source specialized capabilities *outside* the firm that can range from simple contract assembly to quite sophisticated design capabilities. This requires a shift from individual to increasingly collective forms of organization, from the multidivisional (M-form) functional hierarchy (e.g., Williamson 1975, 1985; Chandler 1977) of multinational corporations to the networked global flagship model.

The electronics industry has become the most important breeding ground for this new industrial organization model. Over the last decades, a massive process of vertical specialization has segmented an erstwhile vertically integrated industry into closely interacting horizontal layers (Grove 1996). Until the early 1980s, IBM personified "vertical integration": almost all ingredients necessary to design, produce, and commercialize computers remained internal to the firm. This was true for semiconductors, hardware, operating systems, application software, and sales and distribution. Above all, "IBM was famous (some would say notorious) for the power of its sales force . . . (and distribution system)" (Sobel 1986: 37).

Since the mid-eighties, vertical specialization has become the industry's defining characteristic. Most activities that characterized a computer company were now being farmed out to multiple layers of specialized suppliers, giving rise to rapid market segmentation and an ever-finer specialization within each of the above five main value chain stages. This has given rise to the coexistence of complex, globally organized, product-specific value chains (e.g., for microprocessors, memories, board assembly, PCs, networking equipment, operating systems, applications software, and sales and distribution). In each of these value chains, GFNs compete with each other but may also cooperate (Ernst 2002a). The number of such networks and the intensity of competition vary across sectors, reflecting their different stage of development and their idiosyncratic industry structures.

Information and Communication Technology: Digital Information Systems

The use of DIS to manage these networks has accelerated this process. For the manufacturing of electronics hardware, the use of DIS facilitated geographic dispersion. This is now being mirrored by similar developments for software and electronic design and engineering.

We first need to highlight important transformations in the use of DIS as a management tool. From a machine to automate transaction processing, the focus has shifted to the extraction of value from information resources, and then further to the establishment of Internet-enabled, flexible information infrastructures that can support the extraction and exchange of knowledge across firm boundaries and national borders. A combination of technological and economic developments is responsible for this transformation.

On the technology side, the rapid development and diffusion of cheaper and more powerful information and communication technologies (e.g., Sichel 1997; and Flamm 1999) has considerably reduced transaction costs. In addition, the move toward more open standards in DIS architecture (UNIX, Linux, HTML) and protocols (TCP/IP) enabled firms to integrate their existing intranets and extranets[2] on the Internet, which, by reducing cost and multiplying connectivity, dramatically extended their reach across firm boundaries and national borders.

Compared to earlier generations of DIS, the Internet appears to provide much greater opportunities to share knowledge with a much greater number of people faster, more accurately, and in greater detail, even if they are not permanently colocated (Litan and Rivlin, 2001; Ernst 2001a, 2001b). The most commonly used technologies today facilitate *asynchronous* interaction, such as e-mail or non-real-time database sharing. But as data transfer capacity ("bandwidth") increases, this is creating new opportunities for using technologies that facilitate *synchronous* interaction such as real-time data exchange, video-conferencing, as well as remote control

[2] An "intranet" is defined as a private network contained within an organization (a firm) that consists of many interlinked local-area networks (LANs). Its main purpose is to share company information and computer resources among employees. An "extranet," in turn, is a private network that links the flagship via conventional telecommunications networks with preferred suppliers, customers, and strategic partners.

of manufacturing processes, product quality and inventory, maintenance and repair, and even prototyping. This has created new opportunities for extending knowledge exchange across organizational and national boundaries, hence magnifying the scope for vertical specialization. Equally important, wireless Internet-based technologies have increased the mobility of DIS.

On the economic side, vertical specialization, particularly pronounced in the electronics industry, poses increasingly complex information requirements (e.g., Chen 2002; Macher, Mowery, and Simcoe 2002). As firms now have to deal with constantly changing, large numbers of specialized suppliers, they need flexible and adaptive information systems to support these diverse linkages. These requirements become ever more demanding as flagships attempt to integrate their dispersed production, knowledge, and customer bases into global and regional networks. DIS now need to provide new means to improve global supply chain management and speed-to-market. DIS also need to provide for effective communication between design and manufacturing, and for the exchange of proprietary knowledge. The semiconductor industry provides examples for both developments (e.g., Macher, Mowery, and Simcoe 2002): vertical specialization gives rise to the separation of design ("fabless design") and manufacturing ("silicon foundry"). This creates very demanding requirements for knowledge exchange between multiple actors at distant locations, say, a design house in Silicon Valley and a silicon foundry in Taiwan's Hsinchuh Science Park. Vertical separation of design and production of semiconductor devices in turn has created a vibrant trade in "intellectual property rights" among specialized design firms that create, license, and trade "design modules" for use in integrated circuits.

In addition, far-reaching changes in work organization have fundamentally increased the requirements for information management and for the exchange of knowledge (e.g., Ciborra et al. 2000). The transition from Fordist "mass production" to "mass customization" requires a capacity to constantly adapt products or services to changing customer requirements, "sensing and responding" to individual customer needs in real time (Bradley and Nolan 1998). This necessitates dynamic, interactive information systems and a capacity to rapidly adjust the organization of firms and corporate networks to disruptive changes in markets and technology. Third, real-time resource allocation, performance monitoring, and accounting became necessary, due to the short-term pressures of the financial system (quarterly reports) and the shortening life cycles of products and technologies. Fourth, to cope with ever more demanding competitive requirements, firms have to continuously adapt their organization and strategy, hence the demand for flexible DIS.

Following Brynjolfsson and Hitt (2000), I argue that the impact of DIS on economic performance is mediated by a combination of intangible inputs as well as intangible outputs that act as powerful catalysts for organizational innovation.[3] After a while, these induced organizational changes may lead to productivity growth by reducing the cost of coordination, communications, and information processing. Most importantly, these organizational changes may enable firms "to increase output quality in the form of new products or in improvements in intangible aspects of existing products like convenience, timeliness, quality and variety." (Brynjolfsson and Hitt 2000: 4). In short, we are talking about a complex process that involves a set of interrelated ("systemic") changes (Milgrom and Roberts 1990). By combining DIS with changes in work practices, strategies, and products and services, a firm transforms its organization as well as its relations with suppliers, partners, and customers.

Once we adapt such a framework, it becomes clear that firms that participate in GFNs can reap substantial benefits from using DIS as a management tool. There is ample scope for cost reduction across all stages of the production process, both for the flagship company and for local suppliers. Procurement costs can be reduced by means of expanded markets and increased competition through Internet-enabled online procurement systems. Another cost-reducing option is to shift sales and information dissemination to lower-cost online channels.

The transition to Internet-based information systems can drastically accelerate speed-to-market by reducing the time it takes to transmit, receive, and process routine business communications such as purchase orders, invoices, and shipping notifications. There is much greater scope for knowledge management: documents and technical drawings can be exchanged in real time, legally recognized signatures can be authenticated, browsers can be used to access the information systems of suppliers and customers, and transactions can be completed much more quickly.

A further advantage can be found in the low cost of expanding an Internet-based information system. While establishing a network backbone requires large up-front fixed investment costs (purchasing equipment, laying new cable, training), the cost of adding an additional user to the network is negligible. The value of the network thus increases with the number of participants ("network externalities"). In addition, the Internet and related organizational innovations provide effective mechanisms for constructing flexible infrastructures that can link together and

[3] Intangible inputs include, for instance, the development of new software and databases, the adjustment of existing business processes, and the recruitment and continuous upgrading of specialized human resources. Of equal importance are intangible outputs that would not exist without DIS, like speed of delivery, flexibility of response to abrupt changes in demand and technology, and organizational innovations, such as "just-in-time" (JIT), "mass customization," the built-to-order (BTO) production model, integrated supply chain management (SCM), and customer-relations management (CRM).

coordinate knowledge exchange between distant locations (Hagstrøm 2000; Pedersen, Tølle, and Vesterager 1999; Antonelli 1992).

This has important implications for organizational choices and locational strategies of firms. In essence, Internet-enabled DIS foster the development of leaner, meaner, and more agile production systems that cut across firm boundaries and national borders. The underlying vision is that of *networks of networks* that enable a global network flagship to respond quickly to changing circumstances, even if much of its value chain has been dispersed. DIS, especially the open-ended structure of the Internet, substantially broadens the scope for vertical specialization. It allows global flagships to shift from *partial* outsourcing, covering the nuts and bolts of manufacturing, to *systemic* outsourcing that includes knowledge-intensive support services, such as software production, electronic design services, business process outsourcing, maintenance and repair of information systems, and skill transfer and training (Ernst 2004d).

The Flagship Network Model

Theoretical Foundations

Until recently, these fundamental changes in the organization of international production have been largely neglected in the literature, both in research on knowledge spillovers through FDI and in research on the internationalization of corporate R&D. This is now beginning to change. There is a growing acceptance in the literature that, to capture the impact of globalization on industrial organization and upgrading, the focus of our analysis needs to shift away from the industry and the individual firm to the international dimension of business networks (e.g., Bartlett and Ghoshal 1989; Gereffi and Korzeniewicz 1994; Ernst 1997; Rugman and D'Cruz 2000; Birkinshaw and Hagstrøm 2000; Borrus, Ernst, and Haggard 2000; Pavitt 2003; Ernst and Kim, 2002b; Ernst and Ozawa 2002). Flagship-driven corporate networks are of course only one of diverse complex digital formations that are currently reshaping the international economy (see contributions by Sassen, Garcia, and Latham in this volume).

My model of GFNs emphasizes three essential characteristics: (1) *scope*—GFNs encompass all stages of the value chain, not just production; (2) *asymmetry*—flagships dominate control over network resources and decision making; and (3) *knowledge diffusion*—the sharing of knowledge is the necessary glue that keeps these networks growing.

A focus on international knowledge diffusion through an extension of firm organization across national boundaries distinguishes my concept of GFN from network theories developed by sociologists, economic geographers, and innovation theorists that focus on localized, mostly interpersonal networks (e.g., Powell and Smith-Doerr 1994). The central problem of these theories is that industries now operate in a global rather than a localized setting (Ernst et al. 2001). Important complementarities exist, however, with work on global commodity chains (GCC) (e.g., Gereffi and Korzeniewicz 1994). A primary concern of the GCC literature has been to explore how different value chain stages in an industry (e.g., textiles) are dispersed across borders, and how the position of a particular location in such a GCC affects its development potential through access to economic rents (e.g., Gereffi and Kaplinsky 2001; Henderson et al. 2001).[4] Strong complementarities also exist with research on computer-based flexible information infrastructures that frequently uses the terms "extended enterprise" or "virtual enterprise," where the first stands for more durable network arrangements, while the latter stands for very short-term ones (e.g., Pedersen 1999; Jørgensen and Krogstie 2000; and various issues of the electronic journal www.virtual-organization.net).

As for the dynamics of network evolution, my approach complements the transaction cost approach to networks and vertical disintegration that centers on the presumed efficiency gains from these organizational choices (e.g., Williamson 1985, 1975; Milgrom and Roberts 1990). The latter approach, however, skips some of the more provocative chapters in the economic history of the modern corporation. Chandler's vibrant histories (e.g., 1997) show that the quest for profits and market power via increased throughput and speed of coordination were more important in explaining hierarchy than the traditional emphasis on transaction costs. This implies that the analysis of the determinants of institutional form must move beyond a narrow focus on transaction costs to the broader competitive environment in which firms operate. It is time to bring back into the analysis market structure and competitive dynamics, as well as the role played by knowledge and innovation. Like hierarchies, GFNs not only promise to improve efficiency but can permit flagships to sustain quasi-monopoly positions, generate market power through specialization, and raise entry barriers; they also enhance the network flagships' capacity for innovation (Ernst 1997; Borrus, Ernst, and Haggard 2000: chap. 1).

Network Characteristics

GFNs differ from MNCs in three important ways that need to be taken into account in the study of knowledge diffusion (Ernst 2002a, 2002b, 2004d). First, these networks cover both intrafirm and interfirm transac-

[4] Unfortunately, no one has as yet come up with a convincing and robust set of indicators. How should academic researchers, even with the best possible funding, be able to measure distribution of rents across borders when global flagships like Enron and telecom majors excel in the development of sophisticated off-balance-sheet financial techniques and transfer pricing?

tions and forms of coordination: a GFN links together the flagship's own subsidiaries, affiliates, and joint ventures with its subcontractors, suppliers, service providers, as well as partners in strategic alliances. A network flagship like IBM or Intel breaks down the value chain into a variety of discrete functions and locates them wherever they can be carried out most effectively, where they can improve the flagship's access to resources and capabilities, and where they are needed to facilitate the penetration of important growth markets.

Second, GFNs differ from MNCs in that a great variety of governance structures are possible. These networks range from loose linkages that are formed to implement a particular project and are dissolved after the project is finished, so-called virtual enterprises (e.g., Pedersen et al. 1999: 16), to highly formalized networks, "extended enterprises," with clearly defined rules, common business processes, and shared information infrastructures. What matters is that formalized networks do not require common ownership: these arrangements may or may not involve control of equity stakes.

Third, vertical specialization ("outsourcing" in business parlance) is the main driver of these networks (Ernst 2002b). GFNs help flagships to gain quick access to skills and capabilities at lower cost overseas locations that complement the flagships' core competencies. As the flagship integrates geographically dispersed production, customer, and knowledge bases into GFNs, this may well produce transaction cost savings. Yet the real benefits result from the dissemination, exchange, and outsourcing of knowledge and complementary capabilities.

Increasingly, the focus of outsourcing is shifting from assembly-type manufacturing to knowledge-intensive support services, like supply chain management, engineering services, and new product introduction. Outsourcing may also include design and product development. This indicates that GFNs also differ from traditional forms of subcontracting: much denser interaction between design and production and other stages of the value chain require substantially more intense exchange of information and knowledge. Network flagships increasingly rely on the skills and knowledge of specialized suppliers to enhance their core competencies.

Two distinctive characteristics of GFN that are enhanced by DIS shape the scope for international knowledge diffusion: a rapid yet concentrated dispersion of value chain activities and, simultaneously, their integration into hierarchical networks.

Concentrated Dispersion

GFNs typically combine a rapid geographic dispersion with spatial concentration on a growing but still limited number of specialized clusters. To simplify, I distinguish two types of clusters (Ernst 2002a): "centers of excellence" that combine unique resources, such as R&D and precision mechanical engineering, and "cost and time reduction centers" that thrive on the timely provision of lower-cost services.[5] Different clusters face different constraints to knowledge diffusion, depending on their specialization, and on the product composition of GFNs. The dispersion of clusters differs across the value chain: it increases, the closer one gets to the final product, while dispersion remains concentrated especially for high-precision and design-intensive components.

Let us look at some indicators in the electronics industry, a pace setter of the flagship network model (Ernst 2002b, 2004d). On one end of the spectrum is final PC assembly that is widely dispersed to major growth markets in the United States, Europe, and Asia. Dispersion is still quite extended for standard, commodity-type components, but less so than for final assembly. For instance, flagships can source keyboards, computer mouse devices, and power switch supplies from many different sources, in Asia, Mexico, and the European periphery, with Taiwanese firms playing an important role as intermediate supply chain coordinators. The same is true for printed circuit boards. Concentration of dispersion increases, the more we move toward more complex, capital-intensive precision components: memory devices and displays are sourced primarily from "centers of excellence" in Japan, Korea, Taiwan, and Singapore; and hard disk drives from a Singapore-centered triangle of locations in Southeast Asia. Finally, dispersion becomes most concentrated for high-precision, design-intensive components that pose the most demanding requirements on the mix of capabilities that a firm and its cluster needs to master: microprocessors, for instance, are sourced from a few globally dispersed affiliates of Intel, two American suppliers, and one recent entrant from Taiwan.[6]

In other words, geography continues to matter, even when DIS and high-velocity transportation are used. Rapid cross-border dispersion thus coexists with agglomeration. GFNs extend national clusters across national borders. This implies three things: first, some stages of the value chain are internationally dispersed, while others remain concentrated. Second, the internationally dispersed activities typically congregate in a limited number of overseas clusters. And third, agglomeration economies continue to matter, hence the path-dependent nature of development tra-

[5] "Cost & time reduction centers" include the usual suspects in Asia (Korea, Taiwan, China, Malaysia, Thailand, and now also India for software engineering and web services) but also exist in once peripheral locations in Europe (e.g., Ireland, central and eastern Europe, and Russia), Latin America (Brazil and Mexico), some Caribbean locations (like Costa Rica), and a few spots elsewhere in the so-called rest of the world (RoW).

[6] Ernst (2002a) provides a systematic analysis of the diversity of cluster dispersion, using examples from the semiconductor and hard drive industries.

jectories for individual specialized clusters. In short, the new mobility of knowledge remains constrained in space: while cross-border exchange of knowledge has penetrated new geographic areas, it remains limited to a finite number of specialized clusters.

Integration: Hierarchical Networks

A GFN integrates diverse network participants who differ in their access to and position within such networks and hence face very different opportunities and challenges. These networks do not necessarily give rise to less hierarchical forms of firm organization (as predicted, for instance, in Bartlett and Ghoshal 1989, and in Nohria and Eccles 1992). GFNs typically consist of various hierarchical layers, ranging from network flagships that dominate such networks, due to their capacity for system integration (Pavitt 2003), down to a variety of usually smaller, local specialized network suppliers.

FLAGSHIPS

The flagship is at the heart of the network: it provides strategic and organizational leadership beyond the resources that, from an accounting perspective, lie directly under its management control (Rugman 1997: 182). The strategy of the flagship company thus directly affects the growth, strategic direction, and network position of lower-end participants, like specialized suppliers and subcontractors. The latter, in turn, "have no reciprocal influence over the flagship strategy" (Rugman and D'Cruz 2000: 84).[7] The flagship derives its strength from its control over critical resources and capabilities that facilitate innovation, and from its capacity to coordinate transactions and knowledge exchange between the different network nodes.

Flagships retain in-house activities in which they have a particular strategic advantage; they outsource those in which they do not. It is important to emphasize the diversity of such outsourcing patterns (Ernst 1997). Some flagships focus on design, product development and marketing, outsourcing volume manufacturing, and related support services. Other flagships outsource as well a variety of high-end, knowledge-intensive support services. This includes, for instance, trial production (prototyping and ramping up), tooling and equipment, benchmarking of productivity, testing,

[7] With Rugman's flagship model, we share the emphasis on the hierarchical nature of these networks. However, there are important differences. Rugman and D'Cruz (2000) focus on localized networks within a region; they also include "nonbusiness infrastructure" as "network partners."

process adaptation, product customization, and supply chain coordination. It may also include design and product development.

To move this model a bit closer to reality, I distinguish two types of global flagships: (1) Original equipment manufacturers (OEM) that derive their market power from selling global brands, regardless of whether design and production is done in-house or outsourced; and (2) "contract manufacturers" (CM) that establish their own GFN to provide integrated manufacturing and global supply chain services (often including design) to the OEM.

LOCAL SUPPLIERS

Local suppliers differ substantially in their capacity to benefit from the new mobility of knowledge (Ernst, Ganiatsos, and Mytelka, 1998). Greatly simplifying, we distinguish two types of local suppliers: higher-tier and lower-tier. "Higher-tier" suppliers, like Taiwan's Acer group (Ernst 2000), play an intermediary role between global flagships and local suppliers. They deal directly with global flagships (both OEMs and CMs), possess valuable proprietary assets (including technology), and have sufficient resources to upgrade their absorptive capacities. Some of these higher-tier suppliers have even developed their own mini-GFN (Chen 2002). With the exception of hard-core R&D and strategic marketing, which remain under the control of the OEM, the lead supplier must be able to shoulder all steps in the value chain. It must even take on the coordination functions necessary for global supply chain management.

"Lower-tier" suppliers are the weakest link in the GFNs. Their main competitive advantages are low cost, speed, and flexibility of delivery. They are typically used as "price breakers" and "capacity buffers" and can be dropped at short notice. This second group of local suppliers rarely deals directly with the global flagships; they interact primarily with local higher-tier suppliers. Lower-tier suppliers normally lack proprietary assets; their financial resources are inadequate to invest in training and R&D; and they are highly vulnerable to abrupt changes in markets and technology, and to financial crises.

Contradictions

It is important to emphasize that nothing guarantees the uninterrupted growth of digital formations in the corporate sector. As with other such formations, inherent contradictions may well cause the pendulum to swing in the opposite direction. In this last section, we highlight problems in the efficiency of coordinating GFNs, focusing on recent developments in the electronics industry. In essence, these contradictions reflect a growing tension between increasingly complex interactions between multitier networks of networks and limited organizational capabilities to cope with the resulting coordination requirements.

Networks of Networks: Outsourcing Based on Contract Manufacturing

The "New Economy" boom in the United States has accelerated a long-standing trend toward vertical specialization. Especially in the electronics industry, outsourcing based on contract manufacturing became the "panacea of the '90s"(Lakenan et al. 2001:3), a "New American Model of Industrial Organization" (Sturgeon 2002). Two interrelated transformations need to be distinguished: supply contracts and M&A. Global brand leaders like Dell, the original equipment manufacturers, increasingly subcontract manufacturing and related services to U.S.-based global contract manufacturers, like Flextronics. Equally important, however, is that the very same CMs have acquired existing facilities of OEMs, as the latter are divesting internal manufacturing capacity, seeking to allocate capital to other activities that are expected to generate higher profit margins, such as sales and marketing and product development.

This has created increasingly complex, multitier "networks of networks" that juxtapose global ties between the two large global players (the OEMs and CMs), as well as intense regional ties with smaller firms (the local network suppliers). A focus on complex, multitier networks of networks distinguishes this analysis from Sturgeon's (2002) modular production network model. That model focuses on two actors only: global OEMs and CMs, most of them of American origin. OEMs and CMs are perceived to interact in a virtuous circle where each can only win. In that model, nothing can stop continuous outsourcing through contract manufacturing: "turn-key suppliers and lead firms co-evolve in a recursive cycle of outsourcing and increasing supply-base capability and scale, which makes the prospects for additional outsourcing more attractive" (Sturgeon 2002: 6).

Limitations to the U.S.-Style CM Model

In contrast, my analysis emphasizes serious limitations to the U.S. model of contract manufacturing, forcing both OEMs and CMs to adjust and rationalize the organization of their networks. That model was based on the assumption of uninterrupted demand growth. In reality, however, demand and supply only rarely match. This simple truth was all but forgotten during the heydays of the "New Economy."

Industry observers highlight seven important limitations.[8] First, global contract manufacturing is a highly volatile industry. While powerful forces push for outsourcing, this process is by no means irreversible. Major OEMs retain substantial internal manufacturing operations; they are continuously evaluating the merits of manufacturing products or providing services internally versus the advantages of outsourcing. Second, global CMs are now in a much weaker bargaining position than OEMs, whose number has been reduced by the current downturn and who are now much more demanding. In principle, important long-term customer contracts permit quarterly or other periodic adjustment to pricing based on decreases or increases in component prices. In reality, however, CMs "typically bear the risk of component price increases that occur between any such re-pricings or, if such re-pricing is not permitted, during the balance of the term of the particular customer contract" (Jabil 2001: 49).

A third important limitation of the U.S. CM model represents tradeoffs between specialization advantages and rapid inorganic growth through M&A. In economic theory, vertical specialization is supposed to increase efficiency, that is, to reduce the wastage of scarce resources. It is not clear whether the recent rapid growth of CM has produced this result. The excessive growth and diversification that we have seen during the "New Economy" boom may well truncate the specialization and efficiency advantages of the CM model. The leading CMs have aggressively used M&A to pursue in parallel four objectives that do not easily match: rapid growth; a broadening of the portfolio of services that they can provide; a diversification into new product markets (especially telecom equipment); as well as an expansion of their own production networks, establishing a global presence at record speed. Yet this forced pace of global expansion may well create an increasingly cumbersome organization that could undermine the supposedly primary advantage of the CM model: a capacity for rapid scaling up and scaling down, in line with the requirements of the OEMs.

Fourth, the rapid expansion of GFNs is subject to extreme risks and uncertainty. This reflects the much greater volatility of international operations compared to domestic ones. Managing GFNs thus requires major efforts, in terms of management time and resources, which of course conflicts with the need to keep overheads at very low levels.

Take the assessment of the risks involved in its international operations by a major U.S. global contract manufacturer (Jabil). In its 10K report for 2001 (p. 50), the company emphasizes the following risks:

[8] This section is based on e-mail correspondence with Bill Lakenan, lead author of a recent study by Booz-Allen & Hamilton on global contract manufacturing (Lakenan, Boyd, and Frey 2001); recent 10K reports of the leading U.S. global CMs; and author's interviews at affiliates of global CMs in Malaysia. See also Maltz et al. (2000), and Benson-Armer et al. (2004).

> difficulties in staffing and managing foreign operations; political and economic instability; unexpected changes in regulatory requirements and laws; longer customer payment cycles and difficulty collecting accounts; receivable export duties; import controls and trade barriers (including quotas); government restrictions on the transfer of funds to us from our operations outside the United States; burdens of complying with a wide variety of foreign laws and labor practices; fluctuations in currency exchange rates, which could affect local payroll, utility and other expenses; inability to utilize net operating losses incurred by our foreign operations to reduce our US income taxes; . . . (and, especially in lower-cost locations) . . . currency volatility, negative growth, high inflation, limited availability of foreign exchange.

Fifth, rapid growth, based on the use of stock as a currency for mergers and acquisitions, is extremely risky and contains the seed of future problems. It stretches the already limited financial resources of CMs, which typically have to cope with very low margins. The downturn of the global electronics industry has further increased these financial pressures on leading U.S.-based CMs.[9] This of course raises the question whether this will lead to off-balance sheet financing techniques to hide accumulated debt.

Sixth, in contrast to the original expectation that outsourcing based on contract manufacturing may improve inventory and capacity planning, global brand leaders in the electronics industry who rely heavily on outsourcing, have experienced very serious periodic mismatches between supply and demand. When a product unexpectedly becomes a hit, outsourcing provides these OEMs with only a limited capacity for scaling up. During a recession, on the other hand, OEMs cannot abruptly reduce orders that they had previously placed with CMs.[10]

Lastly, there seems to be a conflict of interest between OEMs, who are looking for flexibility, and CMs, who are looking for predictability and scale. For instance, OEMs focus on early market penetration and rapid growth of market share to sustain comfortable margins. OEMs thus need flexibility in outsourcing arrangements that allows them to divert resources at short notice to a given product, if it becomes a hit. This sharply contrasts with the situation of CMs: with razor-thin margins, they need to focus ruthlessly on cost cutting. CMs need predictability: "they want

[9] Ironically, these pressures are particularly severe for those CMs, like Solectron, that have aggressively diversified beyond the PC sector into telecommunications and networking equipment, the high-growth sectors of the "New Economy" boom.

[10] Take Cisco. During the peak of the "New Economy" boom, from 1999 to 2000, demand for its products grew by 50 percent. Reliance on CMs produced severe component shortages and a massive backlog in customer orders. When demand fell abruptly, starting from the fall of 2000, Cisco found itself saddled with excess capacity of $2.25 billion that it had put in place to meet expected demand growth. Excess capacity of this magnitude is deadly in time-sensitive industries like electronics.

to make commitments in advance to reap benefits like big-lot purchases and decreased overtime." (Lakenan et al. 2001: 10).

These conflicting interests complicate the coordination of CM-based outsourcing arrangements. They also require substantial changes in the organization of both OEMs and CMs, as well as an alignment of incentives through contract terms and agreements. If such alignment does not occur, it may well be that the new mobility of knowledge will face new constraints. The irony is that the more dispersed and digitized these global networks, the more difficult it becomes to coordinate them.

In short, effective outsourcing requires that both flagships and CMs acknowledge their conflicting interests. Further, with complexity comes uncertainty. In industries with rapidly shifting technologies and markets, OEMs have no way to predict with any accuracy the specifications of what they will need, in terms of capacity, design features, and configuration, and in terms of the specific mix of performance requirements. In the electronics industry, all of these variables can change quite drastically and at short notice. Such high uncertainty has important implications for the reorganization of CM-based outsourcing arrangements. Flexibility now becomes the key to success. Proceeding by conjecture ("stochastically") takes over from a deterministic approach. Flagships need adjustable networks to "satisfy a range of possible demand profiles with a portfolio of customizable capacity." They "need access to—and the ability to turn off—big chunks of production more quickly than ever contemplated in order to capture profitability" (Lakenan et al. 2001: 11, 12).

Conclusions

This chapter demonstrates that digital formations in the corporate sector are shaped by the evolution of cross-border forms of corporate networking practices, especially global flagship networks, and the increasing use of digital information systems to manage these networks. These two interrelated transformations in the organization of international business are *gradually* reducing constraints to international knowledge diffusion. GFNs expand interfirm linkages across national boundaries, increasing the need for knowledge diffusion, while DIS not only enhance information exchange, but also provide new opportunities for the creation, sharing, and joint utilization of knowledge. In the emerging global network economy, we thus need to reconsider and amend the "stickiness-of-knowledge" proposition.

The approach that I have chosen focuses on international knowledge diffusion through an extension of firm organization across national boundaries. I explored how two distinctive characteristics of GFNs, which are enhanced by DIS, shape the scope for international knowledge diffusion: a rapid yet concentrated dispersion of value chain activities, and, simultaneously, their integration into hierarchical networks. I demonstrated that the new mobility of knowledge is an *unintended* consequence of the evolution of global flagship networks. The more dispersed and complex these networks, the more demanding their coordination requirements. Hence, knowledge sharing is the necessary glue that keeps these networks growing.

But this occurs in complex ways. Knowledge diffusion has created new "cost-and-time-reduction centers" in lower-income regions that thrive on the timely provision of blue-collar knowledge support services like supply chain management, design services, and detailed engineering. Yet the sources of knowledge creation remain concentrated in a few global "centers of excellence" that combine unique capabilities in research, global branding, standard-setting, and system integration (Ernst 2004e). While reducing the constraints to knowledge diffusion can enhance global development, the critical issue remains the unequal distribution of the sources of innovation that global network flagships are unlikely to relinquish easily. Of global R&D, 86 percent takes place in industrialized countries, with the United States occupying the leading position with 37 percent (Dahlman and Aubert 2001: 34). For instance, the R&D budget of Microsoft, at around $6.2 billion (for 2003), exceeds China's total R&D budget. The United States has raced ahead in the most prized areas of technological innovation, as far as these can be measured by patent statistics. The U.S. "innovation score" measures the number of patents granted by the U.S. Patent Office, multiplied by an index that indicates the value of these patents.[11] Since 1985 the U.S. "innovation score" has more than doubled, a rate far better than in any other country (CHI/MIT 2003). In 2002 all fifteen leading companies with the best record on patent citations were based in the United States, with nine of them in the IT sector.

[11] The citation index measures the frequency of citation of a particular patent. When the US Patent Office publishes patents, each one includes a list of other patents from which it is derived. The more often a patent is cited, the more likely it is a pioneering patent, connected with important inventions and discoveries. An index of more than 1 indicates that patents are cited more often than would be expected for a specific group of technologies, while less than 1 indicates they are cited less often than expected.

References

Antonelli, C. ed. 1992. *The Economics of Information Networks.* North Holland, Amsterdam: Elsevier.

Archibugi, D., and J. Michie. 1995. The Globalisation of Technology: A New Taxonomy. *Cambridge Journal of Economics* 19:121–40.

Bartlett, C. A., and S. Ghoshal. 1989. *Managing across Borders: The Transnational Solution.* Boston: Harvard Business School Press.

Bell, Martin, and K. Pavitt. 1993. Technological Accumulation and Industrial Growth: Contrasts between Developed and Developing Countries. *Industrial and Corporate Change* 2(2).

Benson-Armer, R. J., D. L. Dean, and J. C. Kelleher. 2004. Getting Contract Manufacturers Back on Track. *The McKinsey Quarterly* (July).

Birkinshaw, J., and P. Hagstrøm, eds. 2000. *The Flexible Firm. Capability Management in Network Organizations.* Oxford: Oxford University Press.

Borrus, M., D. Ernst, and S. Haggard, eds. 2000. *International Production Networks in Asia. Rivalry or Riches?* London: Routledge.

Bradley, S. P., and R. L. Nolan, eds. 1998. *Sense and Respond: Capturing Value in the Network Era.* Boston: Harvard Business School Press.

Braudel, F. 1992. *The Perspective of the World, Civilization and Capitalism, 15th–18th Centuries.* Vol. 3. Berkeley: University of California Press.

Breschi, S., and F. Malerba. 2001. The Geography of Innovation and Economic Clustering: Some Introductory Notes. *Industrial and Corporate Change* 10(4): 817–34.

Brynjolfsson, E., and L. M. Hitt. 2000. Beyond Computation: Information Technology, Organizational Transformations and Business Performance. Manuscript, Sloan School of Management, MIT.

Chandler, A. D. 1977. *The Visible Hand: The Managerial Revolution in American Business.* Cambridge: Harvard University Press.

Chandler, A. D., and J. W. Cortada. 2000. The Information Age: Continuities and Differences. In *A Nation Transformed by Information,* edited by A. D. Chandler and J. W. Cortada. Oxford: Oxford University Press.

Chen, S. H. 2002. Global Production Networks and Information Technology: The Case of Taiwan. *Industry and Innovation* 9(3) (special issue on Global Production Networks).

CHI/MIT. 2003. Report on Innovation Scores survey, at www.CHI.com.

Christensen, C. M. 1997. *The Innovator's Dilemma: When New Technologies Cause Great Firms to Fail.* Boston: Harvard Business School Press.

Ciborra, C. U., et al. 2000. *From Control to Drift: The Dynamics of Corporate Information Infrastructures.* Oxford: Oxford University Press.

Dahlman, C. J., and J. E. Aubert. 2001. *China and the Knowledge Economy.* Washington, DC: World Bank Institute, The World Bank.

Dicken, P. 1992. *Global Shift: Transforming the World Economy.* 3d ed. New York: Guilford Press.

Dunning, J. 1981. *International Production and the Multinational Enterprise.* London: George Allen & Unwin.

———, ed. 2000. *Regions, Globalization and the Knowledge-Based Economy.* Oxford: Oxford University Press.

Ernst, D. 1997. *From Partial to Systemic Globalization. International Production Networks in the Electronics Industry.* Report prepared for the Sloan Foundation project on Globalization in the Data Storage Industry, Data Storage Industry Globalization Project Report 97–02, Graduate School of International Relations and Pacific Studies, University of California at San Diego.

———. 1998: High-Tech Competition Puzzles. How Globalization Affects Firm Behavior and Market Structure in the Electronics Industry. *Revue d'Economie Industrielle* 85.

———. 2000. Inter-Organizational Knowledge Outsourcing: What Permits Small Taiwanese Firms to Compete in the Computer Industry? *Asia Pacific Journal of Management,* 17(2): 223–55.

———. 2001a. The Internet, Global Production Networks and Knowledge Diffusion. Challenges and Opportunities for Developing Asia. In *PTC2001: From Convergence to Emergence: Will the User Rule*? Proceedings of the 2001 Annual Pacific Telecommunications Council Conference, January 14–18, Honolulu.

———. 2001b. The Evolution of a Digital Economy. Research Issues and Policy Challenges, *Economia e Politica Industriale* 28 (110) (October): 127–39.

———. 2002a. Global Production Networks and the Changing Geography of Innovation Systems: Implications for Developing Countries. *Journal of the Economics of Innovation and New Technologies* 11(6): 497–523.

———. 2002b. The Economics of Electronics Industry: Competitive Dynamics and Industrial Organization. In *The International Encyclopedia of Business and Management (IEBM), Handbook of Economics,* edited by W. Lazonick. London: International Thomson Business Press.

———. 2004a. Complexity and Internationalisation of Innovation: Why Is Chip Design Moving to Asia? *International Journal of Innovation Management,* special issue in honor of Keith Pavitt. Also available as East-West Center Economics Working Paper #64.

———. 2004b. Pathways to Innovation in Asia's Leading Electronics Exporting Countries—A Framework for Exploring Drivers and Policy implications. *International Journal of Technology Management.*

———. 2004c. Late Innovation Strategies in Asian Electronics Industries—A Conceptual Framework and Illustrative Evidence. Manuscript, East-West Center.

———. 2004d. Global Production Networks in East Asia's Electronics Industry and Upgrading Perspectives in Malaysia. In *Global Production Networking and Technological Change in East Asia,* edited by S. Yusuf, M. A. Altaf, and K. Nabeshima. Washington, DC: World Bank and Oxford University Press.

———. 2004e. Limits to Modularity—A Review of the Literature and Evidence from Chip Design. East-West Center Economics Working Paper #71.

Ernst, D., T. Ganiatsos, and L. Mytelka, eds. 1998: *Technological Capabilities and Export Success: Lessons from East Asia.* London: Routledge.

Ernst, D., P. Guerrieri, S. Iammarino, and C. Pietrobelli. 2001. New Challenges for Industrial Districts: Global Production Networks and Knowledge Diffusion.

In *The Global Challenge to Industrial Districts. Small and Medium-Sized Enterprises in Italy and Taiwan,* edited by P. Guerrieri, S. Iammarino, and C. Pietrobelli. Aldershot: Edward Elgar.

Ernst, D., and Kim, L. 2002a. Global Production Networks, Knowledge Diffusion and Local Capability Formation. *Research Policy* 31(8/9) (special issue in honor of Richard Nelson and Sydney Winter).

———. 2002b. Introduction: Global Production Networks, Information Technology and Knowledge Diffusion. *Industry and Innovation* 9(3) (special issue on Global Production Networks).

Ernst, D., and B. Lundvall. 2004. Information Technology in the Learning Economy—Challenges for Developing Countries. In *Evolutionary Economics and Income Inequality,* edited by E. Reinert. London: Edward Elgar.

Ernst, D., and D. O'Connor. 1992. *Competing in the Electronics Industry: The Experience of Newly Industrialising Economies.* Paris: Development Centre Studies, OECD.

Ernst, D., and T. Ozawa. 2002: National Sovereign Economy, Global Market Economy, and Transnational Corporate Economy. *Journal of Economic Issues* 36(2) (June).

Ernst, D., and J. Ravenhill. 1999. Globalization, Convergence, and the Transformation of International Production Networks in Electronics in East Asia. *Business and Politics* 1(1).

Flaherty, T. 1986. Coordinating International Manufacturing and Technology. In *Competition in Global Industries,* edited by M. Porter. Boston: Harvard Business School Press.

Flamm, K. 1999. Digital Convergence? The Set-Top Box and the Network Computer. In *Competition, Innovation and the Microsoft Monopoly: Antitrust in the Digital Market Place,* edited by J. A. Eisenach and T. M. Lenard. Boston: Kuwer Academic Publishers.

Foray, D., and W. E. Steinmueller. 2001. Replication of Routine, the Domestication of Tacit Knowledge and the Economics of Inscription Technology: A Brave New World? Paper presented at conference in honor of Richard R. Nelson and Sidney Winter, Danish Research Unit in Industrial Dynamics (DRUID), Aalborg, Denmark, June 12–15.

Gereffi, G., and R. Kaplinsky. 2001. The Value of Value Chains. *IDS Bulletin* 32(3) (special issue on Global Value Chains).

Gereffi, G., and M. Korzeniewicz, eds. 1994. *Commodity Chains and Global Capitalism.* Westport, CT: Praeger.

Ghoshal, S., and C. A. Bartlett. 1990. The Multinational Corporation as an Interorganizational Network. *Academy of Management Review* 15(4): 603–25.

Grove, A. S. 1996. *Only the Paranoid Survive. How to Exploit the Crisis Points That Challenge Every Company and Career.* New York: Harper Collins Business.

Hagstrøm, P. 2000. New Wine in Old Bottles: Information Technology Evolution in Firm Strategy and Structure. In *The Flexible Firm: Capability Management in Network Organizations,* edited by J. Birkinshaw and P. Hagstrøm. Oxford: Oxford University Press.

Henderson, J., et al. 2001. Global Production Networks and the Analysis of Economic Development. Manuscript. Manchester Business School, University of Manchester.

Jabil. 2001. Form 10-K Report to the U.S. Securities and Exchange Commission. Washington, DC.

Jørgensen, H. D., and J. Krogstie. 2000. Active Models for Dynamic Networked Organisations." Working Paper #23, Institute of Computer & Information Sciences, Norwegian University of Science and Technology (available at http:// hdj.jok@informatics.sintef.no).

Kim, L. 1997. *Imitation to Innovation. The Dynamics of Korea's Technological Learning*. Boston: Harvard Business School Press.

Kogut, B. 1985. Designing Global Strategies: Profiting from Operational Flexibility. *Sloan Management Review* (Fall).

Kogut, B., and N. Kulatilaka. 1994. Operating Flexibility, Global Manufacturing, and the Option Value of a Multinational Network. *Management Science* 40(1) (January).

Kogut, B., and U. Zander. 1993. Knowledge of the Firm and the Evolutionary Theory of the Multinational Corporation. *Journal of International Business Studies* (fourth quarter).

Lakenan, B., D. Boyd, and E. Frey. 2001. Outsourcing and Its Perils. *Strategy + Business*, Booz-Allen & Hamilton, no. 24 (third quarter).

Langlois, R. N. 1999. External Economies and Economic Progress: The Case of the Microcomputer Industry. *Business History Review* 66 (Spring): 1–50.

Lerner, J., and J. Tirole. 2000. The Simple Economics of Open Source. Manuscript. Harvard Business School.

Litan, R. E., and A. M. Rivlin. 2001. *The Economic Payoff from the Internet Revolution*. Washington, DC: Brookings Institution Press.

Macher, J. T., D. C. Mowery, and T. S. Simcoe. 2002. eBusiness and the Semiconductor Industry Value Chain: Implications for Vertical Specialization and Integrated Semiconductor Manufacturers. *Industry and Innovation* 9(3) (special issue on Global Production Networks).

Maltz, A. B., et al. 2000. Lessons from the Semiconductor Industry. The International Sematech Semiconductor Logistics Forum Study. *Supply Chain Management Review* (November/December).

Markusen, A. 1996. Sticky Places in Slippery Space: A Typology of Industrial Districts, *Economic Geography* 72: 293–313.

Milgrom, P., and J. Roberts. 1990. The Economics of Modern Manufacturing: Technology, Strategy, and Organization. *The American Economic Review* 80(3): 511–28.

Nohria, N., and R. G. Eccles. 1992. *Networks and Organizations: Structure, Form, and Action*. Boston: Harvard Business School Press.

Nolan, R. L. 2000. Information Technology Management since 1960. In *A Nation Transformed by Information*, edited by A. D. Chandler and J. W. Cortada. Oxford: Oxford University Press.

North, D. C. 1996. Institutions, Organizations, and Market Competition. Keynote address to Sixth Conference of the International Joseph Schumpekr Society, Stockholm, June 2–5.

Pavitt, K. 2003. Are Systems Designers & Integrators "Post-Industrial" Firms? In

Systems Integration and Firm Capabilities, edited by A. Prencipe, A. Davies, and M. Hobday. Oxford: Oxford University Press.

Pedersen, J. D., M. Tølle, and J. Vesterager. 1999. Global Manufacturing in the 21st Century. Report prepared for ESPRIT, European Commission, Brussels, November 30.

Porter, M. 1990. *The Competitive Advantage of Nations.* London: Macmillan.

Powell, W., and L. Smith-Doerr. 1994. Networks and Economic Life, In *The Handbook of Economic Sociology,* edited by N. Smelser and R. Swedber. Princeton: Princeton University Press.

Richardson, G. B. 1996. Competition, Innovation and Increasing Returns. DRUID Working Paper no. 96–10, Department of Business Studies, Aalborg University.

Rugman, A. M. 1997. Canada. In *Governments, Globalization and International Business,* edited by J. H. Dunning. London: Oxford University Press.

Rugman, A. M., and J. R. D'Cruz. 2000. *Multinationals as Flagship Firms: Regional Business Networks.* Oxford: Oxford University Press.

Sichel, D. E. 1997. *The Computer Revolution: An Economic Perspective.* Washington, DC: Brookings Institution Press.

Sobel, R. 1986. *IBM vs. Japan. The Struggle for the Future.* New York: Stein & Day.

Sturgeon, T. 2002. Modular Production Networks: A New Model of Industrial organization. *Industrial and Corporate Change* 11(3).

UNCTAD. 1993. *World Investment Report, 1993: Transnational Corporations and Integrated International Production.* Geneva.

Wilkins, M. 1970. *The Emergence of Multinational Enterprise.* Cambridge: Harvard University Press.

Williamson, O. E. 1975. *Markets and Hierarchies: Analysis and Antitrust Implications.* New York: The Free Press.

———. 1985. *The Economic Institutions of Capitalism: Firms, Markets, Relational Contracting.* New York: The Free Press.

Yau, P., and N. Das. 2001. *Growing Trade in Intellectual Property Rights.* New York: Credit Suisse First Boston.

Zander, U., and B. Kogut. 1995. Knowledge and the Speed of the Transfer and Imitation of Organizational Capabilities: An Empirical Test. *Organizational Science* 6(1).

Zanfei, A. 2000. Transnational Firms and the Changing Organisation of Innovative Activities. *Cambridge Journal of Economics* 24:515–42.

NETWORKS OF COOPERATION

Cooperative Networks and the Rural-Urban Divide

D. LINDA GARCIA

THE POSITIVE ROLE that networked information technologies can play in fostering economic development is now widely recognized.[1] These technologies have proven extremely useful not only in promoting and sustaining economic activities of all kinds, but also in enhancing human potential—a key ingredient for the success of any development strategy. The value of these technologies will likely loom even larger in the future, given both their enhanced capabilities as well as a more service-oriented global economy in which production and marketing activities are networked worldwide. In preparation, many developing countries are currently looking to communication and information technologies to help them bypass the long and arduous process of industrialization, allowing them—straight away—to join the information age. Likewise, many development organizations and international NGOs are now focusing their funding and efforts on issues related to the "digital divide." Increasingly aware of the positive externalities associated with global interconnection, even the private sector is joining together in international forums, such as the World Economic Forum and the Global Business Dialogue, to promote worldwide access.

Even the strongest advocates of infrastructure deployment are quick to point out, however, that communication and information technologies, although necessary, are insufficient for sustainable development (Hudson 1997). In fact, more often than not, these technologies have acted as a double-edged sword, giving rise to both positive and negative outcomes. The two-sided nature of network technologies are particularly

[1] There is without doubt a growing body of evidence that shows a significant positive correlation between investment in telecommunications and economic growth. Analyzing thirty-two years of U.S. data, Cronin et al. (1993) found, for example, that causality operates in two directions: telecommunications investments increase to a significant degree with economic growth, while economic growth expands with the investment in telecommunications. Similarly, in their study of the fifty U.S. states, Dholakia and Harlam (1994) not only confirmed this causal relationship but also found that the link between telecommunications infrastructure and economic development is strengthened when other factors such as education and physical infrastructure are simultaneously taken into account.

pronounced in rural areas (Innis 1951). On the positive side, these technologies can overcome the barriers of distance and time, so they allow rural communities to link up to the growth potential of larger, city economies. At the same time, however, because they can foster a net outflow of resources from rural areas, these technologies often serve to undermine the long-term economic viability of rural communities.

Today, rapid advances in communication technologies are once again restructuring and redefining rural communities and markets. Whereas in the past, networking technologies brought rural villages and towns into a larger, national community, now they link communities worldwide. Just as industrialization served to disadvantage rural areas, so too might the global information economy. In the future, for example, profits and growth opportunities will be ever more closely linked to transaction costs. Under such circumstances, cities—which benefit greatly from economies of agglomeration—will have an even greater advantage over nonmetropolitan areas than they have today (Sassen 1989; Castells 1989).

Whether or not advanced networking technologies put rural areas at greater risk will depend not only on their capabilities and accessibility, but also—and perhaps more importantly—on the social, economic, and political context in which these technologies are deployed. Having access to new technologies without the skills to employ them, for example, will yield little, if any, benefits. To achieve the desired results, therefore, technology deployment strategies must be linked to complementary social and economic policies that address other—and often more formidable—developmental barriers.

Thus, if rural areas are not to be left behind in the global economy, new ways must be found to design technology-based networks to meet rural needs. In particular, attention must be focused not simply on the problem of deploying advanced technologies, as has been the case in recent discussions of the "digital divide" (Compaine 2001). Equally, if not more, important is the task of creating the optimal conditions for reaping the benefit of these technologies in a rural setting. With this goal in mind, this chapter explores the role that cooperative institutions might play in addressing the rural-urban divide. The case is made that locally based cooperatives can play a unique role in promoting the diffusion of a networked-based economic infrastructure in rural communities. In particular, such institutions can serve not only to link rural communities to the global economy, but also—and as importantly—to reinforce their local environments. Because cooperatives generate social as well as financial capital, they can help to foster innovation and learning. Equally important, being locally based, networked providers can tailor networks to meet the particular needs of a rural community. Thus, networks can be specifically designed to reinforce local strengths while compensating for local weaknesses. By embedding their economies at the local level, rural communities will be less at risk as well as better positioned to reap the benefits of global markets.

In today's liberal, deregulatory environment, such cooperative solutions are hardly in vogue. To the contrary, competition is the prevailing maxim as well as the criterion according to which all policies are typically assessed. Not surprisingly, therefore, the problem of the global digital divide is typically viewed as one of technology deployment, while its solution is sought—more often than not—by promoting trade liberalization and support for foreign investment. In laying out a cooperative approach to address the "digital divide" in rural areas, this chapter calls for a new epistemological approach. Market strategies are appropriate only to the extent that markets function well. Such is not the case in rural areas. For advanced networking technologies to promote rural economic development, what is needed is not so much competitive market strategies but rather social innovations that can not only serve to foster deployment but also compensate for the multiple market failures typically found in rural communities.

Characterizing Rural Communities

Rural communities are by no means all alike; nonetheless, they share a common set of problems that are associated with the "rural condition." To address these common problems, it is important to have a clear analytical notion of what rural entails. For without such a conceptualization, it is impossible to identify appropriate strategies for coping with these problems. As described by Hoff, Braverman, and Stiglitz (1993: ix):

> [i]n order to design effective policies to remedy a market failure, one has to understand its underlying source. One needs also to recognize that the interactions among markets are not limited to ones of price and income, as modeled in general equilibrium theory. What happens in one sector or market can have repercussions on the nature of transaction costs, risks, and enforcement mechanisms used in other markets. To design effective development policies, one therefore needs a theory of rural organization.

Given the diversity of rural communities, the choice of a conceptualization must depend on the analytical problem at hand. Thus, in the case of telecommunications, policymakers might stress factors such as remoteness and population density, insofar as these are the most important variables determining the cost of network deployment. Others, concerned primarily about the preservation of "places," might look instead at historical longevity and the community-based structures to be found in rural areas. For the purposes of this chapter, it is useful to characterize rural communities in terms of the organizational structure of their markets as well as their location and positioning in relation to the global hierarchy of markets.

Conceptualizing rural areas in terms of the organization of their markets is in keeping with the history of rural communities and the market failures and urban dependencies that traditionally have been associated with rural economies. As described by Jane Jacobs (1994: 124), many rural economies are "passive economies" insofar as they do not create economic change themselves but rather respond to forces unloosed in distant cities. Employing the French hamlet of Bardou as a descriptive metaphor of the rural economic predicament, she notes:

> Time and again like a toy on a string, Bardou has been jerked by some *external* economic energy or other. In ancient times the site was exploited for its iron, then abandoned. In modern times it was depopulated when distant jobs attracted its people, then repopulated by city people. The jerks were not gentle. But when cities and city people left Bardou alone, had no uses for it, the place either had no economy whatever, as when it was wilderness, or else a subsistence economy that remained unchanging.

Not surprisingly, in many of these areas, the obstacles to economic growth and development are found in excess. Infrastructures are poor; markets are subject to numerous failures; investment capital is scarce; and human resources are underdeveloped. Moreover, these problems are highly interdependent, so market failures in one area often spill over and compound those in others. The result is a vicious circle that spirals downward (Hoff, Braverman, and Stiglitz 1993). Thus, for example, one finds a positive relationship between a population's density and organizational formation as well as among a population's literacy level, the degree of urbanization, and the extent of organizational formation (Pennings 1981; Stinchcombe 1965).

Conceptualizing rural communities in terms of the organization of their markets is also necessary if we are to understand how advanced networking technologies might improve the prospects of rural economies. For it is precisely by reordering and reconfiguring organizational and geographical relationships that information-based networking technologies can contribute to economic development. In the past, rural communities suffered because the expansion of markets took place at the expense of local ties, thereby depriving them of economies of scale, scope, and agglomeration. Explaining this tradeoff, Evans and Wurster (2000: 23–24) point out that "the communication of rich information has required proximity and dedicated channels whose costs or physical constraints have limited the size of the audience to which information could be sent. Conversely, the communication of information to a large audience has required compromises in bandwidth (amount), customization, and interactivity." Today, in contrast, given the greater capacity and functional characteristic of networking technologies, this tradeoff is no longer required. Thus, highlighting the ways in which networking technologies can affect the organization of economic activities may point the way to new, more promising, development strategies for the future.

Patterns of technology diffusion and innovation are also related to organizational structures, especially the way in which they determine the availability of information and the effectiveness of agencies of technology diffusion (Brown 1981). One finds, for example, that although markets may do well in performing these information-related functions in urban areas, in rural areas socially based institutions and conventions are often required (Hoff, Braverman, and Stiglitz 1993).

To capture these critical variables while at the same time accommodating the diversity of rural communities, this chapter views "rural" and "urban" as ideal types that are located at opposite ends of a continuum.[2] Accordingly, depending on their organizational structure and geographical positioning, communities will be considered to be more or less rural depending on the extent to which their markets (1) are remote and nonintegrated into urban, national, and global markets; (2) suffer from failures due to information asymmetries; and (3) are unable, given their low population densities, to take advantage of economies of scale and agglomeration associated with city-based economies. Focusing on the "rural" end of the spectrum, the discussion will lay out networking strategies that are specifically designed to address these types of failures.

The Impact of Networking Technologies

Although isolated and remote, rural communities do not exist in a vacuum. They are linked to the world surrounding them through a variety of transportation and communication networks, and the information and commodities that flow across them. The impact of these networking technologies has not always been favorable, however. To anticipate how today's networks might structure rural-urban economic relationships, it is useful to consider the impact that the railroads, the telegraph, and the mass media had on rural American communities more than a century ago.

[2] As Hart (1995: 64) has noted, classifying the extremes is not a problem. As he points out, "the traditional rural-urban divide has become a continuum. The ends of this continuum are not debatable. No one, for example, would argue that mid-town Manhattan is rural, or that a wheat field in North Dakota is urban, but the rural-urban continuum has no unambiguous 'natural' break that is generally recognized and accepted."

Before the advent of these technologies, social life within rural American farm communities was self-contained. The provision of services was unspecialized. The community provided the institutional context in which families organized to worship and educate their children. Members of each community relied on their families and other local institutions to cushion the hardships of rural life.

The advent of the telegraph and the railroad served to undermine this self-sufficiency. By extending their ties and expanding their markets, communication technologies made rural communities more vulnerable to external developments and events over which they had little control. The vast network of transportation and communication technologies not only served to channel resources away from rural communities; they also created conditions for economic success that rural communities were increasingly unable to fulfill.

To compete in the new economy, rural communities required—at a minimum—access to advanced transportation and communication networks. For the shifts in the national economy were not accidental. They were closely associated with the development of regional and national infrastructures, and a rural area's proximity to these trade networks proved to be a critical factor in determining its economic viability. Rural communities, however, typically lagged behind urban areas in the diffusion process. Because of the high fixed costs entailed in constructing networks, networking providers focused on deploying technologies first to high-density urban areas where the costs of deployment were lower and could be shared across a wider, more lucrative, customer base. As a result, favorably situated businesses in high-density, urban corridors usually enjoyed a head start of several decades in utilizing networking technologies, thereby gaining a significant competitive advantage.

This uneven pattern of deployment was clearly evident, for example, in the case of the telephone. First came major trunks linking northeastern cities, followed by lines to smaller towns in their immediate hinterlands, and then connection to midwestern cities. Thus, although the telephone was patented in 1876, it took twelve years before it reached Chicago. Transnational service was not inaugurated until 1915. Reaching rural areas took even longer. So much so, in fact, that by 1950 only 40 percent of all farm residences had telephones (Office of Technology Assessment 1992).

Having access to networking technologies was no guarantee of economic success, however. In fact, as often as not, networking technologies had a deleterious impact, serving to favor urban economies over rural ones. The rise of urban areas at the expense of rural economies resulted in part from the vastly increased scale and scope of the national market, made possible by networking technologies.[3] Using the telephone and telegraph, for example, businesses were able to expand their spheres of operation and centralize decision-making in distant headquarters. As firms extended their research, transaction costs increased, forcing firms to become larger and larger. As described by DuBoff (1983: 257):

> The telegraph dramatically enlarged information networks; it saved time, reduced the need for large inventories, decreased financing requirements, and prompted elimination of middlemen. But "competition" and "monopoly" are not, as neoclassical theory implies, polar opposites. The telegraph improved the functioning of markets and enhanced competition, but it simultaneously strengthened forces making for monopolization. Larger scale business operations, secrecy and control, and spatial concentration were all increased as a result of telegraphic communication.

While urban communities had the resources to support business organization on such a grand scale, rural economies did not.

The advent of the mass media also reinforced the development of a national marketplace, exacerbating the growing disparity between rural and urban areas. The emergence of inexpensive popular magazines such as *The Saturday Evening Post, The Ladies Home Journal,* and *Country Gentleman* intensified competition for advertising among segments of the publishing industry, and the winners in this competition reflected shifts in the nation's marketing system (Peterson 1964). The metropolitan press increasingly tied its fortunes to department stores and chains, and magazines were well positioned to run advertisements for nationally marketed consumer goods that were sold through all kinds of outlets. As a result, the small, local retailers, which had once served their communities with little competition, suddenly faced a succession of new challenges—department stores, mail-order firms, and chain stores (Office of Technology Assessment 1992).

Compounding the problems of rural America, networking technologies served at the same time to undermine the social cohesiveness of local communities—the very attribute that had been their mainstay. Weakening their sense of autonomy and resolve, networking technologies made it more difficult for rural communities to develop strategies necessary to compete in an increasingly national marketplace.

[3] This development is clearly illustrated by a convergence of prices across the nation. As Richard Duboff (1983: 257) notes with respect to the cotton market, "Data on cotton prices in New York show diminishing fluctuations over time. The average spread between lowest and highest prices narrowed steadily, except during the Civil War and its aftermath, and the steepest declines in high-low price ranges and dispersion of prices from decade averages came in the 1850s—'the telegraph decade,' as it might be called."

The impact of the telegraph on rural communities provides a case in point. Because of the high costs, telegraph use was confined largely to businesses and the press; few people used it for social communication, at least in the United States. Among the press, the telegraph fostered the standardization and central processing of news reports, allowing all Americans to read the same national and international news stories for the first time. But standardized content diminished the community's importance in the eyes of its local citizens, while centralization shifted the locus of control from local editors to national press association headquarters and bureaus (North 1884).

To meet the needs of the industrial economy, new towns and trade centers emerged, located at reasonable traveling distance from farm communities. Taking advantage of improved transportation and communication networks, these centers were, in turn, linked more and more to urban areas, leaving rural towns to fend for themselves.[4] The subsequent deployment of modern highway facilities served only to reinforce this uneven pattern of development and its associated impacts. Although road building brought rural and urban areas closer together, it forced many small communities to deal with urban values for the first time. Highways also facilitated massive rural out-migration. Concomitantly, by facilitating specialization in agriculture, highways reduced the need for farm labor, inducing many rural residents to seek urban jobs. Highways also contributed to population decentralization. Nonfarm employment expanded in the hinterlands along freeways and modern roads. Industrial belts grew up in towns and countryside along highways, especially in the southern and border states. The nation's midsized cities linked by freeways also grew at the expense of rural communities (Office of Technology Assessment 1992).

Today, rapid advances in information-based networking technologies are once again restructuring and redefining rural communities and markets. However, whereas in the past communication and information technologies brought rural villages and towns into a larger, national community, now they link communities worldwide. In a global environment such communities will be faced with far more competition in the very sectors—

[4] According to Swanson (1990: 22), rural communities were self-contained production units. However, with industrialization, "previous social formations, such as the rural church or the one-room, six-grade school house, gave way to the demands of new industrial employers and regional and national trade. Rural schools were not expected to prepare children for the financial and technical demands of a rapidly industrializing agriculture and nonfarm sector. Local socioeconomic networks such as cooperative harvesting (and risk taking), and quasi barter exchange systems that mediated local production and consumption under non commercial conditions were gradually subordinated to and/or eclipsed by new institutions."

such as primary products and manufacturing—on which they depend for their livelihoods. At the same time, the terms of trade for these sectors will continue to decline relative to service and knowledge-based industries.

In an information-based global economy, profits and growth opportunities will be ever more closely linked to transaction costs. Under such circumstances, cities—which benefit greatly from economies of agglomeration—will have an even greater economic advantage over nonmetropolitan areas than they have today. In fact, cities will themselves be ranked depending on their size and importance, with those on the top of the hierarchy serving as central hubs and access ramps of the global economy (Gottman 1983; Sassen 1989; Castells 1989).

Equally troubling for rural communities is the possibility that the extension and intensification of global interactions might occur at the expense of local ties (Amin and Thrift 1995; Cox 1997). Describing the basic features entailed in globalization, David Held (1995: 21) notes, for example: "What is new in the modern global system is the stretching of social relations in and through new dimensions of activity—technological, organizational, administrative and legal among others—and the chronic intensification of patterns of interaction mediated by such phenomenon as modern communication networks and information technology." If—in the course of this "stretching"—globalization further undermines the links between rural communities and urban centers upon which their economies depend, the long-term viability of these communities will be at risk.[5]

To reverse this pattern will require a concerted and integrated effort that not only addresses the problems that rural economies have traditionally faced in competing in worldwide markets. Such an effort must also be designed to help local economies develop innovative economic strategies that allow them to link up to global markets, while at the same time reinforcing and replenishing their economies by "reembedding" them in their local environments. To identify such strategies, it is necessary first to consider more closely the forces driving the globally networked economy.

Imperatives of the Globally Networked Economy

The organizational requirements of a networked economy are likely to be quite different from those in the industrial era. In a highly complex and

[5] As Jacobs (1994: 124) has characterized this type of situation, "Economies that have previously served city markets or have sent out people to city jobs or received technology, city transplants, city money, can eventually lose their ties to cities. If they do, their people sink into lives of rural subsistence. But as they adjust to sheer subsistence, they shed or lose many former practices and skills."

rapidly changing global economy, gaining competitive advantage no longer depends solely on achieving efficiency and cost reduction. Increasingly, it depends on the effectiveness of businesses—their ability to innovate, respond just-in-time, focus on quality, and establish more cooperative interfirm and intrafirm relationships. Instead of standardization, flexible production systems are called for, which allow businesses to respond quickly to changing demand, and to customize their products without sacrificing economies of scope.[6]

In this new environment, vertical bureaucracies are pushed to their limits. Businesses everywhere are enhancing their flexibility by downsizing and outsourcing. They are increasingly purchasing in the market what they need, whether preassembled parts, logistical support systems, customized communication services, or packaged business information. As described by Grabher (1993: 16):

> [The] strategy of vertical integration was successful when the pace of technology change was relatively slow, production processes were well understood and standardized and production runs turned out large numbers of similar products. Today, however, such large-scale vertical integration has serious weaknesses: inability to respond quickly to competitive changes in international markets; resistance to process innovations that alter the relation between different stages of the production process; and relative lack of willingness to introduce new products.

Information-based networking technologies are both driving and facilitating the adaptation of business to these structural changes. Configured in a networked architecture, these technologies not only extend the reach of market transactions as in the past; perhaps more importantly, they can also greatly enhance the density and functionality of market transactions, thereby generating the kinds of economies of agglomeration that hitherto were available only in tight-knit urban markets (Garcia 1998). Using these technologies, businesses can integrate and compress the time from product innovation to marketing to drive demand and maximize customer responsiveness. Coupled together loosely, they can rearrange their activities around teams and networks to bring together everyone involved in the life-cycle of a product. Working together and sharing the same information, they can carry out all business processes in parallel. This kind of network structure reduces the time involved in product development and leads to higher-quality products (Garcia 1998).

[6] As characterized by Ayres (1992: 21), "The key to the suggested 'new paradigm' for economic growth is that increasing flexibility progressively reduces the cost differential between customized and standardized products. The smaller this differential, the greater the demand for diversity and, hence, flexibility. But this process, in turn, leads to further improvement in the manufacturing process, traditional cost-driven engine of growth."

As these technologies and their various functions are brought together into integrated and interactive networks, more and more trade will take place electronically, in a virtual environment. Already, companies are moving many of their key activities online (Garcia 2001). Thus, for example, General Motors and Ford Motor Company have set up an electronic market for all goods and services they buy. Likewise, in the energy sector, Royal Commerce One and Royal Dutch/Shell link global buyers of oil, gas, and chemicals. Not surprisingly, given these developments, business-to-business e-commerce is predicted to increase from $50 million in 1998 to $1.3 billion in 2003 (Bruno 2000).

Geographic Implications

How these electronic organizations and markets evolve, and the actual form that they take, will have significant consequences for the functioning of the global economy, and the way in which costs and benefits are distributed among countries and regions within countries. Because electronic markets can reduce the overall costs of doing business, they can greatly enhance efficiency and lead to expanded trade. But the pattern of this expansion will depend on a number of factors, including the rate and evenness of diffusion, the rules for interconnection, and the structure of the network architecture (Garcia 2001).

What can be said for certain, however, is that local places will continue to matter (Sassen 1998). However, these places will not remain untouched by the global expansion of markets; to the contrary, in order to survive, they will have to redefine themselves in relationships to these markets.[7] For, as Massey has argued, territorial places are never static; rather, they are "constructed out of the juxtaposition, the intersection, the articulation of multiple social relations and should be seen as 'shared spaces' riven with internal tensions and conflicts" (Massey 1993: 18–19).

Where the local and global meet, two interrelated forces are likely to be at work. On the one hand, globalization is operating to eliminate the key economic distinctions that are associated with specific places (Storper 1997a: 20). Thus, for example, we are witnessing the standardization of tastes, technologies, and techniques on a global scale. However, and somewhat paradoxically, this aspect of globalization is giving rise to totally new types of production techniques that are embedded in territorial lo-

[7] As Storper (1997a, 26) notes, "A model of the global firm does not so much imply deterritorialization of the economic process as a recasting of the role of territories in complex, intraorganizationally and interorganizationally linked global business flow."

cales. As Storper (1997a: 35; 1997b) has emphasized, the type of globalization we are experiencing today:

> opens up markets to products based on superior forms of 'local knowledge;' it consolidates markets and leads to such fantastic product differentiation possibilities that markets refragment and with them, new specialized and local divisions of labor reemerge; and it in some ways heats up the competitive process albeit among giants creating new premia on technological learning that requires the same firms that become new global supply oligopolists to root themselves in locationally specific relational assets.

In the globally networked economy, old industrial cities, which were designed to accommodate mass production, will continue their decline. However, urbanization will continue apace (Storper 1991; Sassen 1998). Given the breakdown of economic and political boundaries, metropolitan areas will extend their connections by incorporating urban regions into their vastly expanded networks that stretch across the globe (Gereffi and Korzeniewicz 1994). At the same time, new industrial regions are likely to emerge in places that were previously underdeveloped. Describing the new geography of the networked global economy as it is presently unfolding, Scott (1998: 68) notes:

> the developed areas of the world are represented as a system of polarized regional economies each consisting of a central metropolitan area and a surrounding hinterland (of indefinite extent) occupied by ancillary communities, prosperous agricultural zones, smaller tributary centers and the like. . . . Each metropolitan nucleus is the site of intricate networks of specialized but complementary forms of economic activity, together with large, multifaceted local markets, and each is a locus of powerful agglomeration economies and increasing return effects. As such they are not only large in size but also constantly growing yet larger. These entities can be thought of as the *regional motors of the new global economy.*

The result is the rise of the "Galactic" city, which extends from one major metropolis to another. Although such areas appear unplanned and disorderly, they have an inherent logic to them. Describing their emergence in the United States, Lewis (1995: 50) notes:

> this new galactic city is an urban creation different from any sort Americans have ever seen before. And because it does not spread across the rural landscape along a solid front the way cities used to, many people—scholars included—fail to recognize it for what it is, a genuine city. It performs all the functions that American cities have always performed: commercial, industrial, residential, and social. What makes it different is its geographic arrangement, which to many casual observers (and even some of its inhabitants) seems disorderly and even unsettling.

Strategies for Rural Survival

The imperatives of the globally networked economy need not hollow out rural communities. In fact, these forces can be "pinned down" in some places to provide the basis for sustainable economic growth. To reap the benefits of the new economy, however, will not be easy. Characterizing the challenge, Amin and Thrift (1995: 10) note:

> Increasingly, the pressure posed by globalization is to divide and fragment cities and regions, to turn them into arenas of disconnected economic and social processes and groups. Nevertheless, these places continue to embrace singular and common identities in order to live in, or challenge, the global. The critical question which remains then, is whether the politics and policies of place are appropriate or sufficient for securing acceptable levels of social and economic well-being within the global.

To secure a place for themselves in the future, rural communities must reengineer themselves to meet the requirements of a knowledge-based network economy in much the same way that many businesses have had to do. However, in contrast to large-scale firms, which are unbundling their operations, and reconfiguring their operations into loosely coupled networks, rural communities must instead integrate their economic activities and "thicken" their institutions by reinforcing their local and regional ties.

To maintain their places in the networked global economy, rural communities will need to play to their strengths rather than their weaknesses. Communities competing with one another to attract companies and investment are likely to be much less successful than in the past. With shorter product cycles, companies are likely to be less grounded in any particular place. Thus, rural communities must rely on one another (Jacobs 1994). Instead of competing for low-wage, low-skilled jobs, they must find new and complementary ways to add value to the production chain by building their own unique set of assets.[8] Only then can they gain a secure niche in the global production system (Gereffi and Korezeniewicz 1994).

[8] As Storper (1997a: 20) has explained, what makes territories economically distinct and gives them leverage is their asset specificity. As he notes, "Territorialized economic development may be defined as something quite different from mere location or localization of economic activity. It consists . . . in economic activity which is dependent on resources that are territorial specific. These 'resources' can range from asset specificities available only from a certain place, or more importantly, assets that are available only in the context of certain inter-organizational or firm-market relationships that necessarily involve geographic proximity, or where relations of proximity are markedly more efficient than other ways of generating these asset specificities."

One model that might serve rural areas well in this regard is that of the "industrial district." In an industrial district, small and medium-sized firms are networked together in a geographic region (Asheim 1994; Pyke and Sengenberger 1992; Amin 1994; Rabellotti 1997). Each firm within the network specializes in some aspect of a common production system, allowing them to jointly reap many of the benefits of vertical integration, hitherto available only to large firms. As described by Henry, Barkley, and Zhang (1999: 32):

> Proximity between the more specialized firms and their input suppliers and product markets enhances the flow of goods through the production system, an especially important consideration for firms using just-in-time inventory replacement procedures. Ready access to product and input markets is also beneficial to firm survival since shortened product life cycles mandate quicker adaptability to market changes. And a spatial concentration of industry activity provides the pool of skilled labor required by the computer-aided technologies and flexible manufacturing organizations (Henry, Barkley, and Zhang 1999).

To the extent that firms in an industrial district are jointly serviced by the governance structure in the region in which they are located, they can also gain significant "external economies" that—although external to the firm—are internal to the region. These locational economies serve not only to reduce overall costs, but also to allow communities to use their limited resources in the most cost-effective manner. Benefits can accrue to local and regional communities, for example, if there are a wider pool of skilled labor, specialized businesses and financial services, as well as infrastructure investments (Harrison 1992).

Locational economies stem not only from economic factors but from social and cultural factors as well. Characterizing the associational benefits to be derived from industrial districts, Amin (1994: 65) notes, for example:

> Thus, beyond the attributes of individual entrepreneurs, industrial districts act like a collective brain: the product of years of experience and know-how pulsing through every channel of the local economic system (firms, institutions, households, etc.), and thereby enabling the creation, and dissemination of new "stories" [*sic*] innovation and knowledge on a generalized basis. This capability is, as it were, in the "air" and in the "blood" of the inhabitants of an industrial district, transmitted on the basis of intergenerational continuity and face-to face contacts. In possessing such a diffuse innovation capability, Marshallian industrial districts are able to assimilate and transmit new industrial "stores" across the entire system.

Among the benefits that network participants have attributed to greater collaboration and information sharing, for example, are improvements in marketing, competitiveness, profitability, and product development. To capture all of these synergies, organizational as well as economic factors must be taken into account. The most effective strategies, and the ones most likely to preserve the integrity of rural places, will be those that successfully match the structure of rural markets to the opportunities afforded by new technologies.

A Technology-Based Strategy

In a globally networked economy, how networks are designed and configured will be a matter of great import. Increasingly much of the information and knowledge that was once held personally is now embedded in software-based components and networks, where it can be used to support a wide range of economic activities. Depending on the way in which networks are configured, and how they structure relationships and perceptions as well as distribute information, they can be employed either to empower or to weaken the position of rural communities in economic transactions or exchanges.[9]

Fortunately, today's networking technologies are better suited to support rural economies. Defined by software and supporting almost all forms of communication, networking technologies are more flexible, versatile and easy to use than ever before. Moreover, because these networks can be organized on a decentralized basis, they can be more easily customized to the tasks at hand (Garcia 2001).

This flexibility can be a boon to rural communities, helping them to better reap the benefits of the globally networked economy. For the first time, these communities can design networks to support their unique development needs. Moreover, because rural networks can be organized on a decentralized basis, they can more easily be customized to support horizontal relationships and local ties.

Put more concretely, just as businesses are employing networking technologies to establish industry-based portals, so too might rural communities use these technologies to establish regionally based rural portals, which can serve as "virtual industrial districts." However, in contrast to business portals, which are being established along industry sector lines, a regional rural portal would be configured, instead, around geographic boundaries and provide e-business services to cover the needs of an entire

[9] Thus, for example, a CEO might adopt new computer-based manufacturing technologies for the purpose of gaining greater control over job-related knowledge. Similarly, manufacturers might seek to lock in customers and suppliers by controlling database access through proprietary network standards. Likewise, vendors of information and communication services might try to limit competition by restructuring access to the information gateway or the intelligent network switch.

region. Such services might include, for example, an e-commerce platform, community-based information relating to education, health care, and social and governmental services, helping not only to spread costs but also to generate new synergies and positive externalities.

When operating in a "virtual environment," rural and urban areas are more likely to be on equal ground. A regional rural portal, for example, would allow remote communities from across an entire region to link up and cooperate with like communities elsewhere, thereby reinforcing local knowledge, restraining destructive competition among communities, and limiting the drain of resources to more urbanized areas. Instead of the "learning companies" so often touted in business circles, virtual industrial districts would foster "learning regions" (Scott 1998). Moreover, by participating in such regional rural portals, rural communities would benefit not only from greater economies of agglomeration but also from the external economies associated with industrial districts. These locational economies would serve not only to reduce overall costs but also to allow communities to use their limited resources in the most cost-effective manner.

To leverage such advantages will require much more than advanced technologies. To employ networking technologies strategically, rural communities must have some control over network deployment and design.

The stakes in designing rural networks, therefore, are very high. Moreover, the choice made will be irreversible, at least in the short and medium terms. Once a decision is made, technology tends to become fixed on a given trajectory (Arthur 1989). This pattern is especially evident with networked information technologies, which require vast amounts of capital and investment. Thus, periods of rapid technological advances such as we are witnessing today provide rural communities with a rare opportunity to rethink and restructure the way that they interact with the global economy.

Deployment and Diffusion Strategies

Recognition of the need to develop networked-based development strategies is now widespread. To this end, many developing countries have undertaken major efforts to promote the deployment of advanced networking technologies. In keeping with global pressures for deregulation and trade liberalization, most of these countries have adopted supply-driven strategies that aim to encourage foreign investment in telecom deployment. However, if advanced communication technologies and services are to operate to the benefit of remote rural communities, technology deployment alone will not be enough. At the same time, decision-makers in developing countries must create the optimal conditions for the productive use of these technologies and their incorporation into everyday life.

As Hoff, Braverman, and Stightz (1993: ix) have emphasized: "to design effective policies to remedy market failure, one has to understand its underlying source. One needs also to recognize that the interaction among markets are not limited to ones of price and income, as modeled in general equilibrium theory. What happens in one sector or market can have repercussions on the nature of transaction costs, risks, and enforcement mechanisms used in other markets."

To capture the critical variables for success, technology deployment strategies must be linked to diffusion strategies. Whereas technology deployment refers to the physical provision of infrastructure facilities, technology diffusion can be defined as the process by which technologies and technical innovations are extended and adapted over time and space, and integrated into day-to-day social and economic activities (Brown 1981). As Hanna, Guy, and Arnold (1995: xi) have described this process with respect to information technologies: "diffusion involves more than acquiring computerized equipment and microelectronics-based products and related know-how. It involves the development of technical change-generating capabilities to adapt given technology to a widening range of needs."

This process is a cumulative, iterative one; once deployed, new technologies continue to evolve, and they are "reinvented" in response to changing needs and circumstances. The course the diffusion process takes is determined not only by technical and economic factors, such as technology advances and declining costs, but also by social and institutional factors, such as the availability of mechanisms for information learning and information exchange.

A focus on diffusion is especially important in the case of remote rural areas, which are characterized by thin markets and institutional structures. To overcome these obstacles, innovative nonmarket approaches may be required. Cooperative, community—based approaches can be especially effective under such circumstances. In many rural areas, for example, cooperative arrangements have long been employed to promote resource sharing and information pooling. Moreover, when such efforts are community based, organizational arrangements can be tailored to local environments, building on and enhancing existing strengths and resources.

Notwithstanding the critical relationship between deployment and diffusion, rarely are these two strategies effectively combined. Not surprisingly, therefore, the debate today over the digital divide is dominated by a concern about supply (Warschauer 2003).[10] Unfortunately, supply-

[10] In contrast to many others who have such high aspirations for information and communication technologies, Warschauer is not a technology determinist. To the contrary: as he describes it, to promote social inclusion requires much more than overcoming the "digital divide." Because technology and society are "intertwined and co-constitutive" (2003:7),

driven deployment strategies—focusing almost exclusively on the problem of access—work all too often to undermine the very socioeconomic conditions that are required to encourage widespread and sustainable usage (Garcia and Gorenflo 1997).

Designing and implementing technology diffusion strategies to meet social as well as economic criteria is, moreover, becoming increasingly difficult, given today's deregulated, global economic environment, in which the political and economic modus operandi is to let markets take their course. Under such circumstances, developing countries have little leeway to craft holistic communication policies. Lacking capital and finding themselves deeply in debt, many depend on foreign investment to support their telecommunications infrastructure development. When such investment is forthcoming, its provision is typically based on strict market criteria, which serve to reinforce the concentration of infrastructure in urban areas. Sometimes, developing countries can successfully negotiate concessions, requiring foreign operators, for example, to meet basic universal service goals. However, such agreements rarely foster diffusion. By their very nature, they are supply driven.

The Case of U.S. Telephone Cooperatives

The experience of rural telephone cooperatives in the United States provides an example of one strategy that might be pursued by developing countries today. Although countries differ considerably in terms of their social, economic, and political contexts, the rural cooperative model has a universal appeal. Equally important, even in today's deregulatory climate, cooperatives can be employed so that competitive and cooperative markets complement, rather than replace, one another.

U.S. telephone cooperatives were critical not only in providing affordable telephone services to many rural communities but also in linking them up with the national marketplace. Building on and reinforcing the strengths of their own communities, these rural cooperatives played an important role in promoting not only the universal deployment of communication technologies and services, but also their widespread use in support of economic and community development.

In the United States, rural communities first entered the telephone business in 1894, when the original Bell Telephone Company patents expired.

strategies to promote technology access must go hand in hand with those that aim to provide users the wherewithal to participate fully in society. As he states, "What is at stake is not access to ICTs in the narrow sense of having a computer on the premises but rather access in a much wider sense of being able to use ICTs for personal or socially meaningful ends" (p. 32).

Shunned by urban-based telephone companies, rural residents took it upon themselves to provide their own phone service, relying almost exclusively on local capital and labor. In many local villages, doctors and other local professionals took the initiative; whereas in more remote areas it was farmers who set up the first telephone lines.

Rural phone companies organized themselves in a wide variety of ways. Some purely private companies, which functioned as intercom systems, consisted of a single line, which was owned and shared by a small group of people. Others were organized on a profit-seeking basis, taking the form of privately owned and commercial stock companies. Mutual stock companies, in contrast, were owned entirely by users. Organized on an informal basis, their members paid a prorated share of the capital expenditures, maintenance, and improvement fees. Farmer lines were typically set up as purely private or mutually owned systems (Annenberg Washington Program 1994). Thus, for example, to join the Liberty Telephone Company in 1910, one had to pay an up-front fee of $25; provide a telephone, a pole, and some labor; as well as pay a flat annual fee of $7 for service (Meyer 1912).

Rural phone companies were able to make do with such limited resources by sharing what they had and keeping their expenses to a minimum. Local farmers, for example, often built networks using their own materials and tools. When necessary, they purchased equipment from independent manufacturers or through mail order catalogues distributed by such firms as Sears and Roebuck and Montgomery Ward. Having built their own networks, these farmers had little trouble maintaining them. Problems did arise, however, when they resorted to very low-quality equipment and poles, which sometimes included barbed wire and fence posts (Atwood 1984). Overall, however, the model was a great success. By 1920, 39 percent of all farmers had obtained rudimentary service, and in some midwestern states the number of telephones per person exceeded that in the East.

Despite their initial successes and the important service benefits that they provided, rural telephone companies' fates were inextricably linked to those of the communities they served. With industrialization and the onset of the Great Depression, these companies were no longer able to sustain themselves. Many failed. Because urban-based telephone companies were unwilling to serve these thin, unprofitable markets, service in rural communities continued to deteriorate. Thus, by 1940, only 25 percent of all farm residencies in the United States had working telephones (United States Census 1949: 1).

This trend was reversed only when the federal government decided to adopt a less market-oriented, and more community-based, approach to telephone deployment in rural areas. To promote rural telephony, the government turned to the Rural Electrification Administration (REA), which had already proven successful in bringing electricity to rural areas. The model advocated by the REA—the cooperative—was designed to address the problem of market failures in rural economies.

As in the case of the telephone, rural residents had greatly lagged behind urban residents in accessing electricity. By 1935, less than 12 percent of all America's farms were served. Private utilities were unwilling to provide service because demand seemed low and the technical problems high. At first, the federal government sought to assist and encourage private industry rather than displace it. When industry failed to respond, President Roosevelt created the REA, which bypassed municipal and private industry with its own grass-roots, cooperative networks (Garwood and Tuthill 1963). Although the REA's goals were ambitious—universal high-quality service, rapid deployment, and low rates—it was successful in achieving them. Few rural cooperatives defaulted. By 1940, 3 percent of all farmers had electricity; by 1950, 78 percent were receiving service; and by 1959, 96 percent (United States Department of Agriculture 1989).

Rural electric cooperatives also played an important role in economic development. The cooperatives aggressively recruited and served industrial, commercial, and suburban customers, which had the effect of increasing the number of consumers each year, from 5 million in 1960 to 12 million in 1987. In so doing, they greatly facilitated the movement of industrial, commercial, and nonfarm residences to rural areas (United States Department of Agriculture 1989).

Looking for a new mission in the late 1940s, the REA welcomed the task of helping to deploy telephones to rural areas. With its authority expanded by Congress, the REA helped to achieve high-quality, state-of-the art telephone service in rural communities. To serve widely scattered residences, it pioneered technology to reduce the size of wire, its cost of installation, and its vulnerability to lightning and icing. REA borrowers also replaced party lines with one-party service. Rates were standardized, and comprehensive "area" coverage was provided. Attesting to the program's success, 94 percent of all farms were served by telephones in 1990 (United States Department of Agriculture 1989).

Like the electric cooperatives, telephone cooperatives played a central role in promoting community and economic development. Even today, these local cooperatives continue in this tradition. A recent survey of Internet deployment in rural areas shows that telephone cooperatives were instrumental in bringing Internet services to their communities, even when it was not in their immediate financial interest to do so (Garcia 1996).[11]

[11] The prospects of new business opportunities and/or threat of competition were of little concern, for example, to the nineteen early movers who began providing Internet access

Replicating the U.S. Model

The successful U.S. experience with cooperatives is by no means unique. Similar bottom-up community efforts have emerged in a number of European countries, including Norway, Sweden, Finland, and the Netherlands. Although organizational approaches have varied depending on local circumstances and political culture, all these efforts have served—as they did in the United States—to speed the deployment and diffusion of telecommunications services to rural and remote areas. Today, in the developing world cooperative efforts—such as microcredit organizations—have likewise proved successful in promoting both technology diffusion and innovation, as well as in providing new business opportunities.

Notwithstanding such successes, recently a number of factors have limited the broad applicability of the cooperative model for telecommunications. Given, for example, the rising cost and growing size and complexity of networks, it became increasingly difficult to piece together at the local level all the necessary financial and human resources required to create community-based telephone systems. Network interdependencies and the need for interconnection also favored centralized network ownership and administration, essentially foreclosing the option of a decentralized, cooperative approach.

In today's global, more privatized environment, there are fewer such constraints. Taking advantage of the higher performance and enhanced variety of new networking technologies as well as the much greater flexibility that they afford, new networking solutions can be employed to deploy advanced technologies on an ad hoc and customized basis. Already, many countries are deploying less costly communication systems. In Asia-Pacific and Latin America, for example, many countries are using very small aperture terminals (VSATs) to provide both public and private services. Such systems can support voice traffic, facsimile, and low-rate data transmission. In other countries, such as India, fixed cellular radio systems are often used (Blumenthal 2000; Hudson 1997). These radio-based systems are easier to deploy than wireline services, and they have lower up-front investment costs, which can be shared among subscribers. These technologies can, moreover, be deployed on a step-by-step basis, with new

between March 1993 and March 1995. Hardly any of these providers had a business plan when they set out. Nor, for the most part, did they anticipate a profit. Twelve of the nineteen were small or very small in size and had limited resources (less than 20 employees, 3,500 access lines, and $3.5 million in annual revenues). When asked in followup interviews why they had decided to provide Internet service, most said that they had simply wanted to meet their community's needs. As one company manager explained: "We don't expect to make a profit; we would be happy if we could just break even" (Garcia 1996).

cells added in response to growing demand. Because radio technology is more reliable than wireline technology, such systems also have lower maintenance costs.

Given the possibility of technology leapfrogging, there is no reason why developing countries cannot employ bottom-up cooperative approaches to serve users in small communities and in rural areas. The synergies and positive externalities thereby generated would encourage not only technology deployment, but also—and as importantly—diffusion and innovation (Hanna, Guy, and Arnold 1995). At first, small-scale pilot projects might be undertaken, which are customized to the social, economic, and political conditions at hand. Building on local knowledge, and being locally based and embedded in their communities, telecommunication cooperatives might gradually expand their activities to create regional rural portals in the manner described above.

Notwithstanding the benefit of new technologies, cooperative efforts may need initial public support and assistance, as was the case in the United States. At the very least, national governments will need to adopt a proactive stance in gaining the acquiescence and support of incumbent providers, assuring that local providers have equal interconnection rights. If local credit markets are unable to generate adequate seed funding, initial loans may also be required. But their magnitude need not be excessive, especially if the technologies used can be deployed in stages, in response to growing demand. Experience suggests that a need for some technical assistance and technology transfer will also be likely, at least in the short term. Equally important, but more difficult to control at the national level, is the establishment of a global communication regime that—recognizing the differences and increasing divergence between urban and rural markets—is more receptive to such innovative approaches.

The Need for a Rural Telecom Regime

The cooperative movement in the United States benefited from a dual set of regulatory arrangements, one for urban and one for rural areas. Recognizing that the structure of markets in urban and rural areas was significantly different, the government promoted universal deployment not solely through the use of cross-subsidies within the framework of a regulated national monopoly, but also by providing rural telephone cooperatives loan subsidies and technical assistance under the auspices of the Rural Electrification Administration.

Today, developing countries have much less recourse to pursue such options. In an increasingly global economy, many critical issues are now being worked out at the supranational level in a vast array of nongovernmental and governmental organizations that have mushroomed to address these burgeoning tensions. As a result of this intervention, national policy-makers are often constrained in establishing their own agendas (Scott 1998; Biersteker 1995).

This loss of autonomy has been particularly evident in the area of telecommunications. Given the increased global provisioning of telecom equipment and services, as well as tremendous growth in global trade in services, international pressures for deregulation, privatization, and liberalization of this sector continue to mount. Leaders in developing countries are especially susceptible to such pressures. Competing among themselves for foreign direct investment, and in search of mechanisms to pay off their foreign debts, most developing countries have been quick to follow the lead of their cohorts and counterparts in the industrialized West, and to introduce telecom reforms. The choice, however, is not an easy one. Describing the motivations and tensions inherent in these decisions, one observer has noted:

> Perhaps for the first time communications are being recognized as a strategic underpinning of civilization, as important as the provision of clean water. The implicit fear for many countries must be that an inadequate infrastructure will forever keep a national economy out of the world economic structure that is shaping up for the 21st century, in addition to the fear that government relinquishes an important tool. It is into this cauldron that telecom policy is being pushed. (McClelland 1992: 31)

The market forces driving telecom reform are being reinforced by the new international telecommunication regime. Deeply embedded in the world trade regime, communication policymakers no longer view telecommunication as a means of achieving social and economic objectives but rather as an end in and of itself. That is to say, telecommunications is increasingly perceived to be a commodity, to be bought and sold in the marketplace much like any other commodity (Garcia 2002).

Championed by the United States, the commodification of communication has today become the international norm. This transformation from a technologically oriented telecommunication regime to a commodity-based trade regime gained its first momentum in 1994, at the Uruguay Round of the General Agreement on Tariffs and Trade (GATT) negotiations. The United States sought to use these negotiations to open up the world market to domestic services providers, and to foster competition and—with it—reduced prices for telecommunication users. The economic stakes for the United States were very high. For example, in 1994 global telecommunication revenues—totaling $513 billion—constituted more than 2 percent of global GDP, while international traffic grew at a compound annual rate of 15.2 percent over the course of the previous decade (Hudson 1997: 417). So adamant was the United States in its intent, its representatives balked at the 1996 negotiations, refusing to continue discussions until adequate concessions had been made.

The U.S. objective of creating a trade regime for telecommunication is now almost fully realized. Lobbied intensively by the United States and its European allies, GATT participants agreed to negotiate the liberalization of trade in services in accordance with the most-favored nation principles of the GATT. In 1997 the General Agreement on Trade in Services (GATS) went into effect, to be implemented by the World Telecommunications Organization (WTO).[12] According to this agreement, each signatory must file an individual schedule of commitment indicating which services it will bring into compliance with the GATS guidelines. Signatories to the Annex on Telecommunications are obliged to ensure that any service supplier of any other member is accorded access to and use of telecommunication transport networks and services on reasonable and nondiscriminatory terms and conditions. A separate Reference Paper lays out members' interconnection obligations as well as regulatory practices—such as the creation of an independent regulatory agency—deemed most appropriate for achieving competition. By the year 2000, seventy-two countries, representing more than 91 percent of global telecommunication revenues, had become signatories to the Annex (Collins 2000).

Although serving primarily the interests of those developed countries that have a comparative advantage in the telecommunications sector, the goals of the Telecommunications Annex were couched in more universal terms. Thus, for example, in announcing the agreement, the director general of the WTO predicted that gains in global income over the following decades would total close to $1 trillion. Even more important, he said: "this deal goes well beyond trade and economics. It makes access to knowledge easier. It gives nations large and small, rich and poor, better opportunities to prepare for the challenges of the 21st century. Information and knowledge, after all, are the raw material of growth and development in our globalized world" (Mansell and Wehn 1998: 191).

Likewise, the GATS agreement paid tribute to the special needs of the developing world. Thus, it characterizes one of its purposes as facilitating "the increasing participation of developing countries in trade and services and [strengthening] their domestic capacity and its efficiency and competitiveness" (Feltham 2000: 151). In addition, the agreement recognizes the unique needs of many developing countries, and the difficulties they

[12] A second major governance regime—the Information Technology Agreement—was concluded for information technology in January 2000. Signatories constitute 90 percent of the world trade in information technologies (Mansell and Wehn 1998).

may encounter in the transition. Thus, section 5 (g) of the Annex specifically adds protection for them, allowing developing countries to "place reasonable conditions on access to and use of public telecommunication transport networks to strengthen its domestic telecom infrastructure and service capacity and to increase its participation in international trade in telecom services" (Feltham 2000: 151).

Even if the Telecommunications Annex is universal in intent, its impacts are hardly likely to be experienced uniformly. As telecommunications markets are liberalized, the forces of competition will serve to concentrate the provisioning of facilities and services in high-density urban areas. To the extent that global providers serve rural areas, they will be compelled to "cream skim" at the margins, further depleting rural areas of their overall networking resources. The uneven pattern of global development will be reinforced as a result.

To anticipate the results, one need only consider the pattern of infrastructure development as it is evolving in a developed country such as the United States. Most of the deployment of advanced networking technologies is now focused almost exclusively in large cities and urbanized regions, where demand is highest and the customer base the most lucrative. Thus, for example, the top seven metropolitan areas host 62 percent of the nation's Internet backbone capacity; the top 21 metropolitan areas, 87.5 percent. What is worse, a number of large companies are actually abandoning their rural customers. Thus, for example, since 1994, the incumbent U.S. West—which provides service in fourteen states—has sold off more than four hundred of its exchanges. Likewise, GTE has divested itself of many of its exchanges. Explaining the company's behavior, GTE noted: "This repositioning effort is part of an overall corporate plan announced in April 1998 to generate after-tax proceeds of $2–$3 billion to be redeployed into other higher growth initiatives" (Selwyn, Kravtin, and Coleman 1998: 20).

This kind of divestiture makes total economic sense, given the intense competitive characteristics of the global telecommunications market. However, the incentives that drive increasingly globally oriented providers are hardly likely to foster community building at the local and regional levels. As Scott (1998: 37) has emphasized: "competitive contests and rivalries corrode those subtle processes of association, cooperation, and communal solidarity that are critical to much of the economic success and social welfare in the contemporary world."

To reverse such developments, a deliberate rural strategy will be required. Just as in the United States, where the government was forced to adopt a nonmarket strategy to assure that rural communities had equal access to critical infrastructure, so too might decision-makers in local, regional, national, and international arenas today. In a highly complex, globally networked economy in which economies are increasingly interdependent, competition can no longer be viewed as an end, in and of itself. Instead, competition must be viewed as a policy tool, which is more or less appropriate depending upon the circumstances. Given such a radical change in the mindsets of today's policymakers, innovative rural solutions, which draw upon and reinforce local strengths and resources, can surely be found.

References

Amin, Ash. 1994. The Potential for Turning Informational Economies into Marshallian Industrial Districts. In *Technological Dynamism in Industrial Districts: An Alternative Approach to Industrialization in Developing Countries.* New York: United Nations.

Amin, Ash, and Nigel Thrift. 1995. *Globalization, Institutions and Regional Development in Europe.* New York: Oxford University Press.

Annenberg Washington Program in Communications Policy Studies of Northwestern University. 1994. *Speeding Telephone Service to Rural Areas: Lessons from the Experience in the United States.* Washington, DC: The Annenberg Washington Program.

Arthur, Brian. 1989. Competing Technologies, Increasing Returns, and Lock-in by Historical Events. *Economic Journal* 99(394): 116–31.

Asheim, Bjorn T. 1994. Industrial Districts, Inter-Firm Cooperation and Endogenous Technological Development: The Experience of Developed Countries. In *Technological Dynamism in Industrial Districts: An Alternative Approach to Industrialization in Developing Countries.* New York: United Nations.

Atwood, Roy A. 1984. *Telephone and Its Cultural Meaning in Southern Iowa.* Iowa City: University of Iowa.

Ayres, R. U. 1992. CIM: A Challenge to Technology Management. *International Journal of Technology Management* 7 (1–3) (December): 17–39.

Biersteker, Thomas J. 1995. The Triumph of Liberal Economic Ideas in the Developing World. In *Global Change, Regional Response: The New International Context of Development,* edited by Barbara Stallings, 174–197. New York: Cambridge University Press.

Blumenthal, Marjorie. 2000. Architecture and Expectations: Networks of the World—Unite! In *The Promise of Global Networks,* edited by Jorge Reina Schement, 1–52. Washington, DC: Aspen Institute.

Brown, Lawrence A. 1981. *Innovation Diffusion: A New Perspective.* New York: Methuen.

Bruno, Leo. 2000. Briefing: The Broadband Era. *The Red Herring.* February.

Castells, Manuel. 1989. *The Informational City: Restructuring and the Urban Regional Process.* Oxford: Basil Blackwell.

Collins, Katherine. 2000. International Accounting Rate Reform: The Role of International Organizations and Implications for Developing Countries. *Law and Policy in International Business* 31(3): 1077.

Compaine, Benjamin M., ed. 2001. *The Digital Divide: Facing a Crisis or Creating a Myth?* Cambridge: MIT Press.

Cox, Kevin R., ed. 1997. *Spaces of Globalization: Reasserting the Power of the Local.* New York: Guilford Press.

Cronin, Francis J., Elisabeth K. Colleran, and Paul L. Herbert. 1993. Telecommunications and Growth: The Contribution of Telecommunications Infrastructure Investment to Aggregate and Sectoral Productivity. *Telecommunications Policy* 17(9): 677–90.

Dholakia, Ruby Roy, and Bari Harlam. 1994. Telecommunications and Economic Development: Econometric Analysis of the US Experience. *Telecommunications Policy* 18(6): 470–77.

Duboff, Richard. 1983. The Telegraph and the Structure of Markets in the United States, 1845–1890. *Research in Economic History* 8:253–77.

Evans, Philip B. and Thomas S. Wurster. 2000. *Blown to Bits: How the New Economics of Information Transforms Strategy.* Cambridge: Harvard Business School Press.

Feltham, Jennifer Laura. 2000. Polish Telecommunications Law: Telecommunications Takes Off in Transition Countries, But at What Price Are They Becoming Wired? *Vanderbilt Transnational Law Journal.* 33(1): 149–183.

Galston, William A. and Karen J. Baehler. 1995. *Rural Development in the United States: Connecting Theory, Practice, and Possibilities.* Washington DC: Island Press.

Garcia, D. Linda. 1996. Who? What? Where? A Look at Internet Deployment in Rural America. *Rural Telecommunications* 15(6) (November/December).

———. 1998. *Global Electronic Commerce and Transaction Costs.* Washington, DC: Institute for Technology Assessment.

———. 2001. The Architecture of Global Networking Technologies. In *Global Networks, Linked Cities; Urban Connections in a Globalizing World,* edited by Saskia Sassen. New York: Routledge.

———. 2002. Crafting Communication Policy in a Competitive Environment. *Encyclopedia of Library and Information Science* 72(35): 61–90.

Garcia, D. Linda, and Neil Gorenflo. 1997. Best Practices for Rural Internet Deployment: The Implications for Universal Service Policy. Paper presented to the Telecommunications Policy Research Conference, Alexandria, VA.

Garwood, John D., and W. C. Tuthill. 1963. *The Rural Electrification Administration: An Evaluation.* Washington, DC: American Enterprise Institute.

Gereffi, Gary, and Miguel Korzeniewicz, eds. 1994. *Commodity Chains and Global Capitalism.* Westport, CT: Greenwood Press.

Gottman, J. 1983. *The Coming of the Transactional City.* College Park: University of Maryland Press.

Grabher, Gernot, ed. 1993. *The Embedded Firm: On the Socioeconomics of Industrial Districts.* New York: Routledge.

Hanna, Nagy, Ken Guy, and Erik Arnold. 1995. *The Diffusion of Information Technology: Experience of Industrial Countries and Lessons for Developing Countries.* Washington, DC: World Bank.

Harrison, B. 1992. Industrial Districts: Old Wine in New Bottles. *Regional Studies* 26(5): 469–83.

Hart, John Fraser. 1995. "Rural" and "Farm" No Longer Mean the Same. In *The Changing American Countryside: Rural People and Places,* edited by Emery N. Castle, 63–76. Lawrence: University of Kansas Press.

Held, David. 1995. *Democracy and the Global Order: From the Modern State to Cosmopolitan Governance.* Stanford: Stanford University Press.

Henry, Mark S., David J. Barkley, and Yibin Zhang. 1999. *Industrial Clusters in the TVA Region: Do They Affect Development of Rural Areas.* Lexington: TVA Rural Studies Program, University of Kentucky.

Hoff, Karla, Avishay Braverman, and Joseph E. Stiglitz. 1993. *The Economics of Rural Organization: Theory, Practice, and Policy.* New York: Oxford University Press, for the World Bank.

Hudson, Heather. 1997. *Global Connections: International Telecommunications Infrastructure and Policy.* New York: John Wiley & Sons.

Innis, Harold. 1951. *The Bias of Communication.* Toronto: University of Toronto Press.

Jacobs, Jane. 1994. *Cities and the Wealth of Nations: Principles of Economic Life.* New York: Vintage Books.

Lewis, Pierce. 1995. The Urban Invasion of Rural America: The Emergence of the Galactic City. In *The Changing American Countryside: Rural People and Places,* edited by Emery Castle, 39–62. Lawrence: University of Kansas Press.

Mansell, Robin, and Uta Wehn. 1998. *Knowledge Societies: Information Technology for Sustainable Development.* Oxford: Oxford University Press, for the United Nations.

Massey, D. 1993. Global Sense of Place. In *Space, Place, and Gender,* 146–56. Minneapolis: University of Minnesota Press.

McClelland, Stephen. 1992. The International Dimensions: The PTTs. *Telecommunications.* June.

Meyer, C. W. 1912. How We Built a Home-Owned Farmers' Telephone Line. *Telephony.* November 16.

North, S.N.D. 1884. *History and Present Condition of the Newspaper and Periodical Press of the United States.* Washington, DC: U.S. Government Printing Office.

Office of Technology Assessment. 1992. *Rural America at the Crossroads: Networking for the Future.* Washington, DC: U.S. Government Printing Office.

Pennings, J. M. 1981. Environmental Influences on the Creation Process. In *The Organizational Life Cycle,* edited by J. R. Kimberly and R. H. Miles, 135–60. San Francisco: Jossey-Bass.

Peterson, Theodore. 1964. *Magazines in the Twentieth Century.* 2d ed. Urbana: University of Illinois Press.

Pyke, F., and W. Sengenberger. 1992. *Industrial Districts and Local Regional Economic Regeneration.* Geneva: International Institute for Labor Studies.

Rabellotti, Roberta. 1997. *External Economies and Cooperation in Industrial Districts: A Comparison of Italy and Mexico.* New York: St. Martins Press.

Sassen, Saskia. 1989. *Cities in a World Economy.* London: Basil Blackwell.

———. 1998. *Globalization and Its Discontents: Essays on the New Mobility of People and Money.* New York: New Press.

Scott, Allen. 1998. *Regions of the World Economy: The Coming Shape of Global*

Production, Competition and Political Order. New York: Oxford University Press.

Selwyn, Lee, Patricia D. Kravtin, and Scott A. Coleman. 1998. *Building a Broadband America: The Competitive Key to the Future of the Internet.* Boston: Economics and Technology.

Storper, Michael. 1991. *Industrialization, Economic Development and the Regional Question in the Third World: From Import Substitution to Flexible Production.* London: Pion Limited.

———. 1997a. Territorial Flows and Hierarchies in the Global Economy. In *Spaces of Globalization: Reasserting the Power of the Local,* edited by Kevin R. Cox, 19–44. New York: Guilford Press.

———. 1997b. *The Regional World: Territorial Development in a Global Economy.* New York: Guilford Press.

Stinchcombe, A. L. 1965. Social Structure and Organizations. In *Handbook of Organizations,* edited by J. G. March, 142–93. Chicago: Rand McNally.

Swanson, Louis. 1990. Rethinking Assumptions about Farm and Community. In *American Rural Communities,* edited by. A. E. Luloff and Louis E. Swanson, 19–33. Boulder: Westview Special Studies in Contemporary Social Issues.

United States Census. 1949. Compiled by the U.S. Congress, House Committee on Agriculture.

United States Department of Agriculture, Rural Electrification Administration. 1989. *A Brief History of Rural Electric and Telephone Programs.* Washington, DC: USDA, REA.

Warschauer, Mark. 2003. *Technology and Social Inclusion: Rethinking the Digital Divide.* Cambridge: MIT Press.

Networks, Information, and the Rise of the Global Internet

ROBERT LATHAM

SHOULD FIELDS LIKE international relations provide theories for the formation of global infrastructures? Although such infrastructures, from communication and transport to financial systems, are fundamental to globalization and the nature of the global realm, such theories hardly exist. Such theories may help us understand how global systems emerge and take a certain form. This chapter will probe the outlines of one for a system essential to our world, the Internet.

Why have such theories not come into being?[1] In the first place, the long history of infrastructure, such as communication systems, has mostly been a national one. Industrialized states have been able to establish within their borders relatively well-integrated networks with uniform standards, such as is commonly seen in a national telephone system.[2] The very success of such national systems since the nineteenth century has meant that global communication systems have typically taken form as a set of connections among discrete national communication networks (an internetwork). This form describes, for example, the matrix of transboundary links among national telephone networks that makes international calling possible. Standards can vary, but the internetwork will work as long as the links compensate for different standards.

A global network, in contrast, is composed of links across national boundaries that do not join gateways to other networks (as in an inter-

[1] There have been some outstanding exceptions, such as Hughes (1983) and David and Bunn (1988).

[2] The term system is being used to describe the overall array of technologies, applications, regulations, and connections that constitute a form of communication (like telephony) that can be national or international. By network is meant the web of connections (both human and machine)—and the technologies that support those connections—that make a system possible. Networks and systems overlap considerably and perhaps differ ultimately only in emphasis and comprehensiveness (with networks, the emphasis is on connections and what makes them possible; with systems, the emphasis is on total configurations of relevant elements and forces).

network) but constituent nodes that are administered integrally, with uniform standards of operation. The historically successful development of separate national communication systems—and the tendency toward internetworking that follows that success—has not prevented the emergence of global networks. A good example is a global network of orbiting satellites and ground stations.[3]

A second reason for the lack of theories of infrastructure formation might be that we have an easy time answering some questions about infrastructure, such as how internetworks come into being (thus, no theory is necessary). Internetworks have historically flowed from agreements among states to connect their national communications systems based on accepted technologies of linkage: a precedent established by telegraphy in the nineteenth century as country after country agreed, first bilaterally, to link up and, then multilaterally, to establish a regime of wire connection (see, for example, Zacher and Sutton 1996).[4] The formation of a single globe-spanning network is also easily accounted for. Such networks have typically taken form as the possession or projection of the interests of one country or a condominium of countries. In the nineteenth century Britain constructed a global telegraphic network spanning its empire, which the British state directly and—through British firms—indirectly controlled (Headrick 1991; Hugill 1999). Twentieth-century satellite networks were controlled—after being initially a U.S. possession (Comsat)—by Intelsat, a condominium (Cowhey 1990: 176–82).[5]

Although it involves both national networks and transboundary internetworks, the emergence of the Internet is not easily accounted for in this way. The Internet was not established nor is it maintained via international agreement, even though there are instances where bilateral agreements produced international connections, and international organizations such as the International Telecommunications Union (ITU) and the European Union (EU) played a role in its development.[6] It is also not the

[3] Any given communications system can comprise both global networks and global internetworks. For example, global telephony involves global satellite networks that facilitate internetworking among national telephone systems. Of course, any set of gateways linking discrete networks can be viewed as a simple network in its own right (hence the use of the term "*network* of networks" to describe the Internet, for example). However, I will reserve the distinction between an internetwork (a set of links via gateways among networks) and a network (a set of links facilitating communication among users based on a uniform set of technologies and protocols, administered by some common authority).

[4] On the history of connecting telegraphic systems, see Headrick (1991); Codding, (1952: 5–34); and Standage (1998: 68–104).

[5] Of course, the United States continued to exercise great influence.

[6] I am thinking of some of the early agreements to establish ARPANET nodes in Europe in the 1970s. Note also that regime-like agreements—in the broad sense of the term—can certainly be found within the Internet via voluntary organizations like the World Wide Web

product or possession, directly or indirectly, of any single state or group of states, however important one takes its origins in the United States and actions by the U.S. state to be. Most importantly, the very nature of the Internet is relatively exceptional in telecommunications history because, as an internetwork, it directly and indirectly links a diverse range of not just national but subnational, regional, and global networks. The question is, how does an internetwork comprising such varying network types and scales come into being to become the primary global computer communications system?

It is natural to turn to international regime analysis to provide an answer to this question. As suggested above, regimes do bear on infrastructure. Their analysis can help us understand how agreements can form about an infrastructure that is in the interest of one group or another or is efficient and economical. However, while the Internet's development was noticeably aided and shaped by the relations and interactions among sets of agents and institutions (experts, states, corporations, international organizations), these relations *on their own* do not explain the Internet's success.[7] However important the development of various transnational communities and coalitions of IT experts and organizations has been, the emergence of the Internet as a global communication system cannot be explained in itself by the success of a given coalition of transnational or transgovernmental actors and institutions (coalitions—like state actions and international agreements—were a necessary but not sufficient factor). That is because coalitions broad enough to affect IT policy around the world emerged as the Internet itself emerged (in other words, the collaboration occasioned by the construction of the Internet produced the coalitions that could contribute to its continuation and robustness).[8]

The only other contending starting points for explaining the emergence of the Internet are network economics (emphasizing the power of "network effects" or "externalities") and folk theories from the IT world (emphasizing the overwhelming attraction of Internet technology). I will suggest that even if network effects can explain the large-scale growth of networks, it cannot explain their initial emergence. IT folk theories in turn suffer from assuming away the crucial incentive to interconnect that makes the use of technology attractive in the first place.

Consortium (W3C). However, this use of regime is distinct from the classic use in IR where states establish and abide by various rules and norms.

[7] There is no single, comprehensive history of such relations and interactions. A good start is Abbate (1999).

[8] The formation of these coalitions is crucial for understanding the development of a technology, as Bruno Latour (1987) shows, but not necessarily for explaining the emergence of a communication system, which itself, as an infrastructure, is the very thing that makes coalescing possible in the first place.

What I am after in a theory of infrastructure formation is some framework for understanding the logics of global infrastructure emergence—that is, how an emerging organization of infrastructure impacts its development and prospects for successful evolution.[9] For me, the crucial factor in thinking about the infrastructural logic of the Internet is the *relations among networks.* I will argue that the crucial motor of success was the basic logic whereby computer networks would form and then connect or not connect (and the consequences in the aggregate of such formation and connection). Why did an emerging Internet expand from a relatively small number of interconnecting networks to thousands of networks and connections? What logic made such interconnecting of interest to the organizations administering computer networks? What implications does such a logic hold for the character of the emerging communications system as it scaled up into a web of thousands, then millions, of users?

Like global infrastructure formation, internetwork relations are understudied by social science. Despite the recent popularity of the term "Internet" and the phrase "network of networks," relations among networks (rather than relations in networks) fails to gain serious attention, even while the study of networks grows.

I will concentrate mostly on constructing a model of that logic rather than on the politics of pursuing agendas or the trajectories of cooperation and conflict among relevant organizations. If the Internet is a network of networks, then it is essential to identify the mechanism through which this internetwork formed. This chapter should be viewed as a first step in this direction. While I attempt, for heuristic purposes, to model a process, I continually endeavor to place that model in its historical context.[10]

[9] Karl Deutsch has come the closest to this question, but he focused less on why infrastructures are organized a particular way per se and more on the flows across them. Another, less internationally focused, exception is Ithiel de Sola Pool. See also the recent work of Ronald Deibert (1997).

[10] In contrast, the recently published fine book by Susanne Schmidt and Raymund Werle (1998), *Coordinating Technology,* emphasizes the play of actors, institutions and their interests, perceptions, and models to explain the technical solutions that become standards (the process of standardization). But by starting with a fixed process and set of actors (the International Telecommunications Union and International Organization for Standardization, or ISO) the authors are forced to focus on the process that produced standards (the PTT model, as I grossly call it) that did not prevail as the mode of internetworking for the world (of course the European protocols are used and thriving in places, but in the context of an Internet environment rather than the reverse). Internet formation essentially becomes an anomaly for Schmidt and Werle, or a sideshow to the legitimate process of standardization. What I am asking is that we turn this on its head and designate a facet of infrastructural logics—internetworking—to drive analysis. This does not rule out, however, the possibility of analysis that gets at the same dynamics discussed below by starting instead with a close analysis of the relevant politics and conflict.

Obviously, behind and in every network there is an organization. I therefore do not seek to privilege infrastructural logics in order to displace political processes as explanatory forces (the importance of which is underlined in Guice 1998). Regimes matter, as do political and economic agendas, and interorganizational conflict. To argue that logics were agents of change per se is sheer technological determinism. Indeed, without ultimately taking social and political processes into consideration, the very notion of system success is rendered meaningless (success among whom? for what?). Rather, my position is that the logic of internetworking is what shaped the terms of and possibilities for the success or failure of a computer communication system in the worlds where such system formation mattered.[11] One could have imagined national systems of computer communications emerging along the lines of earlier communications systems such as the telephone: that is, with robust national development unfolding first, followed after a span of time by a state-led process of international interconnection. As I will discuss below, there was one model of computer networking of this sort on the table that was of interest especially to the national telecommunications agencies in Europe. It was a serious alternative and failed to take hold because its advocates ignored the compelling social logic of interconnection that the Internet represented. Thus, to the extent that I do consider specific political processes, it is around the conflict between the two basic alternatives to network formation, the homogeneous interconnection of national networks versus the heterogeneous interconnection of diverse networks.

What Needs to Be Explained?

I seek to provide an answer to the question of how it was possible for the Internet to defy the typical pattern of international interconnection and become the primary global system of computer communication. The emphasis on global is important because what distinguishes the trajectory of the Internet as a communications system is that it did reach relatively quickly across borders, especially to Europe and then Asia. Although much of the development of the Internet was located in the work of U.S.

[11] In this respect I do not reject the social constructivist approach to the study of technology, which emphasizes the play of power, perspectives, and agendas among contending groups, as exemplified by Bijker (1995) and adopted by Schmidt and Werle (1998). More accurately, I share important assumptions with the approach by taking contingency seriously, treating the success of a technology as an explanandum rather than an explanans, and recognizing that social groups do not necessarily precede the emergence of a technological system but can emerge with it. Where I depart from that perspective is in my emphasis on the influence of the logic of the emerging system as a crucial shaper of outcomes.

agencies and networks, international interconnections were there from the start and grew in significance rapidly, so that now by one meaningful measure the Internet is at best only 55 percent U.S. based—a predominance that continually drops.[12] Thus, we should understand global to mean that the Internet spans the globe, however unevenly.[13]

However, given that its developmental origins were in the United States, a question arises that is corollary to the basic question about system formation: why did it successfully spread from a predominantly U.S. base to other industrialized countries and on to the rest of the world?

That the Internet was initially anchored in the United States does not mean that my task is to explain why the United States was the crucial, but by far not the only, source of innovation.[14] Rather I seek to explain why such innovation became the basis for a global communications system. Success for those innovations required that others take them up and join in. In other words, innovative, efficient, and competitive services (i.e., cheaper lines) might explain why the Internet emerged from the United States, but not necessarily why it spread from there. Efficient and competitive U.S. services, especially early on, did not really aid non-U.S. network-builders who had to rely on their own systems and services. And superior technology is in the eye of the beholder, as Europeans expended a great deal of energy developing an alternative set of networking protocols that even the U.S. government endorsed.

These points raise the issue of which period of Internet development I am concerned with. There are three distinguishable periods. Phase one covers the early 1970s through mid-1980s. This is the period of initial development and experimentation, when the very first interconnections, including international ones, occur and begin to advance. A second phase is marked by the proliferation of interconnections, especially transboundary ones, from the mid-1980s to the early 1990s. NSF figures prominently as a supporter and foundation for such interconnections. Finally, the phase of real take-off occurs as commercialization sets in from the early to mid-1990s up to the present, marked by the advent of the World Wide Web.[15] My analysis focuses on periods one and two, although I will

[12] That basis is domain names, which Zook (2000) argues is a good representation of distribution relative to other measures, such as hosts or network infrastructure.

[13] I will also use the terms transboundary (any connection across national boundaries) and international (transboundary connections tied to some sort of national or state endeavor, such as intermilitary networking).

[14] For an analysis of exactly this issue, see Mowery and Simcoe (2001). It should not be forgotten that Europe has been an important source of innovation as well. The most famous innovation to come from there is the World Wide Web, developed by the scientific research organization headquartered in Switzerland, CERN.

[15] For roughly the same periodization, see Guice (1998) and Mowery and Simcoe (2001).

draw out in the conclusion some implications of the analysis for the current period.[16]

When trying to understand what the Internet is as a system, one can take the name literally: an interconnection of disparate networks that allows communication among users of computers (via, for example, e-mail, file transfer and storage, remote login). Based on that definition, there are three levels from which to view the Internet: the information available and circulating; the web of links among individuals and organizations producing, maintaining, and accessing such information; and the ensemble of software and hardware that makes such links and communication possible, such as TCP/IP. Success for the Internet required that growth occur on all three levels (after all, greater connectivity has no meaning unless it involves greater communication). The process that was able to bring all three levels together and yield global growth was the interconnection of networks, each of which had content, users, and hard/software. The mechanism driving that interconnection, I will argue, involved a potent mix of information and links across such networks, within which an emerging technology of connection served as a catalyst.

I will conclude by considering how the logics of Internet formation help us better understand the possibilities for establishing the social purpose of this communication system.

Two Approaches to Transboundary Internetworking

Because the Internet developed as an internetwork of disparate networks, any network in a given country, in principle, could connect directly to any other network in or outside that same country. Networks were not required to send data through a national data network, which then forwarded the data to another national data network, which in turn forwarded the data to the ultimate destination inside that country. Of course, as the Internet matured, the average user and network needed first to get access to Internet service providers (ISPs) and rely on large-scale data backbones to make connections to other networks far away. But these providers are really networks themselves connecting directly in a patchwork of links—across national boundaries—that is global in scale. And while national data networks did emerge and persist today, they operate within the context of the Internet patchwork rather than an internetwork of national networks.

[16] Is this a late phase of the Internet, or are we merely in the beginning stages of a much deeper and longer trajectory of development for digital technologies of connection, within which the Internet is merely an initial social experiment? My hunch is that the latter is right.

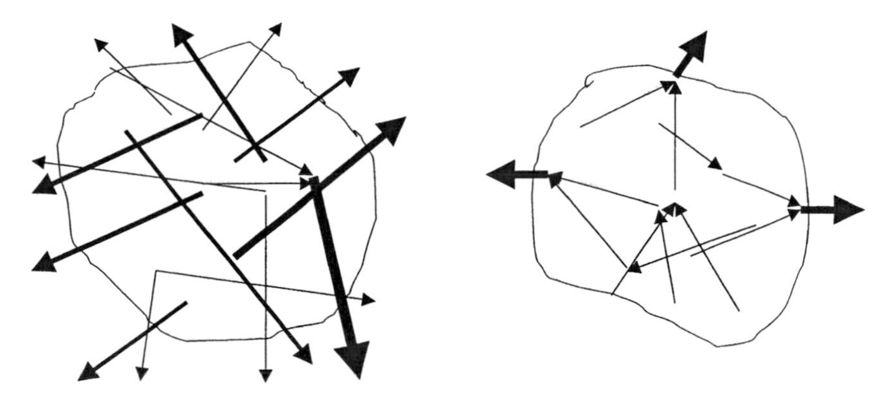

Figure 1. Transboundary internetworking styles

We can crudely visualize the difference between the two approaches to transboundary internetworking in the following figure meant to depict national boundaries (fig. 1). The form on the left represents the many lateral interconnections emerging from a given country initiated by discrete networks (the variation in line thickness should convey that networks differ in size, particularly bandwidth). The form on the right represents the interconnections of a national-level public network that takes place at the national border through a national gateway. I label the left side lateral internetworking and the right vertical internetworking (to convey that it is necessary that subnational networks move up and down through national networks).

It is all too easy to fall into the trap of assuming that lateral internetworking was an overwhelmingly attractive approach, the success of which was made inevitable by U.S. state support, and that the Internet was some sort of juggernaut emerging from inside the United States as a U.S. entity extending outward with unstoppable momentum.[17] This view is mistaken. At the most basic level, it should be recognized that the United States was not the only site of computer network formation. Europe invested considerably in networks. These networks—which had their own momentum—were interconnected within Europe as well as to the United States, and the process started in the early 1970s, as internetworking began to become an issue in the computer-networking world.

While some of the emerging networks and interconnections in Europe were based on U.S. networking approaches (e.g., Eunet, which was the

[17] Guice (1998) also emphasizes the importance of considering alternative routes to computer networking.

European UNIX Network), others followed an approach to vertical networking that was centered in Europe but was embraced around the world, including in the United States.

In and of itself, European computer networking was not a force that represented an alternative to the Internet. These networks across the 1980s were linking up to networks in the Internet constellation and thereby became a part of that constellation.[18] In this respect, they were important to the Internet's success as a global system.

What did offer an alternative was the vertical networking vision that was advanced especially in Europe (and to the extent that the European networks would have been drawn up into that networking project—rather than the Internet—they would have been a challenge). That vision was strongly associated with the large national telecom agencies, Post, Telegraph, and Telephone or PTTs, and was aimed at establishing a national public data network in each country along the lines of all previous international telecommunications systems. The point was to make sure data was channeled through the already existing telephone system in a manner that allowed the agencies to charge for access and usage (Libicki 1995: 80–81, 87; Gillies and Cailliau 2000: 65). Private networks—especially in the process of proliferation—were frowned on and viewed as a nuisance and sometimes even a threat to the project of one large public data network per country (Abbate 1999: 160–66; Hirsch 1975a, 1975b). But such networks existed and were flourishing, ironically, often because of the very efforts of European governments. For the PTTs, the trick would be to get them and all comers to accept the large public networks as the basic infrastructure and ultimate global context of (inter)networking.

The conflict between the two visions of internetworking—represented in figure 1—was fought out on the terrain of standards for networking and internetworking (standards to be taken up and applied by all networkers as protocols for making connections). Since the international standards bodies were made up of national government memberships, these organizations (specifically the International Standards Organization, or ISO, and the ITU) were naturally oriented toward the PTT approach and its associated protocols, called the Open Systems Interconnection (OSI). Basically, the OSI suite of networking protocols was oriented toward establishing control over the movement of data: as data went from one point to another, the route could be traced. This would allow charges to be levied, and limits to be set on who could carry data, not unlike the way international telephony works. In the Internet suite of

[18] These interconnections are documented in Quarterman (1990). For an interesting survey of the thinking and people behind them, see Malamud (1992).

protocols (Transmission Control Protocol/Internet Protocol, or TCP/IP) the data could go any available route, which made cost capture and control impossible at the time.[19]

We know now that this model failed to capture the destiny of computer networking, along with its protocols. At the time, however, there was considerable momentum for it as public networks were established (which still exist), as the protocols were embraced by networks and organizations worldwide, including U.S. corporations eager to sell OSI products and the U.S. state, which adopted them for its own networking needs and accepted them as the future of global computer networking.[20] While the United States promoted lateral internetworking (as will be discussed below), it did so to advance the work of Arpa and the possibility of scientific and technical collaboration, not to establish an Arpa-based Internet as the global computer communications system.[21] Putting this in historical perspective, it was not until the end of the 1980s that one could say that the chances for the OSI suite to triumph were fully wiped away as the scales in favor of TCP/IP tipped decidedly in its direction, as just about all networks accepted the latter as the primary means of interconnection.

It is not possible here to explore in any substantial fashion why the PTT model failed. My purpose in this chapter is to explain how it was possible for a lateral internetworking system to come into being. But to the extent that the dynamic between the two models helps us understand better how the Internet model succeeded and what was at stake in that success—a success that was achieved ultimately at the expense of the alternative—it deserves consideration.[22]

[19] It is not possible in this space to discuss the complex history of this conflict (TCP/IP vs. OSI). My interest is to draw attention to its existence and relevance to the success of the Internet. For analysis and portraits of the conflict, see Abbate (1999); Salus (1995); Malamud (1992); Drake (1993); and Tannenbaum (1989).

[20] This adoption is associated with the Government Open Systems Interconnection Profile (GOSIP). See the Department of Commerce report, *Standards in Process* (Cerni 1984), to understand some of the ways this choice was framed in the United States. See also Cerf and Lyon, n.d. (approx. early 1980s, internal agency document), for a glimpse of how the Arpa networking community tried to fight back against the implications of the U.S. embrace of the international standards and convince the U.S. government to advance the TCP/IP cause.

[21] This may seem to contradict the points made below about the promotion of the Internet from the United States. However, one needs to separate the attempt to advance international scientific collaboration and to support Arpa and its particular networking approach (to say nothing of the military's overall endeavors) from the decisions made regarding protocols for other agencies and more generally the face of the U.S. government in international fora.

[22] I will lay aside the counterfactual of what would have been the fate of the PTT model if the Internet alternative did not exist or failed for other reasons.

The momentum in Europe for developing networks and internetworks was helped along by the prospect of establishing an international approach that was viewed to be clearly in the interest of European telecom agencies. Success for the PTT approach was anticipated, and TCP/IP was at best tolerated temporarily and at worst shunned.[23] It is not clear how networking in Europe would have proceeded without this prospect for success, but anticipation of success did create a sense that the investment was worthwhile.

But the dynamic between systems was not zero-sum in that the development of the PTT model directly contributed to the success of the Internet. First, the large number of networks that were put in place through European initiative helped—upon integration into the Internet constellation—make the Internet a global system. Second, the implementation of OSI technologies, when added to TCP/IP technologies, helped produce a more diverse networking environment overall. Such diversity supported the Internet model because its protocols were well suited to operate in such diversity (Abbate 1999: 178).[24]

However, a zero-sum view was prevalent among Internet-oriented technologists who have argued the PTT model was inferior; an inferiority that has generally been cast as a function of the positive attributes associated with the Internet.[25] That is, what is wrong with the PTT model is that it is not the successful emerging Internet approach. Mostly it is argued that one approach was superior to the other for achieving the ends of global internetworking—more technically effective (accommodates diversity, can be flexibly applied to interconnections of one's choice), economical (available as open source for application and based on a more simple design), timely (was on the scene early in the cycle of network formation), and attractive regarding organizational culture (emerging out of the bottom-up heterogeneous culture of networkers rather than the top-down approach of bureaucrats).

These qualities are generally held out as explanations for the success of the Internet and, in turn, the failure of the PTT model that lacked these qualities (OSI undermined diversity as mentioned above; interconnection was complicated and poorly thought through; the protocols—especially the manuals—had to be purchased; the suite was complicated and not eas-

[23] The use of the TCP/IP protocol is not necessary to establish interconnections to Internet constellation networks. One explicit case of avoiding TCP/IP was the UK network called the Joint Network Team (JNT), one of heads of which claimed that "the feeling was that we should align our networking program with UK developments rather than something from across the Atlantic" (cited in Gillies and Cailliau 2000: 66).

[24] Note that OSI protocols are still used. But they are applied in the context of the Internet system.

[25] For overviews of the debates and thinking, see the references in note 19 above.

ily implemented and was released publicly over time; and finally, there never emerged a "networking culture" around it—as there did around, say, Unix systems—even though plenty of networking experts around the world worked with and supported it). But in and of themselves, these attributes do not explain the success of the Internet because they assume the crucial incentive to interconnect.[26] It is more accurate to view the attributes as important supports making possible the network proliferation, maintenance, and interconnection that I will argue is the key to Internet success. The employment of TCP/IP required only minimal changes to the workings of each discrete network.[27] This allowed networks to preserve the identity of their networks in the face of interconnection. Flexibility allowed choices regarding the pursuit of connection to other networks (which could be made in a timely fashion), accessing what was considered of value of in other networks. While the OSI world, as mentioned, increased diversity and networking overall, it was not consistent with the trends toward lateral interconnection described below.

If there was such a marked difference between the two approaches, then how was it possible for the Internet to get a serious toehold in the networking world of Europe? After all, despite all the factors just reviewed that worked in favor of TCP/IP, there were considerable institutional forces arrayed against it (whole governments, international organizations, and seasoned experts). Even if a system is viewed by some actors (in this case, its developers) as working better, this fact does not guarantee its success. Who else views it that way? How is efficiency defined and evaluated and by whom? Many experts and the majority of institutions supported OSI not just because it was consistent with PTT interests but because they believed its design would *in time* be optimal (even if, in the meantime, the Internet approach was more efficaciously applied).

There are, of course, historical precedents for the triumph of "less efficient" technologies. The QWERTY keyboard is held out typically as an example of how less efficient systems can prevail based on the institutional relationships and structures that are entrenched from the past that create path dependencies (David 1985).[28] But—whether or not a technology is

[26] I have put aside the question of whether they are adequate explanations for the failure of the PTT model.

[27] This point about the Internet approach is also made by David (2001: 160, 166–67). David and Bunn (1988: 170, 181) offer this observation about any gateways that link heterogeneous systems. The gateway mediates between the varying systems and allows for their preservation. Some call the Internet a dumb network (Isenberg 1998) because the guidance, channeling, and checking of data movement are not done in the network but at the end points of transmissions (hosts).

[28] Liebowitz and Margolis (1994) question this example by pointing to evidence that the QWERTY technology was as least as efficient as the alternatives. Even if that is the case,

more or less efficient—there must be a process or mechanism allowing entrenchments to be overcome. In my view, what helped undermine the support for OSI was the fact that a connection into the emerging Internet constellation or adoption of Internet protocols (TCP/IP) did not force networks using it to endorse the entire Internet model, forcing a rejection of the PTT model. A network could even use OSI protocols to interconnect to other networks in the Internet constellation. The point is that active engagement with the Internet world could occur without having to make a commitment to it with regard to networking policy overall. As long as the Internet protocols remained simple and minimal (so that networks did not need to reconfigure themselves internally around them), the political costs to European networks of involvement were low. Carl Malamud's book *Exploring the Internet* (1992) documents well how many decisions to link up to the Internet constellation were made on a temporary, interim basis.[29] Interestingly, even Internet-constellation networks, such as NSFnet, had as part of their original policy the promise to migrate ultimately to OSI (Gillies and Cailliau 2000: 79). However, as I will discuss below, over time interim decisions aggregated into a large-scale Internet constellation of global proportions that produced success for that system even in the midst of worldwide support for OSI.

The Basic Model

All the factors just discussed, which stand in favor of the Internet approach, are meaningless unless there is an explanation for the formation of the Internet. In other words, the strongest argument in favor of the success of the Internet is one that articulates a powerful dynamic of formation that is consistent with the historical context of that formation. The rest of this chapter will be devoted to this argument.

Since the Internet is at its heart a set of interconnections among networks, any explanation of internetwork formation must show why such interconnection would come about. I have a very simple model in mind. It starts with the assumption that any given network has value to those who are in it and to those who evaluate it from outside. That is, in straightforward economic terms, it has utility for users or potential users (satisfies wants and is desirable), it is relatively scarce in that it has its own unique qualities, and it involves considerable costs to produce and main-

there is no shortage of other examples, including the one already referred to above: the high-cost telephone systems established in Europe versus the much lower-cost U.S. system.

[29] This pattern was made possible by the slow development of OSI protocols for internetworking, which helped justify promises of temporary use and ultimate migration to full OSI systems.

tain.[30] By interconnecting to a given network you get to access its value. The question is, what is the value of a network?

The most famous depiction of network value is popularly known as "Metcalfe's Law," named for Robert Metcalfe, the inventor of the Ethernet.[31] It quite elegantly states that the value of a network (V) is equal to the square of the number of computers (N) connected to it, $V = N^2$. The squaring of N is meant to depict the potential interaction—$N(N)$[32]—and the various potential activities—such as e-mail communication or content production—that network participation might yield. The original articulation of the law is associated with Metcalfe's development of the Ethernet in the 1970s. It was an argument used to convince Xerox that the growth associated with Ethernets would be polynomial rather than just additive (all Ns). In effect, Metcalfe's Law is a case of what is commonly called network effects or externalities in economics: the utility of something to a user increases as the number of other users grows (e.g., telephones are more valuable to you if you have lots of other people also with telephones to call).

Following from the law, the value of interconnection of one network (N_1) to another (N_2) is $N_1 N_2$ (Varian 1999). In effect, as Varian (1999: 8) puts it, "each network gets equal value from interconnecting" (although from an individual's standpoint, if you are in the smaller of the two networks you get more value because you are linking up to a larger network).

I understand the attraction of Metcalfe's Law: you are dealing with real numbers, as long as you know how many computers or users are connected to networks. But what you are doing is assuming away the messy vagaries of how much content and communication—ultimately value—is actually being generated by users. You are dealing in pure potentialities. As one "web authoring company" observed, although a network of many users may have value as such, "it doesn't mean that simply increasing numbers will add more value" because some "content" is noise, not information. As they see it, "it is the application of the network which really matters."[33] In other words, $V = N^2$ makes too strong an assumption about the use and purposes to which a network is put. There is also the prospect of passive users and the possibility that production is concentrated in a few hands. Further, once networks get very large, the no-

[30] See the discussion of value by Hicks (1946) and Weber's chapter in this volume, which also builds his arguments on the relationship among technology, networks, and value.

[31] One brief commentary on it is in Metcalfe (1995). It is discussed in Shapiro and Varian (1998).

[32] It is of course more accurate to subtract out self-interaction ($n^2 - n$), as Shapiro and Varian (1998: 184) do.

[33] Quoted in Windrum and Swann (1999: 3).

tion that all users might potentially communicate with all others is absurd, as Metcalfe himself recognizes.[34]

Instead of $N(N)$, I would prefer to view the value of a network as equal to the interaction of N, users, and I, information so that $V = N(I)$. Network value is thus a function of the number of users, the amount of information they produce, and its circulation through interactions and activities. Circulation, of course, itself will be a function of network performance in that large amounts of information can only succeed in circulating—among many and few users—if there are high performance levels (e.g., bandwidth to allow file transfer). Thus, by interconnecting, two networks gain $N_1 (I_1) \cdot N_2 (I_2)$ value as users and information are thereby able to mix (via communication and circulation) across the two networks. The attraction of interconnecting in this formula is greater than the one proffered by Varian (N_1N_2) in that you get not only the potential value of interaction across the networks but also the value of access to existing bodies of information.

I chose the category of information—rather than content—to describe the valuable substance that can be generated and circulated on a network among users. Content can be anything that is communicated, without reference it its meaning and ultimately its value per se (cute missives or banner ads). Information, a subset of content, is different. I follow a line of thought developed by Niklas Luhmann (1995), which emerged out of the tradition of Information Theory associated with Claude Shannon, but which departed significantly by squarely linking information and semantic value.[35] Information has value because it is "an event that selects system states" (Luhmann 1995: 67). In other words, information is communication that establishes that a given social system—or some aspect of a system (or a subsystem)—is in one state or another, is this or that way (e.g., the price of gold or the movement of troops).

Of course, there is relative worth to any information—the weather versus a tip on an intended crime or a description of a new computing networking approach. Consistent with Luhmann's conception, I believe, is

[34] Metcalfe (1995); see also Windrum and Swann (1999: 10).

[35] Information in the Shannon framework is a function of the number of choices and the selection of messages among possibilities, such that one would select symbols ultimately to construct a sentence. Shannon wholly divorced semantics from information, which is better conceived in his terms as an engineering concept: namely, bits of o's and 1's that can represent as little as one letter or as much as a whole book. See Shannon and Weaver (1962, especially pp. 95–117). This is very different from the ordinary use of the term. Luhmann (1995: 40, 67–69, 140) is inspired by this conception but brings it back into our more ordinary use of the term by linking possibilities to meaning. Interestingly, Windrum and Swann (1999: 3), in their departure from Metcalfe, also refer to Luhmann with regard to his distinction between meaningful and useful information and "mere information processing," which can involve noisy unwanted messages.

the claim that the salience or worth of information is a function of its effects. That is, it can change to varying degrees the state of things (knowledge of, perception of, or approach to something). These effects can vary according to their scope (what they bear on—from traffic to urban economics) and impact (reverberations across time and space, relating to how robust the information is and how many users/uses it is relevant to).

I understand that I have injected ambiguity back directly into the equation, $V = N(I)$. Information, like utility, is not subject to any straightforward metric. However, other networks of users can judge the relative worth of the information that circulates on another network with regard to its specific substance (what it bears on) and effect (its scope and impact).[36]

Separating N from I has the advantage of not assuming that all potential interconnections yield value in some hypothetical future. It is more anchored in the immediate past and present of a network in that real information must have been or is produced and circulated by users in order for I to yield value. In addition, $N(I)$ allows for a network to have a relatively large value, while also having a small number of users (N). This is particularly relevant to the first phase of Internet development, when networks were small but the information—especially that pertaining to new developments in network technology—could be of great worth (where the network quotient, I/N is $>$ 1 by various magnitudes). Indeed, the proliferation of bounded, specialized networks was prevalent in the early stages of Internet development, which suggests that it is access to the specific mix of a group of users and their information that is of value to outsiders who might want to interconnect, rather than just the prospect of forming a bigger N through internetworking.[37] Particularly in the early, developmental stages, access to networks such as the U.S.-sponsored Arpanet (as discussed below) had a unique, high value in that experimentation was occurring on them producing information that could be applied in the formation of networks elsewhere in the world.

My departure from Metcalfe's approach does not constitute a repudiation of the theory of network effects as an explanation of how networks can explode upward in growth. It should be underscored that there is embedded in $N(I)$ a network effect: as networks interconnect, the rising number of users and uses (information) means more value to any given user. In addition, the combined value of two networks surely exceeds the value

[36] To a degree the search engine Google does that with its weighing of sites according to their prestige.

[37] Notwithstanding, the model still allows for greater value as N grows. Larger-sized networks did develop in the initial phases of the Internet, most famously Usenet, a cooperative network known for its many news groups.

of any one of them alone. I am concerned that the network effects literature assumes the existence of networks, which either ramp up and generate positive feedback effects or fail to do so as they remain too small to achieve expectations among potential joiners that critical mass and thus take-off will obtain (called the start-up problem—see Economides and Himmelberg 1995). In the case of the Internet, we need to explain the initial formation of a network—or more accurately an internetwork. Other factors besides feedback about other users need to be brought in, such as the salience of information mentioned above and the proliferation of discrete networks discussed below. The stakes are distinctive for Internet formation because, as networks interconnect, they are actually building the (inter)network rather than endorsing a particular technology, standard, service, or product associated with an existing—albeit emerging—network of users. In other words, the growth via interconnection that was critical to the formation of the Internet was essentially a process of invention and basic formation (of the internetworking infrastructure) rather than a "bringing to market" of an invention. Once the Internet was established as a global infrastructure of interconnections, its further development—or take-off—might have indeed been driven powerfully by the logic of network effects, toward the end of period two onward into period three.[38]

The Proliferation of Computer Networks

To explain the formation of the Internet, attention must be shifted away from the endpoint of (inter)network development (a large functioning network) toward the starting point. At that starting point, one finds the formation of discrete computer networks in the United States and Europe. Network growth went from a handful of operational networks in the early 1970s to a veritable explosion of networks forming in the tens of thousands during the 1980s as corporations, universities, research organizations, governments, and communities of technology enthusiasts set up networks of various kinds based on diverse technologies.[39]

In a context where there is a growing set of discrete networks, internetworking becomes that much more useful. Where there are only a few

[38] I suspect but cannot argue persuasively that this is the case. Note that the literature on the economics of interconnection that builds on the theory of network effects, as illustrated by Varian's (1999) formula discussed above, is also limited in application to the later Internet because of the assumption of an extant communication system, the very thing I seek to explain.

[39] The single best way to gain a sense of the scope, operations, history, and purposes of these networks around the world as of the late 1980s is to look at Quarterman (1990).

very large networks (e.g., one per country), the salience of internetworking is lower: each large network contains its own built-in value (both users and information) and follows the logic of network effects. Minitel, the huge French national network established in 1983, had little incentive to interconnect to other networks and did not do so until the recent past when keeping unconnected to the burgeoning Internet was impossible. Likewise, the longstanding commercial service Compuserve did not interconnect until the Internet reached more mature stages.

But in the total universe of networks, the large networks of this sort stood out as exceptions. There were notable reasons—some quite obvious, others less so—why smaller networks proliferated and took form as discrete, bounded entities that then would interconnect. Since Internet development depends so much on interconnection between such networks, these reasons are worth considering. I will begin with the conditions underlying network proliferation.

Perhaps the most basic and obvious factor was the spread of computers themselves into modern organizational life, from mainframes initially, to minicomputers, on to personal computers (PCs). Linking individual computers is obviously beneficial so that resources can be shared and electronic communication can occur.[40] Readers of this volume, if old enough, might recall dimly the early efforts of their universities to supply first staff and then students with personal computers, which increasingly were linked up in local area networks (LANs) to facilitate intracampus communication and then connected to the outside world through networks like Bitnet. Among the pioneering corporations in networking is General Motors (Tannenbaum 1989: 36–40), which ambitiously connected its myriad offices and factories in one large corporate network (something we take for granted today).[41]

A second factor was the sponsorship of computer network research and system development by states to ensure that their nation had such capability and could enjoy the benefits of networking inside and outside of government—or pay the price of being left behind by other states.[42] Taking the lead was the United States with the development of its Arpanet (Advanced Research Projects Agency) along initially modest lines in 1968. Other such networks (distinct because they involved sending messages broken down into discrete packets of information) emerged around the same time in Britain at the National Physical Laboratories (NPL). France

[40] See Nolan (2000) for a portrait of the computerization of U.S. business. A classic from the time is of course Bell (1973).

[41] See the chapter by Ernst in this volume for an understanding of the ramifications today of corporate networking.

[42] A classic statement of these concerns in France is Nora and Minc (1980).

also got in the game, as did Germany and the Scandanavian countries.[43] The United States followed its typical pattern and witnessed the establishment of such a network (Telenet) by a private concern.

Within each country (at the national and subnational levels) and across Europe (as a regional endeavor of the European Community), networks were funded and founded. By the mid-1970s the first experiments with public networks emerged as various countries—through their PTTs—across the Atlantic from Spain to Canada set up networks that were meant to be accessible to any user, at a fee of course.

In the private sector, firms—especially Xerox, IBM, AT&T, and Digital Equipment Corporation—promoted the formation of networks based on their own proprietary protocols. The protocols for Bitnet, it should be noted, came out of IBM, which also invested considerably in promoting Bitnet networks in Europe through the European Academic and Research Network (EARN) established in 1983. All together, these various network formations led to a variety of networking protocols and technologies, which expanded the sense of diversity and proliferation.

Such diversity and proliferation would have been impossible without the emergence of computer science and network engineering as fields of study, research, and practice around the industrialized world. This is a third factor. This provided the requisite expertise and human capital on which the various sectors forming networks could draw. Obviously, one of the important payoffs of sponsoring networks—inside and outside of government—was the nurturing of such intellectual capital, as the spirit of experimentation and research occurred simultaneously with implementation.

A fourth factor is the existence of a telecommunications infrastructure—within and across organizational sites—that could be applied to computer networking (e.g., the leasing of telephone lines for data transfers). Imagine how much more difficult it would have been for organizations to set up their own networks if they required the initial laying of lines—locally, nationally, regionally, and globally—that was required with the development of telephony across the nineteenth and twentieth centuries. It would likely have happened, but much more slowly, which would have changed the stakes and patterns of the process of network proliferation.

These four factors were conditions necessary for the formation of discrete networks. They may explain why computer networks emerged initially, and why they were heterogeneous in form and purpose. They do not explain why networking would spread to nonsponsoring or nonexperimenting organizations and be maintained by any organization beyond

[43] It should be noted that the Societé Internationale de Telecommunications Aeronautiques (SITA), which provides communications for air carriers worldwide, was developing an innovative packet switching network.

the point of initial sponsorship. Why would organizations in the United States and Europe establish their own networks rather than wait for access to some large public data network? Organizations must have an incentive to apply or extend the networking experiments and invest in the building and maintenance of their own discrete networks. In other words, what needs to be explained is the success and spread of computer networking, not just its initial testbed development.

Of course, each organization (research and education, business, and government, including the military) had its own individualized incentives based on the specific benefits to it of networking. My interest in developing a simple model is to approach this at a general level. From that angle, what they all shared was the ability to transfer much of their organizational life to a discrete, bounded computer network, including basic communications, institutional memory, and working documents. Consider the U.S. military. Computer networks were seen to be central to command and control, communications, data collection and processing, weapons management, logistics, and so on (Norberg and O'Neill 1996: 1–23). That a network could be bounded was critical, as it allowed for a digital reflection of an organization (even if that organization was nothing more than an emerging research network or consortium). Large-scale public networks do not allow for that, except to the degree that they accommodate boundedness, as is the case today with virtual private networks (VPNs) on the Internet.[44] Referring back to our simple nomenclature, organizations are interested in generating their own network value, $N(I)$, just as other organizations are interested in that value through interconnection.

The preference for highly specialized networks sometimes ran headlong into the preference for more general networks. When the NSF began organizing its networking efforts in the mid-1980s, there ensued a debate between those researchers who preferred a "general-purpose network" and those in physics and chemistry who preferred a network for connections between supercomputers (Rogers 1998: 219–21).

While the NSF ultimately was able to accommodate both sides in its efforts, it should be noted well that the general network was really an internetwork connecting the specialized networks of campuses and research programs. Indeed, internetworking sits right at the tension between specialization and generality in that internetworks must be general enough to accommodate the various bounded networks that seek interconnection.

The double-sided coin of network formation—the tendency toward network differentiation and network interconnection—was a powerful

[44] These operate on the Internet sort of as intranets, with access restricted to employees and invited others.

dynamic, as more and more networks were created along with more and more interconnections among them. That research networks were so prominent early on was of considerable importance to this dynamic. Science does not lend itself easily to serious barriers of access between communities of researchers. Scientists are also very good at producing and circulating information in the form of research, data, and commentary (*I* is likely to be high in research networks). In contrast, firms have concerns about proprietary information that are far more constraining than is typical of the research world. On the other hand, the interest of a firm—as Ernst's chapter shows—in linking to other networks (to coordinate production or distribution) or expanding their own network (to build market share) is obvious. This open/closed duality is manifest in the strong embrace by firms of both intranets (networks internal to a firm) and extranets (the links of those networks externally).

Transboundary Interconnections

Success for the Internet was dependent on the patterns associated with lateral networking. Transboundary internetworking—and thus Internet formation—could proceed apace without waiting for the development of national networks and their interconnection. What was the driver for that formation? The formula $N_1 (I_1) \cdot N_2 (I_2)$ was introduced above to depict the value of interconnection between two networks. Lateral internetworking among discrete, bounded networks produces a great deal of value, more than large-scale network growth does even with the same overall increases in users and information—that is, as N and I in a single large network grow. This certainly is arithmetically true, as long as $N_1 + N_2 \geq N$ and $I_1 + I_2 \geq I$, where N and I are associated with a large network.[45] Even if one assumed that Metcalfe's Law, discussed above, applied to the single large network (N^2), the value of interconnecting two separate networks—rather than growing one network—would still be greater as long as $I > 1$ and the duplication (self-interaction) is subtracted out ($N^2 - N$).[46]

This extra value to interconnection makes substantive sense because lateral links across networks allow for the interaction of discrete trajectories of information production and circulation, organizational culture, and the history of interactions and relations among users on each network. While adding users and information to a single network certainly

[45] There are many instances where $N_1 + N_2 \leq N$ and $I_1 + I_2 \leq I$ as well.

[46] This is on the assumption—following Metcalfe's model—that I is not specified as a value.

increases value (as both *N* and *I* grow), such expansion does not enjoy the level of benefits that a mix of cultures, knowledge, and interactions across two or more separate networks enjoys.

Of course, ultimately, subgroups and cultures can emerge in a single, large network (which some associate with the term "virtual communities"). However, these groups and subnetworks do not entail gaining access to the organizational life that has been transposed to a discrete network to facilitate such activities as sustained research collaboration. This is especially true when such access means learning about the technologies of networking being applied, as was the case in the early formation of the Internet.

A good example of the special value of interconnection is found very early on in the history of computer internetworking. Among the first international connections of networks was the one between the United States and Britain set up in 1973, specifically to the University College, London (UCL). This connection opened up access between Arpanet and emerging British computer research networks. In the words of Peter Kirstein (1999a: 10, 11), a leading figure in networking since then:

> From the outset of the project, we aimed not only to carry out innovative research, but also to provide network services to UK and US groups who wished to cooperate. As early as 1975, there was firm collaboration between many groups in the UK and the US. From the UK viewpoint, the collaborative usage was one of the primary reasons for . . . support of the UCL infrastructure activity. . . . It allowed the British developments to proceed along their own directions, while allowing continued interconnection between the communities on both sides of the Atlantic. As a result, there was no perceived threat of transatlantic dominance.

In other words, the two research networks could, via internetworking collaboration, develop the value of their own networks and contribute to the collective body of information on networking technology.

While the specific value of interconnections explains why networks would pursue them, it does not in itself indicate the conditions and environment that make such actions possible. As was the case with the proliferation of computer networks, the growth of transboundary interconnections depended on a number of underlying factors (recognizing, as argued above, that the proliferation of networks was itself an essential factor). One basic condition was the relatively minimal restrictions in the law and policy of states to the act of interconnecting.[47] Restrictions and constraints did exist: Arpanet could not, in principle, be connected to just

[47] See the discussion by Guthrie in this volume of the minimal restrictions in various IT areas by the Chinese state.

any network. Access was restricted and did not begin to loosen until the 1980s. It was a Department of Defense program, after all. And while continental European policymakers allowed interconnection to U.S.—based networks—and certainly to other European networks—they limited the latter for fear of the very U.S. dominance that Kirstein (1999a: 11) claims was not a British problem. Nonetheless, connections abounded and snowballed across the 1970s and 1980s. There was a culture of scientific research collaboration, international security cooperation, and transactlantic business development in the West, despite a conflict (discussed below) involving different visions of how computer networking should ultimately proceed. Bear in mind that the political context within which the Internet developed was NATO and the OECD in a time of cold war.

In that context, the U.S. state tolerated or even promoted interconnections. On the one hand were efforts to experiment with linking sites across the Atlantic involving research that was directly related to security, such as the very early connection to Norway where the monitoring of Soviet nuclear tests was being undertaken (Cerf et al., n.d.).[48] On the other hand were the efforts associated with NSF to facilitate scientific research of all kinds, but above all on computing. In the words of one 1979 report from the NSF's Computer Science and Engineering Advisory Panel that helped set the terms for NSF's significant networking promotion efforts in the 1980s: "We recommend that NSF provide to qualified computing researchers easy access to an international computer network. This access would create a frontier environment which would offer enhanced communication, collaboration, and the sharing of resources among geographically separated or isolated researchers" (cited in Comer 1983: 748). NSF-established Csnet (1981), linking computer science departments not connected to the Arpanet, had by the mid-1980s been developing over a dozen international interconnections (including Isreal, Switzerland, and Italy). NSFnet, which started running in the mid-1980s, also took international linkage serious. In the early 1990s it even started an initiative (the International Connections Program) to promote them (Goldstein 1995).

All the promotion in the world, however, would have been fruitless without something of value to connect to. Kirstein (1999b) observes that one reason the U.S. networks were so attractive to outsiders was that they were important testbeds for networking developments that could be applied outside the United States. U.S. networks were incredibly able to produce information (*I*) of a high value. It was a potent process: the very ef-

[48] Interestingly, the first instance of computer-like conferencing took place in 1951 in a NATO attempt to link teletype channels to coordinate response to the Berlin Crisis (Vallee 1982: 117–18).

fort to develop infrastructure (the experiments and innovations) became the basis for the global success of the infrastructure. That is, others want access to the process of development, which helps ensure the success of the process. As Kirstein (1999b) makes clear, European networks were not given the necessary support to counter this dynamic.[49]

The flip side of this value production was the question of the sheer number of U.S. networks that were formed. In figure 1 above, it is obvious that if you have more networks with more backbone reaching laterally across borders, you have a great advantage in building a global network centered in that original geographical base. This certainly was an advantage enjoyed by the United States in the history of internetworking. This advantage was a function of the size and unity (political federation) of the United States, which made possible both a wide diversity of discrete network formation (fragmentation) and a national-level program of network promotion (integration).

It also helped that the system, the Internet, which emerged around a constellation of interconnections centered initially in the work of Arpa, was in many ways made for export. In particular, the addressing system (which is now called the domain name system) was able to incorporate an expanding number of networks and was relatively decentralized (e.g., individual networks can decide who has what subdomains).[50] It also helped that much of the new networking equipment associated with Ethernet and the new, relatively cheap routers from Cisco came with the protocols for the Internet, TCP/IP, built-in (Gillies and Cailliau 2000: 68–69). As networks in Europe and elsewhere were set up, existing ones reconfigured, and interconnections sought after and made, the available technologies associated with the Internet were conveniently available and easily applied.

Aggregation and System Formation

These last two conditions draw attention to the question of how an infrastructure system such as the Internet takes form at the macro level as a system, moving squarely into phase two (proliferation) and onward toward phase three (maturation). Historians who think about systems of infrastructure, such as Paul David (David and Bunn 1988: 166) and Thomas Hughes (1983), argue that systems of this sort are not designed de novo but evolve contingently via the play of politics, technology, accident, and

[49] Cf. Mowery and Simcoe (2001: 27).

[50] It became a "globally administered address space" (Malamud 1992: 54). Of course, the space had to be expanded because no one could anticipate the incredible explosion of network formation (ibid. 57).

social relations. The evolutionary perspective is consistent with the model of Internet formation outlined in this chapter, where micro decisions to connect and factors such as discrete, interim technical choices and the (U.S.) advantages of size worked in concert to supply an emerging Internet constellation with a discernable form and a momentum of significant growth. In contrast, the PTT approach involved the design upfront of a comprehensive system from top—macro-internetwork—to bottom—individual network (Vallee 1982: 81–84; Gillies and Cailliau 2000: 65).

In and of itself, the manner of system formation does not explain its success. In addition, it is important not to idealize patterns of development: most systems are combinations of design and contingent evolution, as was the Internet itself (for instance, a considerable amount of design went into the original Arpanet development, as described by Hafner and Lyon [1996] and Norberg and O'Neill [1996]). Moreover, evolutionary systems can fail. What is crucial to the question of success is whether the ways a system scales up, aggregating myriad local actions and decisions, lead to a robust structure (that is, a form whose basic contours reproduce through time). The logic of aggregation has to be consistent with successful but stable growth. Without that stability, the system could morph into something inconsistent with its original "genetic" code (in the case of the Internet, that code comprises the logic of the interconnection of networks plus the specific character of infrastructure formation).

So far the model has been focused on discrete decisions at the micro level (network-to-network). The implication has been that the sum of these decisions produces a set of interconnections that I label the Internet constellation. However, what needs to be established is what the aggregation of interconnections implies for the pursuit of value that I argue is the primary driver of Internet success. In other words, does the cumulation of connections feedback to influence the very process of interconnection itself?[51]

To arrive at an understanding of aggregation, we need to introduce the concept of indirect connections. These are connections a network gains by connecting to a network that is itself connected to other networks. In the history of interconnection, this was a rather common prospect (e.g., the French research network, Aristote, connected to the French Eunet, Fnet, which in turn was connected to Csnet based in the United States and thus the Internet constellation [Quarterman 1990: 446]). To the extent

[51] I am aware of the possibility of falling into a genetic fallacy here (assuming one can explain the operation of something by the same factors that explain its genesis). However, I am training my attention on the end point of genesis, system formation, rather than on the mature development of the system. I am also interested in what implications system formation and maturation (e.g., the World Wide Web) have for the logic of system generation.

that a network has an array of interconnections associated with it, the value of that network is considerably greater than its own $N(I)$. Let's say a given network, N_1I_1, is connected to three other networks. The value of that network can be depicted as $V = N_1I_1\ (N_2I_2 + N_3I_3 + N_4I_4)$.[52] Thus, even if the intrinsic value of network, N_1I_1, is limited, the incentive for any other network to connect to it can be high (based on the indirect connections it offers). This contributed to the process of Internet formation for two reasons: the set of possible (indirect) interconnections among networks mushroomed with each direct connection (direct connections bring in tow indirect connections); and a layer of incentive to interconnect was added to the system (interconnecting to another network to gain access to the emerging Internet constellation via indirect connections).[53]

Where does stability come in? If an explosive number of networks enter the constellation, they have the potential of pushing the system away from its genetic origins: in the Internet's case, the U.S. network infrastructure and approach. This did not happen. If anything, the explosive growth reinforced the centrality of the U.S. backbones and protocols. The very logic of interconnection, which originally, as described above, was so consistent with diversity, in the context of aggregation paradoxically reinforces concentration and stability. It works as follows: once indirect connections emerge, as the number of interconnections increases, so does the redundancy of interconnections (that is, overlapping interconnections that offer any given network multiple paths of access to other networks and the entire constellation). Ultimately, the incentive to establish any additional interconnections decreases (you already have connection and can get to a given network some other, likely indirect, way). This favors the existing base of interconnections (the constellation) because once you have entry into it, you gain access to many other networks. Since any given network has finite resources to establish interconnections (limited number of gateways and administrators of those gateways), this decreases the incentive to connect to any network outside the constellation (or which may be part of some other emergent constellation). So the damper on incentive is twofold: lower incentive to connect to non-Internet networks, and lower incentive to connect to Internet networks with which you already have an indirect connection via the constellation.

[52] The networks indirectly connected aggregate in this additive—rather than multiplicative—fashion because the primary reference remains the network directly connected to, which then interacts with these other networks (no assumptions of interaction can reasonably be made about the indirect connections—except that they are connected). The general equation is $V = N_1I_1 \bullet \sum_{i=2}^{x} N_iI_i$.

[53] This can be understood to be a "small world" network effect.

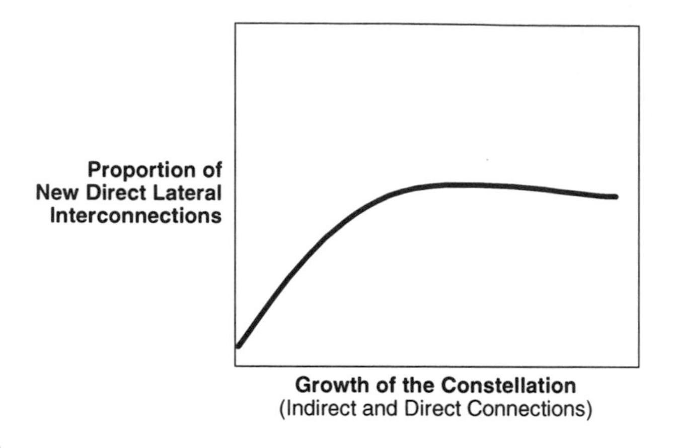

Figure 2

As a result of this dynamic, as indirect connections (redundant, overlapping) expanded (as a percent of total connections in the constellation), the proportion—relative to all connections—of new direct lateral interconnections would slow, level out, or even decrease (fig. 2).[54]

The incentives to join the Internet constellation were of course reinforced by the lower costs and greater efficiency (via bandwidth) of U.S. infrastructure compared with what Europe could offer (Cukier 1995). However, I view this as a facilitating factor (permissive variable) like the efficiency of TCP/IP. That is, what makes efficiency and economy matter is that it gives you access to what you want: connection to the Internet constellation (the information and users associated with the networks). If one did not gain such access, all the lower cost and higher efficiency in the world would be for naught in that it would not deliver the goods demanded.[55]

[54] The upper and lower limits on the number of interconnections (n) ranges from $n(n-1)/2$ at the upper limit, which is the total number of interconnections if all networks were directly connected to one another, down to the lower limit of just n, where all the networks are connected indirectly in a star structure through one juncture point (as though they were spokes on a wheel). The Internet is a mix of direct and indirect connection structures, with backbones and exchanges taking on the role of junctures throughout the system. The evidence for this pattern is the growth of large backbones that serve as the aggregative junctures among networks that are thereby connected only indirectly to one another (in this scheme a network connects to an ISP, which then connects to a backbone). Direct connections are increasingly something only for backbones (as peers) or ISPs that connect to the backbones (as customers of the backbones). See Gareiss (1999) for a description of how even smaller ISPs are suffering under this increasing concentration of bandwidth.

[55] As the purveyors of an incredible surplus of bandwidth in the late 1990s have learned.

Conclusion

Little has been said in this chapter about the implications this model and the analysis holds for current and future developments of the Internet.[56] A few points stand out as suggestions of possible future research in this area.

The formula *N*(*I*) is really a halfway house between a wired past and wireless future. As wireless data networking develops in the twenty-first century, the notion that a network is comprised of *N* sets of users may become meaningless, except for the most restricted networks. At any given time, users will connect to and disconnect from networks on a minute-by-minute basis not unlike, ironically, television audiences do. In this environment, the value of a network will depend even more on the information that flows across it (and which can draw users to it directly). If one cares, as I do, about the quality and nature of information on global data networks (its scope and effect), then it is critical to begin to pursue serious questions about the social purposes of such information and networks.

But there is a serious road block for such pursuit. A system that aggregates up from lateral and flexible interconnections that cumulate leaves relatively little space for designing social purpose into the system as a whole. That would require the production of information at the macro level about the public purposes to which the system is or could be put (that is, various claims about the state or potential state of the system). Discourse in Washington and Brussels and among activist networks on the broader social purposes of the Internet (e.g., education, economic development, distributing resources) did emerge in the early to mid-1990s but fell away quickly as the system became commercialized. These discussions are effectively relegated to the network rather than the internetwork level.[57] To what degree the PTT approach would have opened a space for system-level social purpose design is a question worth researching. However, we should note that the very thinness of the Internet system at the internetwork level was an important factor supporting its success as networks interconnected at a minimum cost to their own operations and identity (allowing thickness at the network level). The tension between

[56] The analysis stops prior to the formation of the World Wide Web and global system of addressing that we associate with today's Internet. To explore in any substantial or rigorous fashion what today's Internet formation means for the model of interconnection described above would require a whole chapter in its own right. However, we can still draw out some implications of the analysis for current developments.

[57] Activists and some developing countries are trying to revive this discussion at the system level. See, for example, www.apc.org.

the desires for unit-level autonomy and system-level political purpose is an old one in the history of political thought. I suggest that internetworking is yet one more terrain within which this tension will shape the development of human interaction.

This chapter suggests that if purposes are not defined at the system level (through law and norms), the form and character of the Internet will be a function of (1) the interaction of the information and purposes of the networks and organizations that compose it; and (2) the patterns of aggregation that I have argued favor concentration and power via backbones and hubs. While the whole (the Internet) may be greater than the sum of its parts (the networks), the question is, how much greater? When research, system development, and interpersonal communications were predominant at the network level, this shaped the culture and substance of the internetwork. It helped that the NSF had a policy of "acceptable use" limiting commercial pursuits on the Internet.[58] However, once that negative wall was lifted, the Internet's form and character could hardly withstand the impact of the burgeoning number of connecting corporate networks, since no positive structure of social purpose was in place (i.e., no model of what the social purposes to which the Internet should be directed). It is naive to call for limits on commercialization unless one has a serious understanding of how to make the Internet something much greater than the sum of its parts. For whom? To what end? By whose authority? Based on what information?[59] The latter is crucial. If information shapes what we know about the state of a system, the question is, who determines which system or subsystem is of concern (a market for handheld computers or life in a third world city) or what aspect of that system we are informed about (stock prices or primary education)?[60]

Social science—and IR in particular—has done little to address the question of how global systems scale up from transboundary interactions. Even in its traditional terrain—the interstate system—no such theory of system scale-up exists. How do bilateral relations aggregate into a macro system? Most analysis takes a system's existence as an assumption and attempts to explain why and how it operates as it does. One notable exception is Mattingly's (1955) study of the Italian city-state system, which shows how fixed embassies emerged among the states; how some states like Rome were centers of exchange and connection in emerging networks

[58] A well-researched paper on the commercialization process is Kesan and Shah (2001).

[59] There are two sides to commercialization: private control and provision of the infrastructure and of the content. For a recent consideration of what is at stake in the latter, see Lessig (2001); on the former, see Kesan and Shah (2001).

[60] These sorts of questions move us toward concerns about the power inherent in shaping the very terms of knowledge and discourse that emerged in the 1960s and 1970s. See, for example, Lukes (1974).

among diplomats and agents from various states. With time an inter-"state" diplomatic communication system emerged across the peninsula. As the attention of IR moves out beyond the interstate system to study transboundary infrastructures and networks, it will have to attend to the genesis of the objects it studies.

References

Abbate, Janet. 1999. *Inventing the Internet.* Cambridge: MIT Press.

Bell, Daniel. 1973. *The Coming of Post-Industrial Society.* New York: Basic Books.

Bijker, Wiebe E. 1995. *Of Bicycles, Bakelites, and Bulbs: Towards a Theory of Sociotechnical Change.* Cambridge: MIT Press.

Cerf, Vincent G., et al. n.d. Cooperative U.S./European Research on Command and Control System Interoperability. Internal Document, Department of Defense.

Cerf, Vincent G., and Robert E. Lyon. n.d (circa early 1980s). Military Requirements for Packet-Switching Networks and Their Implications for Protocol Standardization. Internal Document, Deparment of Defense.

Cerni, Dorothy M. 1984. *Standards in Process: Foundations and Profiles of ISDN and OSI Studies.* Washington, DC: Department of Commerce (NTIA Report 84–170).

Codding, Jr., George A. 1952. *The International Telecommunication Union: An Experiment in International Cooperation.* Leiden: E. J. Brill.

Comer, Douglas. 1983. The Computer Science Research Network Csnet: A History and Status Report. *Communications of the ACM* 26 (10) (October): 747–53.

Cowhey, Peter F. 1990. The International Telecommunications Regime: The Political Roots of Regimes for High Technology. *International Organization* 44 (2) (Spring): 169–99.

Cukier, Neil. 1995. Bandwidth Colonialism? The Implications of Internet Infrastructure on International E-Commerce. Annual Conference of the *Internet Society,* San Jose, CA. Available at http://www.isoc.org/inet99/ie/ie_2.htm (accessed November 2001).

David, Paul A. 1985. Clio and the Economics of QWERTY. *American Economic Review* 75(2) (May): 332–37.

———. 2001. The Evolving Accidental Information Super-Highway. *Oxford Review of Economic Policy* 17(2): 159–87.

David, Paul A., and Julie Ann Bunn. 1988. The Economics of Gateway Technologies and Network Evolution. *Information Economics and Policy* 3(2): 165–202.

Deibert, Ronald J. 1997. *Parchment, Printing and Hypermedia: Communication in World Order Transformation.* New York: Columbia University Press.

Drake, William J. 1993. The Internet Religious War. *Telecommunications Policy* 17 (December): 643–49.

Economides, Nicholas, and Charles Himmelberg. 1995. Critical Mass and Network Evolution in Telecommunications. In *Toward a Competitive Telecommunications Industry,* edited by Gerard Brock, 31–42. College Park: University of Maryland.

Gareiss, Robin. 1999. Old Boys' Network (posted October 7, 1999). Available at http://www.networkmagazine.com/article/NMG20010918S0001 (accessed November 2001).

Gillies, James, and Robert Cailliau. 2000. *How the Web Was Born: The Story of the World Wide Web.* Oxford: Oxford University Press.

Goldstein, Steven N. 1995. Future Prospects for NSF's International Connections Program. Available at http://www.isoc.org/HMP/PAPER/178/html/paper.html (accessed October 2001).

Guice, Jon. 1998. Looking Backward and Forward at the Internet. *The Information Society* 14(3) (July/September): 201–11.

Hafner, Katie, and Matthew Lyon. 1996. *Where Wizards Stay Up Late: The Origins of the Internet.* New York: Simon & Schuster.

Headrick, Daniel R. 1991. *The Invisible Weapon: Telecommunications and International Politics, 1851–1945.* New York: Oxford University Press.

Hicks, John R. 1946. *Value and Capital.* 2d edition. Oxford University Press.

Hirsch, Phil. 1975a. Protocol Control: Carriers or Users? *Datamation* (March): 188–89.

———. 1975b. Protocol for Packet Networks: The Question of Implementation. *Datamation* (May): 189–90.

Hughes, Thomas P. 1983. *Networks of Power: Electrification in Western Society, 1880–1930.* Baltimore: Johns Hopkins University Press.

Hugill, Peter J. 1999. *Global Communications since 1844: Geopolitics and Technology.* Baltimore: Johns Hopkins University Press.

Isenberg, David, S. 1998. The Dawn of the Stupid Network. *ACM Networker 2.1* (February/March): 24–31.

Kesan, Jay P., and Rajiv C. Shah. 2001. Fool Us Once Shame on You—Fool Us Twice Shame on Us: What We Can Learn From the Privitizations of the Internet Backbone Network and the Domain Name System. *Washington University Law Quarterly* 79:89–220.

Kirstein, Peter T. 1999a. Early Experiences with the ARPANET and INTERNET in the UK. *IEEE Annals of the History of Computing* 21(1) (January/March): 38–44.

———. 1999b. Research on Networks versus Networks for Research: The Need for International Internet Testbeds. Keynote paper presented at the ACM SIGCOMM, Boston.

Latour, Bruno. 1987. *Science in Action.* Cambridge: Harvard University Press.

Lessig, Lawrence. 2001. *The Future of Ideas.* New York: Random House.

Libicki, Martin C. 1995. *Information Technology Standards: Quest for the Common Byte.* Boston: Digital Press.

Liebowitz, S. J., and Stephen E. Margolis. 1994. Network Externality: An Uncommon Tragedy. *The Journal of Economic Perspectives* 8(2) (Spring): 133–50.

Luhmann, Niklas. 1995. *Social Systems.* Stanford: Stanford University Press.

Lukes, Steven. 1974. *Power: A Radical View.* London. Macmillan.

Malamud, Carl. 1992. *Exploring the Internet.* Englewood Cliffs: Prentice Hall.

Mattingly, Garrett. 1955. *Renaissance Diplomacy.* New York: Dover Publications.

Metcalfe, Robert M. 1995. From the Ether: A Network Becomes More Valuable as It Reaches More Users. *InfoWorld Magazine* 17(4) (October 2). Available at http://www.Infoworld.com.

Mowery, David C., and Timothy Simcoe. 2001. Is the Internet a U.S. Invention?—An Economic and Technical History of Computer Networking. Paper prepared for the Nelson and Winter Conference, Aalborg, Denmark. June. Available at http://www.business.nuc.dk/dwid/conferences/nw/paper1/mowery.pdf (accessed November 2001).

Nolan, Richard L. 2000. Information Technology Management since 1960. In *A Nation Transformed by Information,* edited by Alfred D. Chandler, Jr., and James W. Cortada, 217–56. Oxford: Oxford University Press.

Nora, Simon, and Alain Minc. 1980. *The Computerization of Society: A Report to the President of France.* Cambridge: MIT Press.

Norberg, Arthur L., and Judy E. O'Neill, 1996. *Transforming Computer Technology: Information Processing for the Pentagon, 1962–1986.* Baltimore: Johns Hopkins University Press.

Quarterman, John S. 1990. *The Matrix: Computer Networks and Conferencing Systems Worldwide.* Bedford, MA: Digital Press.

Rogers, Juan D. 1998. Internetworking and the Politics of Science: NSFNET in Internet History. *The Information Society* 14(3) (July/September): 213–28.

Salus, Peter H. 1995. *Casting the Net: From ARPANET to Internet and Beyond.* Reading, MA: Addison-Wesley.

Schmidt, Susanne, and Raymund Werle. 1998. *Coordinating Technology: Studies in the International Standardization of Telecommunications.* Cambridge: MIT Press.

Shannon, Claude E., and Warren Weaver. 1962. *The Mathematical Theory of Communication.* Urbana: University of Illinois Press.

Shapiro, Carl, and Hal R. Varian. 1998. *Information Rules: A Strategic Guide to the Network Economy.* Cambridge: Harvard Business School Press.

Standage, Tom. 1998. *The Victorian Internet.* New York: Berkley Books.

Tannenbaum, Andrew S. 1989. *Computer Networks.* 2d edition. Englewood Cliffs: Prentice Hall.

Vallee, Jacques. 1982. *The Network Revolution.* Berkeley: And/Or Press.

Varian, Hal. 1999. Market Structure in the Network Age. Paper prepared for *Understanding the Digital Economy* conference, May 1999. Available at http://www.sims.berkeley.edu/~hal/people/hal/papers.html (accessed October 2001).

Windrum, Paul, and G. M. Peter Swann. 1999. Networks, Noise and Web Navigation: Sustaining Metcalfe's Law through Technological Innovation. Available at Userpage.fu-berlin.de/~jmueller/its/conf/torino99/papers/windrun.html (accessed December 2001).

Zacher, Mark W., and Brent A. Sutton. 1996. *Governing Global Networks: International Regimes for Transportation and Communications.* Cambridge: Cambridge University Press.

Zook, Matthew. 2000. Internet Metrics: Using Host and Domain Counts to Map the Internet. *Telecommunications Policy* 24(6/7) (July/August): 613–20.

The Political Economy of Open Source Software and Why It Matters

STEVEN WEBER

OPEN SOURCE IS an experiment in building a political economy—that is, a system of sustainable value creation and a set of governance mechanisms tied to it. It is a system that holds together a community of producers around a counterintuitive notion of property. I mean property in a broad sense—not only who owns what, but what it means to "own" something, what rights and responsibilities property confers. The conventional notion of property is built around variations of a simple claim, the right to exclude you from using something that "belongs" to me. Property in open source is configured fundamentally around the right to distribute, not the right to exclude. (If that sentence feels awkward on first reading, it is a testimony to just how deeply embedded in our intuitions and institutions the exclusion view of property really is.) The open source model is also a political economy that taps into a broad range of human motivations and relies on a creative and evolving set of organizational structures to coordinate behavior.

What would this political economy really look like? The answer to that question is still evolving. Understanding what can now be understood and tracking it forward yields a provocative story about how social organization linked to technology can change the meaning of property, and conversely, how shifting notions of property can alter the possibilities of social organization. The way into that huge agenda is to answer two more immediate questions about open source. How is it that groups of computer programmers (sometimes very large groups) made up of individuals separated by geography, corporate boundaries, culture, language, and other characteristics, and connected mainly via telecommunications bandwidth, manage to work together over time and build complex, sophisticated software systems outside the boundaries of a corporate structure and for no direct monetary compensation? And why does the answer to that question matter to anyone who is not a computer programmer?

The open source model is partly a story about technology, because the success of open source rests ultimately on computer code, code that people find in many cases to be more reliable and faster to evolve than proprietary software built inside a conventional corporate organization. It is also a business and legal story. Open source code does not obliterate profit, capitalism, or the general concept of intellectual property rights. Companies and individuals are creating intellectual products and making money from open source software code, inventing new business models and notions about property along the way.

Ultimately the success of open source is a political story. The open source software process is not a chaotic free-for-all where everyone has equal power and influence. And it is certainly not an idyllic community of like-minded friends where consensus reigns and agreement is easy. In fact, conflict is not unusual in this community; it's endemic and in a real sense inherent to the open source process. The management of conflict is politics, and indeed there is a political organization at work here, with the standard accoutrements of power, interests, rules, behavioral norms, decision-making procedures, and sanctioning mechanisms. But it is not a political organization that looks familiar to the logic of industrial era political economy.

The Analytic Problem of Open Source

The concept of "free" software is not new. In the 1960s and 1970s, the idea of making source code freely available was standard research practice. It was mostly taken for granted in leading computer science departments (such as at MIT and UC Berkeley) and corporate research facilities (particularly Bell Labs and Xerox PARC). Today, however, the majority of software production is organized under the economic logic imposed by a fairly standard intellectual property rights system. Patents, copyrights, licensing schemes, and other means of "protecting" computer software ensure that users cannot reverse-engineer, modify or resell code developed by others. Maintaining control over source code forms the cornerstone of profitability in this model. Indeed, source code is probably the most valuable asset of a firm like Microsoft.

Open source software is fundamentally different, by definition "free"—that is, public and nonproprietary. The Open Source Initiative specifies that software must share three essential characteristics to be considered "open source." Specifically, it must permit the free redistribution of the software, require that the full source code be distributed with any binaries, and allow anyone to modify and redistribute their own versions under these same terms.[1]

[1] Most open source licenses also require that the software itself be made available to others for no more than the cost of distribution. The terms of this definition originated in the Debian Social Contract developed in the mid-1990s by Bruce Perens. See www.debian.org.

There exist today several thousand open source "projects," ranging from small utilities and device drivers to more robust programs such as the e-mail transfer program Sendmail, the HTTP Server Apache, and of course the best known, the operating system Linux. These projects are driven forward by contributions from many, and in a few cases thousands of, developers, who work around the world in a seemingly unorganized fashion and receive neither direct pay nor other compensation for their contributions. Thwarting the conventional economic logic of collective action, these collaborative open source projects demonstrate empirically that *large, complex systems of code* can be built, maintained, developed, and extended in *nonproprietary settings* in which *many developers* work in highly *parallel, relatively unstructured ways* and without *direct monetary compensation.*

Perhaps because the strength of this movement is so counterintuitive, there remains tremendous uncertainty about what drives the open source model. Some observers have thought of the phenomenon in broadly political or sociological terms, trying to understand the internal logic and external consequences of a geographically widespread community capable of producing complex knowledge goods without direct monetary compensation. In early writings and analyses, mostly done by computer hackers who are part of one or another open source project (and are often "true believers"), open source has been characterized variously as:

- A methodology for research and development
- A new business model (requiring new mechanisms for compensation and profit)
- The "defining nexus" of a community geared toward the development of common goods
- A new "production structure" unique to "knowledge economies"
- Even a political philosophy

In part as a result, open source software has suddenly become the repository of extraordinarily diverse hopes and fears about the social and economic consequences of the information revolution. Libertarians see in open source a tool to emancipate individuals from governmental and corporate tyranny. Proponents of free markets see open source as the ultimate low barrier to entry market where only quality counts. Communitarians visualize a cross-national, cross-ethnic, and cross–just about every other traditional boundary community that is working together to advance a shared agenda. Economists see a market in reputation evolving naturally and almost automatically in a space with massively reduced transaction costs. The question is, why has open source software taken on the mantle of the Internet era's Rohrshach test?

The answer is that open source challenges much of what economists, lawyers, and business people believe they know about how intellectual property rights, production, and value-added together are transformed into profit in a modern economy. At a minimum, the arguments and theories that explain why firms exist, why some knowledge is kept private and sold for a price, why some people earn higher salaries than others, and why groups of people often find it hard to work together to produce something that will serve the common good, need to be reinterpreted in light of the success of open source. Some of these arguments may need to be substantially rewritten. This chapter takes preliminary steps in that direction. Building an explanation for open source requires a compound argument capable of reconciling the microfoundations of traditional economic logic with the social and political structures that replace standard notions of "property rights" as the ordering constraints on the organization of software production—and possibly other kinds of knowledge goods as well. The conclusion explores some implications of the success of the open source model, including its generalizability as an example of a community that has been empowered, even in a sense created, by the Internet.

The Economic Foundations—Traditional Approaches to Open Source

MACROECONOMIC APPROACHES

The starting point for most economic analyses of open source is a standard collective action type approach.[2] In this context the economic puzzle is straightforward. For well-known reasons, nonexcludable public goods tend to be underprovided in nonauthoritative social settings. Open source products such as Linux ought to exist at the worst end of this spectrum since they also depend on "collective provision." Recognizing this, Mark Smith and Peter Kollock (1999: 230) go so far as to call Linux "the impossible public good." While projects like it require contributions from a large number of developers, each developer has little incentive to contribute—voluntarily—to a good that he or she can partake of unchecked as a free rider. Simple logic dictates that the system ought to unravel backward, ensuring that no one makes any contributions, and there is no public good to begin with.

Previous attempts to grapple with this paradox have focused on redefining the structural logic of economic exchange. Rishab Aiyer Ghosh, for example, introduces the notion of "nonrival" goods in order to circumvent the "free-rider" trap. Using the image of a cooking pot capable

[2] See the summary and intelligent, if sometimes polemical, critique by Moglen (1999).

of "cloning" all food placed in it, Ghosh suggests that trade in nonrival goods is not plagued by the free-rider problem, as the supply of these goods is inexhaustible. His analogy is of course to the digital Internet, where—once created—software can be downloaded and copied an infinite number of times at essentially zero cost. The individual in this setting faces a different cost-calculus. As Ghosh (1998: 16) explains, "you never lose from letting your product free in the cooking pot, as long as you are compensated for its creation." As long as even one other person contributes an item of some value, trade becomes utility-enhancing for all actors in the system. As Ghosh puts it, "if a sufficient number of people put in free goods, the cooking pot clones them for everyone, so that everyone gets far more value than was put in."

The missing piece in this argument is that it does not actually explain the "trade." What underlying story accounts for the exchange relationship? Strictly speaking, it is still a narrowly rational act for any single individual to take from the pot without contributing—and free ride on the contributions of others. The collective action dilemma remains unsolved. In its traditional form, after all, the system unravels not because free-riders use up the stock of the collective good or somehow devalue it, but because there is no real incentive to contribute to that stock in the first place. The cooking pot starts—and remains—empty.

A solution to this paradox lies in pushing the concept of nonrivalness one step further. Software in some circumstances is more than simply nonrival. Most software, and particularly complex interdependent programs such as operating systems, actually is subject to positive network externalities. Whether it is called a network good or an antirival good (an awkward but nicely descriptive term), the point is that the value of any software increases as more people download and use it. The traditional benefits of standardization and network compatibility provide one explanation for why this is so. As more computers in the world run Linux, for example, it becomes easier for all users of that operating system to share applications and files (as well as gain knowledge useful in solving others' problems).

Perhaps more important, open source software makes an additional and very important use of network externalities, in altering the development process itself. The more individuals actively use a piece of software, the easier debugging becomes as errors are more quickly found and eliminated. Software development also speeds up as the user-base grows. Individuals have more incentive to expend time building plug-ins and coding new features. In practice, most software development takes place in precisely this way—people inside of organizations write code to do things and solve problems that need to be solved within their own organizations. The open source process essentially leverages this huge untapped energy that is usually closeted *within* organizations, by creating a structured process where it can be shared in a coordinated fashion *across* organizational boundaries.

This is the key point recognized by a high-level Microsoft memorandum of summer 1998. Known as the "Halloween Memo," this directive pointed to open source software as a direct threat to Microsoft's revenues and to its quasi-monopolistic position in some markets. As the author recognized, open source software represents a long-term strategic threat to Microsoft because "the intrinsic parallelism and free idea exchange in OSS [open source software] has benefits that are not replicable with our current licensing model." The point is not that open source software is simply able to accommodate free riders. It is actually *antirival* in the sense that *the system positively benefits from what are typically thought of as free riders in a collective good.* Some small percentage of users will provide something of value to the system, even if it is just reporting a bug out of frustration, or encouraging greater commercial support for the platform in general.

MICROECONOMIC APPROACHES

The logic outlined above constitutes a piece of a structural explanation for the success of open source projects. The problem is it provides no explanation for why "core" groups undertake the initial development costs. It remains unclear why these groups arise, and which projects are likely to succeed. A closer look at microeconomic incentives helps to address these questions. Here, Lerner and Tirole in "The Simple Economics of Open Source" (2000), make what is probably the most forceful argument.[3] They portray individual programmers, regardless of whether they work in open source or as employees of a proprietary software firm, as rational actors engaged in straightforward cost-benefit analysis. The immediate benefits to a programmer are private: creating a fix for the specific problem that the programmer faces or leading to direct monetary benefit. The primary cost is the opportunity cost of the time and effort that the programmer expends on the project.

Open source modifies this standard cost-benefit calculus in two significant ways. First, the "alumni effect" should lower the cost of working on open source relative to proprietary code (Lerner and Tirole 2000: 11). Since Unix syntax and open source tools are a standard part of most programmers' educational training, the costs of simply extending the functionality of these existing tools *should* be lower than building proprietary

[3] This is an important paper that draws usefully on others' analyses, while recognizing its own limitations as a "preliminary exploration" that invites further research.

solutions from scratch. Second, there are "delayed" benefits to developing open source programs that create a strong "signaling incentive" (ibid.: 15). These benefits accrue to the programmer's career in ways that are ultimately transformed—or can be transformed—into money. The logic is as follows: ego gratification for solving difficult programming problems is important because it stems from peer recognition. Peer recognition is important because it creates a reputation. And a reputation as a great programmer is monetizable—in the form of job offers, privileged access to venture capital, and the like.

The key point in this story is the signaling argument. As is true in many technical and artistic disciplines, the quality of a programmer's mind and work is not easy to judge in standardized and easily comparable metrics. To be able to assess the talent of a particular programmer takes a reasonable investment of time. The best programmers have a clear incentive to reduce the energy that it takes for others to see and understand just how good they are. Hence the importance of signaling. The programmer participates in an open source project as a strategic act of credentialism—to demonstrate the quality of his or her work. Reputation within a well-informed, committed, and self-critical community is one proxy measure for that quality. Lerner and Tirole argue that the signaling incentive will be stronger when the performance is visible to the audience; when effort expended has a high impact on performance; and when performance yields good information about talent. Open source projects maximize the incentive along these dimensions, in several ways.

With open source, a software user can see not only how well a program performs. He or she can also look to see how clever and elegant is the underlying code—a much more fine-grained measure of the quality of the programmer. And since no one is forcing anyone to work on any particular problem in open source, the performance can be assumed to represent a voluntary act on the part of the programmer, which makes it all that much more informative about that programmer. The signaling incentive should be strongest in settings with sophisticated users, tough bugs, and an audience that can appreciate effort and artistry, and thus distinguish between merely good and excellent solutions to problems. As Lerner and Tirole note, this argument seems consistent with the observation that open source has developed more quickly in more technical settings like operating systems and not in end-user applications. But it is not consistent with much of the more granular behavior seen in the open source community. In fact, reputations built in the open source community have not been a prevalent means to career advancement—if reputation economics is the driving force, it is working for only a few people. A big reason is that most working code does not get looked at in great detail. If the code fails it gets disassembled, but if it runs and runs well, very few people will ever take the time to put it under the microscope. And there is little strategic behavior around reputation in the open source community, as you would expect there to be if monetizable reputations were an important driving force.

Alternate microlevel arguments exist that paint individual incentives as a product of existing social institutions. One of the more interesting is that proposed by Ko Kuwabara (2000). Kuwabara uses a metaphor of complex adaptive systems and evolutionary change to describe the software development process. His account boils down to a series of causal steps. Programmers are motivated by a "reputation game" similar to what Lerner and Tirole depict. But he argues that the social structure alters individual incentives, not vice versa. Because online communities live in a situation of abundance, not scarcity, Kuwabara suggests that they are apt to develop "gift cultures" where social status depends on what you give away, rather than what you control.[4] Expand this into an evolutionary setting over time, and the community will self-organize a set of ownership customs along lines that resemble a Lockean regime of property rights. These ownership customs constitute a sufficient framework for successful and productive collaboration, even if they do not involve explicit legal control over property.

The gift culture idea is an important hypothesis. Gift economies—where social status depends more on what you give away than what you keep—are reasonable adaptations to conditions of abundance. They are often seen among aboriginal cultures living in mild climates and ecosystems with abundant food, as well as among the extremely wealthy in modern industrial societies (Raymond 1998: 99). And the culture of gift economies shares some notable characteristics with that of open source communities: gifts bind people together, encourage diffuse reciprocity, and support a concept of property that resembles "stewardship" more than "ownership" per se. Interestingly, this cultural argument is strongly evident in the writing of Eric S. Raymond, the unofficial ethnographer of the open source movement. In his piece "Homesteading the Noosphere" (1998: 103), Raymond suggests that the gift culture logic works particularly well in software, since the value of the gift (in this case a complex technical artifact) cannot be easily measured except by other members of the software community, who have the expertise to evaluate its technological sophistication. Naturally, therefore, "the success of a giver's bid for status is delicately dependent on the critical judgment of peers."

The culture of open source communities shares some of the characteristics of a gift economy. But there is a key flaw in focusing exclusively on

[4] The "gift culture" argument is taken principally from Raymond (1998). See also Baird (1997).

social constraints when attempting to define individual incentives. Doing this makes it difficult to analyze these incentives in common terms. The gift-culture hypothesis misses the point about the nature of abundance in this setting. Of course the physical tools of programming—bandwidth, disk-space, and processing power—are plentiful and cheap. From a technological standpoint it is certain that each of these will grow more abundant and less expensive over time. Yet when anyone can have a supercomputer on his or her desk, there is little status associated with that "property"—the very abundance of computing power should devalue it. The things that add value in this setting *depend on human mind space and the commitment of time and intellectual energy by very smart people to a creative enterprise.* It is the time and brainspace of smart, creative people that is scarce, and probably becoming more scarce as demand for their talents increases in proportion to the computing power available. Great programming skills are extremely rare. Nor is a *reputation* for greatness typically abundant, because only a certain number of people can really maintain a reputation for being "great writers" at any given point in time.[5]

The Search for New Institutions: Case Studies in Open Source

Macroeconomic approaches do not fully explain the motivations of individual programmers. Microlevel arguments about utility functions do not follow directly from exogenous social structures. A static conception of property rights has traditionally allowed analysts to bridge this gap. Under its logic, the level of analysis problem can be sidestepped because pressures on both levels are expressed in the terms of a common independent variable (money).

But because open source development is not structured around a traditional logic of property rights, bridging the gap between macro- and microlevel approaches is no longer automatic. A successful explanation needs to identify the logic of the particular software licenses and other social constraints that effectively replace standard systems of property rights as the fundamental ordering principles. This section takes on this task by examining in greater depth how two open source communities—the Free Software Foundation and the Linux development community—actually

[5] I say this because standards of "greatness" are themselves endogenous to the quality of work that is produced in a particular population. If there is a normal distribution of quality and the bell curve shifts to the right, what would have been thought excellent in the past is now merely good. The tails of the distribution define excellence in any setting, and they remain small.

work and then drawing general conclusions about the nature of open source development.[6]

The Free Software Foundation

Steven Levy's book *Hackers* (1984) gives a compelling account of the impact the growing importance intellectual property rights in software production had on the programming community, particularly at MIT. With the unbundling of software from hardware in the mid-1970s, many of the best programmers at MIT were hired away into lucrative positions in spinoff software firms. Simultaneously, MIT began to demand that its employees sign nondisclosure agreements in order to use university computing facilities. The newest mainframes, such as the VAX or the 68020, came with operating systems that did not distribute source code—in fact researchers had to sign nondisclosure agreements simply to get an executable copy.

MIT researcher Richard Stallman led the backlash. Driven by moral fervor as well as simple frustration at not being able easily to modify software for his particular needs (such as fixing a printer driver), Stallman in 1984 founded a project to revive the "hacker ethic" by creating a complete set of "free software" utilities and programming tools.[7] Called the Free Software Foundation (FSF), this project aimed to develop and distribute software under what he called the General Public License (GPL), also known in a clever word-play as "copyleft."

The central idea of GPL is to prevent cooperatively developed software or any part of that software from being turned into proprietary software. Users are permitted to run the program, copy the program, modify the program through its source code, and distribute modified versions to others. What they may *not* do is add restrictions of their own. This is the "viral clause" of GPL—it compels anyone releasing software that incorporates copylefted code to use the GPL in their new release. The Free Software Foundation says: "You must cause any work that you distribute or publish, that in whole or in part contains or is derived from the Program [any program covered by this license] or any part thereof, to be licensed as a whole at no charge to all third parties *under the terms of this li-*

[6] These two examples attract a great deal of public attention but are by no means the only important examples; my forthcoming book examines these and others in much more detail.

[7] In Stallman's view, "the sharing of recipes is as old as cooking," but proprietary software meant "that the first step in using a computer was a promise not to help your neighbor." He saw this as "dividing the public and keeping users helpless." See Stallman (1999: 54). For a fuller statement, see http://www.gnu.org/philosophy/why-free.html.

cense."[8] In practice, then, GPLed software can be used and modified freely, with the key restriction that the user cannot restrict the freedom of others to do what they want to do with the software.

Stallman and the Free Software Foundation have created some of the most widely used pieces of UNIX software, including the text editor EMACS, the GCC compiler, and the GDB debugger. As these popular programs were adapted to run on almost every version of UNIX, their availability and efficiency helped to cement UNIX as the operating system of choice for "free software" advocates. But the FSF's success was in some sense self-limiting. Partly this is because of the moral fervor underlying Stallman's approach—not all programmers found his strident libertarian attitude to be practical or helpful. Partly it was a marketing problem. "Free software" turned out to be an unfortunate label, despite FSF's vehement attempts to convey the message that free was about freedom, not price—as in the slogan "think free speech, not free beer."

But there was also a deeper problem in the all-encompassing nature of the GPL and particularly its "viral" clause. Stallman's moral stance against proprietary software clashed with the utilitarian view of many programmers, who wanted to use pieces of proprietary code along with free code when it made sense to do that, simply because the proprietary code was technically good and useful for a task. The GPL did not permit this kind of flexibility (or made it difficult to achieve, requiring a vague distinction between static and dynamic linking of code) and thus posed inconvenient constraints to developers looking for pragmatic solutions to problems.

The Linux Operating System

The history of Linux provides more insight into this phenomenon. Linus Torvalds, in 1990 a computer science student at University of Helsinki, strongly preferred the technical approach of UNIX-style operating systems over the DOS system commercialized by Microsoft.[9] But Torvalds did not like waiting on long lines for access to a limited number of university machines that ran UNIX for student use. And it simply wasn't practical to run a commercial version of UNIX on a personal computer—the software was too expensive and also much too demanding for the limited PCs of the time.

[8] Free Software Foundation (1991). Emphasis added. There are several different modifications to these specific provisions, but the general principle is clear.

[9] "Task-switching" is one major difference between the two systems that was of interest to Torvalds. UNIX allows the computer to switch between multiple processes running simultaneously.

In late 1990 Torvalds came across Minix, a simplified UNIX clone that was being used for teaching purposes at Vrije University in Amsterdam. Minix ran on PCs, and the source code was available. Torvalds installed this system on his IBM AT, a machine with a 386 processor and 4 megabytes of memory, and went to work building the kernel of a UNIX-like operating system with Minix as the scaffolding. In autumn 1991 Torvalds let go of the Minix scaffold and released the source code for the kernel of his new operating system, which he called Linux, onto an Internet newsgroup, along with the following note:

> I'm working on a free version of a Minix look-alike for AT-386 computers. It has finally reached the stage where it's even usable (though it may not be, depending on what you want), and I am willing to put out the sources for wider distribution. . . . This is a program for hackers by a hacker. I've enjoyed doing it, and somebody might enjoy looking at it and even modifying it for their own needs. It is still small enough to understand, use and modify, and I'm looking forward to any comments you might have. I'm also interested in hearing from anybody who has written any of the utilities/library functions for Minix. If your efforts are freely distributable (under copyright or even public domain) I'd like to hear from you so I can add them to the system. (Torvalds 1999b)

The response was extraordinary (and, according to Torvalds, mostly unexpected). By the end of the year, nearly one hundred people worldwide had joined the Linux newsgroup. Through 1992 and 1993 the community grew slowly. New users downloaded it, played with it, tested it in various settings, and attempted to extend and refine it. Flaws surfaced in the form of bugs and security holes, while new features were continually added. Users submitted reports of problems they found, or proposed a fix and sent a patch on to Torvalds. Gradually, the process iterated and scaled up to a degree that just about everyone, including its ardent proponents, found startling. In 1994 Torvalds finally released the first official version of Linux (versio 1.0). The pace of development accelerated, with updates to the system being released on a weekly, or sometimes even daily, basis; today the operating system consists of more than three million lines of code.

This rapid growth is attributable to an extremely large and geographically far-flung community. Indeed, the credits file for the original release names contributors from at least thirty-one different countries. In both the Free Software Foundation and Linux circles, as in most open source communities, there exist a large number of moderately committed individuals who contribute relatively modest amounts of work and participate irregularly, as well as a smaller but much more highly committed group that forms an informal core. A July 2000 survey of the open source community identified approximately twelve thousand developers working on open source projects. Although the survey recognizes difficulties with measurement, it reports that the top 10 percent of the developers are credited with about 72 percent of the code—loosely parallel to the apocryphal "80/20 rule" (where 80 percent of the work is done by 20 percent of the people).[10] Linux user/developers come from all walks of life: hobbyists, people who use Linux or related software tools in their work, and committed "hackers"—some with full-time jobs, and some without.

In both cases, the logic behind the process is both functional and behavioral. Development occurred largely through a game of trial-and-error by people embedded in the culture of a self-aware community. Over time, observers and participants (particularly Eric Raymond) analyzed the emergent process and tried to characterize (inductively for the most part) the key features that made the process work. Drawing largely on Raymond's analysis as well as my own set of interviews, I propose seven key features common to development in successful open source projects.

1. *People pick important problems and make them interesting.* Open source user-developers tend to work on projects that they judge to be important, significant additions to software. There is also a premium for what in the computer science vernacular is called "cool," which roughly means creating a new and exciting function, or doing something in an newly elegant way. There seems to be an important and somewhat delicate balance around how much and what kind of work is done up-front by the project leader(s). User-developers look for signals that any particular project will actually generate a significant product, not turn out to be an evolutionary dead end, but also contain interesting challenges along the way.

2. *Developers look for solutions to their own most pressing problems.* Raymond emphasizes that since there is no formal division of labor imposed on the process, open source developers are free to pick and choose exactly what it is they want to work on. This means that they will tend to focus on an immediate and tangible problem (the "itch that needs to be scratched")—a problem that they themselves want to solve. The Cisco enterprise printing system (an older open source style–project) evolved directly out of an immediate problem—system administrators at Cisco were spending an inordinate amount of time (in some cases half their time) working on printing problems.[11] Torvalds (and others as well) sometimes put out a form of request, as in "isn't there somebody out there who would want to work on 'X' or try to fix 'y' problem?") The

[10] Ghosh and Prakash (2000). Specifically regarding Linux, as of spring 2000, there were approximately 90,000 registered Linux users, a large proportion of whom have programmed at least some minor applications or bug fixes, as well as a core of over 300 central developers who have made major and substantial contributions to the kernel. See http://www.linux.org/info/index.html.

[11] http://ceps.sourceforge.net/index.shtml.

underlying notion is that in a large community, someone will find any particular problem of this sort to be an itch they actually do want to scratch.

3. *Reuse whatever you can.* Open source user-developers are always searching for efficiencies: put simply, because they are not paid directly for contributions, there is a strong incentive never to reinvent the wheel. An important additional point is that there is less pressure on them to do so. This is simply because in the open source environment they know with certainty that they will always have access to the source code and thus do not need to re-create any tools or modules that are already available in open source.

4. *Use a parallel process to solve problems.* If there is an important problem in the project, a significant bug, or a feature that has become widely desired, many different people or perhaps teams of people will be working on it—in many different places, at the same time. They will likely produce a number of different potential solutions. It is then possible for Linux to incorporate the best solution and refine it further.

Is this inefficient and wasteful? That depends. The relative efficiency of massively parallel problem solving depends on lots of parameters, most of which cannot be measured in a realistic fashion. Evolution is messy, and this process recapitulates much of what happens in an evolutionary setting. What is clear is that the stark alternative—a nearly omniscient authority that can predict what route is the most promising to take toward a solution, without actually traveling some distance down at least some of those routes—is not a realistic counterfactual for complex software systems (and many other complex knowledge goods).

5. *Leverage numbers.* The Linux process relies on a kind of law of large numbers to generate and identify software bugs, and then to fix them. Software testing is a messy process. Even a moderately complex program has a functionally infinite number of paths through the code. Only some tiny proportion of these paths will be generated by any particular user or testing program. As Paul Vixie (1999: 98) puts it, "the essence of field testing is *lack* of rigor" (emphasis added). The key is to generate patterns of use—the real world experiences of real users—that are inherently unpredictable by developers. In the Linux process, a huge group of users constitutes what is essentially an ongoing huge group of beta testers.

Eric Raymond (1999: 41) says, "Given enough eyeballs, all bugs are shallow. Implied in this often-quoted aphorism is a prior point: given enough eyeballs and hands doing different things with a piece of software, more bugs will appear, and that is a good thing, because a bug must appear and be characterized before it can be fixed. Torvalds reflects on his experience over time that the person who runs into and characterizes a particular bug and the person who later fixes it are usually not the same person—an observational piece of evidence for the efficacy of a parallel debugging process.

6. *Write code that others can understand and document it well.* In a suffi-

ciently complex program, even open source code may not necessarily be transparent in terms of precisely what the writer was trying to achieve and why. The Linux process depends upon making these intentions and meanings clear, so that future user-developers understand (without having to reverse-engineer) what functions a particular piece of code plays in the larger scheme of things. Documentation is a time-consuming and sometimes boring process. But it is considered essential in any scientific research enterprise, in part because replicability is a key criterion.

7. *"Release early, release often."* User-developers need to see and work with iterations of software code in order to leverage their debugging potential. Commercial software developers understand just as well as do open source developers that users are often the best testers, so the principle of release early release often makes sense in that setting as well. The countervailing force in the commercial setting are expectations: customers who are paying a great deal of money for software may not like buggy beta releases and may like even less the process of updating or installing new releases on a frequent basis.

Open source user-developers have a very different set of expectations. In this setting, bugs are more an opportunity and less a nuisance. Working with new releases is more an experiment and less a burden.[12] The result is that open source projects typically have a feedback and update cycle that is an order of magnitude faster than commercial projects. In the early days of Linux, there were often new releases of the kernel on a weekly basis—and sometimes even daily.

Building the Open Source Economic Logic

The open source process is precisely that—a *process* of production. The software it generates, useful and elegant as it may be in some cases, is ultimately a technical artifact that is an outcome of the production process. The obvious analogy is to the MIT study of lean production in auto manufacturing, *The Machine That Changed the World.* The point is that the machine that changed the world was not a Toyota or any other kind of machine. It was a way of making things. And because its logic was essentially generic, the argument was that it would soon extend beyond Toyota, beyond the auto industry, and beyond Japan.

To understand the open source logic of production, this section examines how open source communities grapple with two fundamental problems in this setting: how to solve coordination problems and how to man-

[12] Linux kernel releases are typically divided into "stable" and "developmental" paths. This gives users a clear choice: download a stable release that is more reliable, or a developmental release where new features and functionality are being introduced and tested.

age complexity. On a less abstract level, this touches on a series of issues that range from who has the right to make decisions about the development of code, to who gets credited for what work, and how conflicts are resolved when they arise.

Coordination Problems

Authority within a firm and the price mechanism across firms are standard ways to coordinate specialized knowledge in a conventional system of property rights. Neither exists in an open source community, where legal ownership is extremely fluid. A simple analogy from ecology suggests what might happen over time as modifications accumulate along different branching chains of software. Speciation—or what computer scientists call code-forking—seems likely. Lacking any constraints of formal ownership or copyright, and given the explicit freedom to modify software code in any way that a user finds desirable, the software should be expected to evolve into incompatible versions, and synergies in development should be lost over time. This of course is very much what happened to UNIX in the 1980s.[13] And if ego is a primary determinant of individual behavior, this coordination problem is made even more acute. When egos get damaged, why don't the owners of those egos walk away from—or even worse, try to undermine—the collective project?

The explanation is not exclusively cultural/structural. Macroeconomic incentives connected to positive network externalities are part of the answer. If developers *think of themselves* as trading innovation for others' innovation, they will want to do their trading in the most liquid market possible.[14] Forking would only reduce the size and thus the liquidity of the market. Viewing software as an "antirival" good creates a similar dynamic: the more open a project is and the larger the existing community of developers, the less tendency there will be to fork. This is because the potential forker faces a difficult problem: it becomes very hard for the renegade to credibly claim that he or she could accumulate a more talented and effective base of developers than already exists in the main code base. Operating with diminished resources, this forked development community could also never promise credibly to match the rate of innovation taking place in the primary code base. It could not use, test, and debug software as quickly. And as a result it could not provide as attractive a payoff in reputation to its developers, even if reputation were shared out

[13] The next section explores this history in more detail.

[14] There is no trade in a formal economic sense, since anyone can withdraw resources from the common pool without being compelled to create anything of value and/or donate to the pool.

more evenly within the forked community.[15] This is why "strategic forking," in contrast to the expectations of the Lerner and Tirole argument, makes little sense. And it is probably why there are no prominent examples of this behavior in the historical record of the community.

Cultural and social norms do play a key role in influencing how these macro and microeconomic pressures play out.[16] A prevalent norm assigns decision-making authority within the community. The key element of this norm is that *authority follows and derives from responsibility.* The more an individual contributes to a project and takes responsibility for pieces of software, the more decision-making authority that individual is granted by the community. In the case of Linux, Torvalds typically validates the grant of authority to "lieutenants" by consulting closely with them on an ongoing basis, particularly when it comes to key decisions on how subsystems are to work together in the software package.

While relatively high levels of trust may reduce the amount of conflict in the system, complicated and informal arrangements of this kind are certain to generate disagreements. There is an additional, auxiliary norm that gets called into play: *seniority rules.* As Raymond (1998: 127) explains: "If two contributors or groups of contributors have a dispute, and the dispute cannot be resolved objectively, and neither owns the territory of the dispute, the side that has put the most work into the project as a whole . . . wins."[17]

But what does it mean to resolve a dispute "objectively"? The notion of objectivity draws on its own, deeper normative base. The open source developer community shares a general conception of technical rationality. Like all technical rationalities, this one exists inside a cultural frame. The cultural frame is based on shared experience in UNIX programming. UNIX was born in the notion of compatibility between platforms, ease of networking, and positive network effects.[18] UNIX programmers have a set of common standards for what is "good code" and what is not-so-

[15] Clearly there are parameters within which this argument is true. Outside of those parameters it could be false. It would be possible to construct a simple model to capture the logic, but it is hard to know—other than by observing the behavior of developers in the open source community—how to attach values to those parameters.

[16] Robert C. Ellickson (1991: 270) provides a compelling argument about the falsifiability of normative explanations.

[17] One interesting additional piece of evidence for these norms is what has happened when the two norms pointed in different directions. Raymond (1998: 128) recalls one such fight of this kind and says "it was ugly, painful, protracted, only resolved when all parties became exhausted enough . . . I devoutly hope I am never anywhere near anything of the kind again".

[18] Indeed, UNIX was developed in part to replace ITS (incompatible time sharing system). The idea in 1969 was that hardware and compiler technology were getting good enough that it would now be possible to write portable software—to create a common software environment for many different types of machines.

good code (Gancarz 1995). These standards draw on pragmatism and experience—the UNIX "philosophy" is a philosophy of what works and what has been shown to work in practical settings over time. From a macro perspective, this is at least as important as the technical "alumni effect" that Lerner and Tirole emphasize on the microeconomic side.

The Open Source Initiative codified this cultural frame by establishing a clear priority for pragmatic technical excellence over ideology or zealotry (more characteristic of the Free Software Foundation). A cultural frame based in engineering principles (not anticommercial ideology) and focused on high reliability and high performance products gained much wider traction within the developer community. It also underscored the rationality of technical decisions driven at least in part by the need to sustain successful collaboration—hence legitimating concerns about "maintainability" of code, "cleanness" of interfaces, and clear and distinct modularity (Tuomi 2000). The mastery of technical rationality in this setting is made clear in the creed that developers say they rely on—"let the code decide."

Leadership matters in setting a focal point and maintaining coordination on it. Torvalds started the Linux process by providing a core piece of code. This was the original focal point. It functioned that way because—simplistic and imperfect as it was—it established a plausible promise of creativity and productivity: that it *could* develop into something elegant and useful. The code contained interesting challenges and programming puzzles to be solved. Together, these characteristics attracted developers who, by investing time and effort on this project, placed a smart bet that their contributions would be efficacious and that there would eventually be a valuable outcome.

In the longer term, leadership matters by reinforcing the cultural norms. Torvalds does, in fact, have many characteristics of a charismatic leader in the Weberian sense. Importantly, he provides a convincing example of how to manage the potential for ego-driven conflicts among very smart developers. Torvalds (1998) downplays his own importance in the story of Linux: while he acknowledges that his decision to release the code was an important one, he does not claim to have planned the whole thing or to have foreseen the significance of what he was doing or what would happen: "The act of making Linux freely available wasn't some agonizing decision that I took from thinking long and hard on it; it was a natural decision within the community that I felt I wanted to be a part of."

When it comes to reputation and fame, Torvalds is not shy and does not deny his status in any way. But he does make a compelling case that he was not motivated by fame and reputation—these are things that simply came his way as a result of doing what he believed in.[19] He continues

[19] The documented history, particularly the archived email lists, supports Torvalds on this point.

to emphasize the fun and opportunities for self-expression in the context of "the feeling of belonging to a group that does something interesting" as his principle motivation. And he continues to invest huge effort in maintaining his reputation as a fair, capable, and thoughtful manager. It is striking how much effort Torvalds puts into justifying to the community his decisions about Linux and documenting the reasons for decisions—in the language of technical rationality that is currency for this community. Would a different leader with a more imperious attitude who took advantage of his or her status to make decisions by fiat have undermined the Linux community? Many in the community believe so (or believe that developers would exit and create a new community along more favorable lines).[20] The logic of the argument to this point supports that belief.

There do exist sanctioning mechanisms to support the nexus of incentives, cultural norms, and leadership roles that maintain coordination. In principle, the GPL and other licenses could be enforced through legal remedies (this of course may lurk and constrain behavior even if it is not invoked). In practice, precisely how enforceable in the courts some aspects of these licenses are remains unclear (McGowan 2003; Merges 1997: 115–36). The sanctioning mechanisms that are visibly practiced within the open source community are two: "flaming" and "shunning" (Raymond 1998: 129). Flaming is "public" condemnation (usually over e-mail lists) of people who violate norms. "Flamefests" can be quite fierce in language and intensity but tend ultimately to be self-limiting.[21]

Shunning is the more functionally important sanction. To shun someone—refusing to cooperate with them after they have broken a norm—cuts them off from the benefits that the community offers. It is not the same as excludability: someone who is shunned can still use Linux. But that person will suffer substantial reputational costs. They will find it hard to gain cooperation from others. They will have to incorporate on an ongoing basis their own work into a code base that they are no longer contributing to in an active way. The threat essentially is to be left on your own to solve problems, while the community can and does draw on its collective experience and knowledge to do the same. This is clearly a strong disincentive to strategic forking, for example, but it also constrains other, less egregious forms of counternormative behavior (such as aggressive ego self-promotion).

[20] Examples of this process are in my forthcoming book.

[21] The intensity seems to be self-limiting, in part because developers understand very well the old adage about sticks and stones vs. words.

Problems of Complexity

To design robust, complex software is a formidable task. Testing, debugging, and maintaining code is generally even harder. As this is a task that needs be divided, the standard industrial response to increasing complexity has been to organize labor within a centralized, hierarchical structure—namely, that of a firm. The firm then manages complexity through formal organization and explicit decisional authority.[22]

Certainly with complex knowledge goods in particular, this is a very imperfect solution. In *The Mythical Man-Month* (1975), a classic study of the social and industrial organization of programming, Frederick Brooks noted that when large organizations add staff resources to a software project that is behind schedule, the project typically falls even further behind schedule. He explained this with an argument that is now known colloquially as Brooks' Law. As you raise the number of programmers on a project, work performed scales linearly (by a factor *n*), but complexity, communication costs, and vulnerability to error scales geometrically by a factor of *n* squared. This (following Becker and Murphy 1992) inheres in the logic of the division of labor for complex knowledge goods. In software the practical manifestation is simple: the number of potential communications paths and interfaces between developers (just as between the pieces of code they write) increases exponentially as the number of developers increases linearly. How does the open source process manage the implications of this "law" among a geographically dispersed community that is not subject to hierarchical command and control?

Eric Raymond (1999: 30) draws a useful but too stark contrast between "cathedrals" and "bazaars" as icons of organizational structure. Cathedrals are designed from the top down, built by coordinated teams who are tasked by and answer to a central authority. Open source projects seem to confound this hierarchical model. Linux appears, at least on first glance, to be much more like a "great babbling bazaar of different agendas and approaches." But there has evolved in Linux a clear hierarchy of decision-making authority, where a decision pyramid leads from the dispersed developer base up through trusted lieutenants who have authority over particular parts of the code, and ultimately to Linus Torvalds, whose decisions are in a sense "final." This hierarchy was put in place in the mid-1990s, precisely in response to the growth of the project beyond the point where Torvalds could realistically manage the complexity on his own.

[22] Of course organization theorists know that a lot of management goes on in the interstices of this structure, but the structure is still there to make it possible.

Programmers explain this with the sly phrase, "Linus doesn't scale." In practice, a great deal of his authority is now devolved down rungs of the hierarchy and decisions made at those levels in effect bear his imprimatur. There is more hierarchical authority here than the popular image of a bazaar captures, although it is remains authority that rests on something other than corporate command and control or the power of money.

The Linux decision-making system is just one example of pragmatic, experimental adaptations to this problem. In fact, open source communities manage complexity in diverse ways. Consider the case of the Berkeley Software Distribution (BSD) model.[23] In BSD, typically, a relatively small committed team of developers writes code. Users may modify the source code for their own purposes, but the development team does not generally take "check-ins" from the public users, and there is no regularized process for doing that. Apache, in contrast, takes in contributions from a wider swathe of developers who rely on a decision-making committee that is constituted according to formal rules, a de facto constitution. The Perl scripting language relies on a "rotating dictatorship" where control of the core software is passed from one member to another inside an inner circle of key developers.

These cases differ from Linux, where the public or general user base can and does propose check-ins, modifications, bug fixes, new features, and so on. There is no formal distinction between users and developers on Linux archive sites. There are essentially no institutional barriers to entry to the debugging and development process. This is true in part because of a common debugging methodology and in part because when a user installs Linux, the debugging/developing environment comes with it (along with the source code, of course). Some users engage in "impulsive debugging"—fixing a little problem (shallow bug) that they encounter in daily use; while others make debugging and developing Linux a hobby or vocation. The key to managing the level of complexity within the software itself, is *modular design.* A major tenet of the UNIX philosophy, passed down to Linux, is to keep programs small and unifunctional ("do one thing simply and well"). A small program will have far fewer features than a large one, but small programs are easy to understand, are easy to maintain, consume fewer hardware system resources, and—most importantly—can be combined with other small programs to enable more complex functionalities.

The technical term for this development strategy is "source code modularization." A large program works by calling on relatively small and self-contained modules. Good design and engineering is about limiting the interdependencies and interactions between modules. Programmers

[23] There are now several BSD projects, which I discuss in detail in my forthcoming book.

working on one module know two things: that the output of their module must communicate successfully with other modules, and that (ideally) they can make changes in their own module to debug it or improve its functionality without requiring changes in other modules, as long as they get the communication interfaces right.

This reduces the complexity of the system overall because it limits the reverberations that might spread out from a code change. Obviously, it is a powerful way to facilitate working in parallel on many different parts of the software at once, since a programmer can control the development of a specific module of code without creating problems for other programmers working on other modules. It is notable that one of the major improvements in Linux release 2.0 was moving from a monolithic kernel to one made up on independently loadable modules. The advantage, according to Torvalds (1999b: 108), was that "programmers could work on different modules without risk of interference . . . managing people and managing code led to the same design decision. To keep the number of people working on Linux coordinated, we needed something like kernel modules."

Torvalds' implicit point is simple: these engineering principles are important because they reduce organizational demands on the social/political structure. In no case, however, are those demands reduced to zero. This is simply another way of saying that libertarian and self-organization accounts of open source software are frankly naive. The formal organization of authority is quite structured for larger open source projects. Torvalds, as noted, sits atop a decision pyramid as a de facto benevolent dictator. Apache is governed by a committee.

How Do They Resolve Conflicts?

Anyone who has dabbled in the software community recognizes that a large number of very smart, highly motivated, self-confident, and deeply committed developers trying to work together creates an explosive mix. Conflict is common, even customary in a sense. It is not the lack of conflict in the open source process but rather the successful management of substantial conflict that needs to be explained—conflict that is sometimes highly personal and emotional as well as intellectual and organizational.[24]

Eric Raymond (1998: 79–137) observes that conflicts center for the

[24] Indeed, this has been true from the earliest days of Linux. See, for example, the e-mail debate between Linus Torvalds and Andrew Tanenbaum (1992: 221–51). Torvalds opens the discussion by telling Tanenbaum, "You lose," "linux still beats the pants off minix in almost all areas," "your job is being a professor and a researcher: That's one hell of a good excuse for some of the brain-damages of minix."

most part on three kinds of issues: who makes the final decision if there is a disagreement about a piece of code; who gets credited for precisely what contributions to the software; and who can credibly and appropriately choose to fork the code. Similar issues of course arise when software development is organized in a corporate setting. Standard theories of the firm explain various ways in which potential conflicts are settled or at least managed by formal authoritative organizations.[25]

The open source community prefers to settle major conflicts through a "battle to consensus." Programmers devote an extraordinary amount of time and energy to this process, trying to convince each other that there are firm technical grounds for preferring one solution or development path to another. It doesn't always succeed—in part because the technical criteria often are not definitive, and in part because personalities get in the way. At that point, leadership takes on a much more important role in conflict resolution. Of course it is a style of leadership that has to justify itself and its decisions to skeptical, independent-minded followers who are free to break away if they so choose.

Linux, in its earliest days, was run unilaterally by Linus Torvalds. Torvald's decisions were essentially authoritative. As the program and the community of developers grew, Torvalds delegated responsibility for subsystems and components to other developers, who are known as "lieutenants." Some of the lieutenants onward-delegate to "area" owners who have smaller regions of responsibility. The organic result is what looks and functions very much like a hierarchical organization where decision making follows fairly structured lines of communication. Torvalds sits atop the hierarchy as a benevolent dictator with final responsibility for managing conflicts that cannot be resolved at lower levels.

Torvald's authority rests on a complicated mix. History is a part of this—as the originator of Linux, Torvalds has a presumptive claim to leadership that is deeply respected by others. Charisma in the Weberian sense is also important. It is notably limited in the sense that Torvalds goes to great lengths to document and justify his decisions about controversial matters. He makes admissions that he was wrong. It is a kind of charisma that has to be continuously re-created through consistent patterns of behavior. Linux developers will also say that Torvald's authority rests on its "evolutionary success." The fact is, the "system" that has grown up under his leadership worked to produce a first-class outcome, and this in itself is a strong incentive not to fix what is clearly not broken.

Ultimately, decisions to accept Torvalds' authority can be traced back to definable incentives—but the incentives themselves depend heavily on the social structure created by the GPL license and by the constructed au-

[25] McGowan (2003) provides a good summary discussion.

thority of the leader. Conflict is expected; indeed it is normatively sanctioned. When an argument ends, as it is also expected to do at some point, the loser has essentially three options. He or she can accept the decision and move on, drop her involvement in the project, or fork the code.

If the decision is to drop out of the project, the opportunity to accrue reputation and affect future decisions about the evolution of the project is lost. The community may lose the involvement of a particular individual, but not more than that (if it is an important individual, obviously the leader has strong incentives to try to heal the wound). The central decision between the other alternatives, to accept the decision or to fork the code, depends in some final sense upon the calculations that I discussed under the subheading "Coordination Problems" above. The open source development process builds momentum as it grows. The larger and more open a project, the higher the threshold for a rational decision to fork the code. The network externalities in the technology have essentially been implanted into the social structure that surrounds it.

Conclusion: Some General Lessons about Political Economy on the Net

Ultimately the intriguing question about open source is how this distinctive process of knowledge production and coordination impacts other realms of the twenty-first-century political economy. The key concepts—user—driven innovation that takes place in a parallel distributed setting, distinct forms and mechanisms of cooperative behavior regulated by norms and governance structures, and the economic logic of "antirival" goods—are generic enough to suggest that software is not the only place where experiments with open source–style systems could flourish. To get to some of the more general implications, there are two myths that first need to be discarded.

The first myth, a surprisingly common conception in the general media, sees open source basically as amusement for enthusiasts; a game of effortless fun among like-minded hobbyists. Some people like to write code: give them a neutral and high bandwidth pipe to communicate and they will get together with other people who feel the same and write code together. Imagine a simple analogy: if all the model train enthusiasts in the world could join their tracks together through the Internet, they would surely build a train set just as elaborate as Linux. Nobody has to tell them how to do this, and surely nobody has to pay them; it's a labor of love. And since everyone basically feels the same way, there is really nothing to argue about.

The macro part of this story is either unarticulated or naively wrong.

Like-minded or not, participants in the open source process argue more or less continuously, and about both technical and organizational issues. These arguments are often intense and emotional. Conflict is not unusual; it's endemic and in a real sense inherent to the open source process. If open source software were simply the collective creation of like-minded individuals who cooperate easily because they are bound together by semireligious beliefs, there would be little disagreement in the process and little need for conflict resolution among developers.[26]

The micro part of this story then shades off into assumptions about altruism. If there is no real mechanism for conflict resolution, then the "hobbyists" must not only be acting for personal satisfaction. They would have to be acting in an explicitly prosocial way, and doing things that others want for the sake of the act itself. This would be altruism. To act selflessly in this setting would be to write and contribute code for no apparent compensation of any sort, other than the personal gratification that comes from doing something that helps someone else.

But the evidence confounds any straightforward version of this argument. If altruism were the primary driving force behind open source software, no one would care very much about who was credited for particular contributions. It wouldn't matter who was able to license what code under what conditions.[27] Certainly people help each other in open source for the sake of helping—as elsewhere in human life, one of the ways people express their values and identities is in the act of providing help. Richard Stallman's original manifesto likened the act of sharing code to neighbors helping each other to put out a fire. But neighbors know each other; they live next to each other over time; a fire next door to my house threatens my house directly; and I have reason to expect reciprocity from my neighbor at some time in the future. The geographically distributed and relative anonymity of the Internet makes altruism a dicier proposition. There is in fact important evidence against the prevalence of altruistic behavior on the Internet, even in settings (such as Peer to Peer networking) where there are zero or very small possible costs to making contributions (Adar and Huberman 2000).

There is another, essentially pragmatic reason to steer clear of altruism as a principal explanation. This has more to with the current discourse in particular segments of social science, where altruism is a highly loaded term. For better or for worse, arguments about altruism invoke an intellectual apparatus that places altruistic behavior *in opposition* to self-

[26] For example, see "After the Microsoft Verdict" in *The Economist* (2000).

[27] Popular media often portray the open source community in this light but fail to account for the fact that many "beneficiaries" of this altruism (apart from the developers themselves) are major corporations that use Linux software.

interest. This can quickly become an unproductive discussion, where people argue about whether or when it makes sense to redefine self-interest so that it can accommodate a desire to do something solely for someone else. In other words, should individual utility functions include a term for the welfare of another? These are important issues, but they don't need to settled or even really engaged for the purpose of understanding open source. By sidestepping this particular aspect of the debate, we can focus more cleanly and directly on what mix of individual motivations are at play, including (but clearly not limited to) a desire to do something that increases the utility or benefit to others.

The second myth goes under the heading "self-organization," a phrase that has become trendy in both popular and scholarly studies of Internet communities. In the context of political community I don't like this term and the reasoning it sometimes represents, for two reasons. The first reason is that self-organization is used too often as a placeholder for an undetermined or underspecified mechanism. When used this way, self-organization becomes a euphemism for "I don't really understand the mechanism that holds the system together." Better-specified notions of self-organization build on the proposition that order arises endogenously out of the local interactions among individuals. Here self-organization simply is being used in contrast to overarching authority or governance—a useful comparison. But that does not relieve the obligation to explain *how* those local interactions actually add up to 'global' order. They do not always do so. We know from simple observation that not all groups of programmers "self-organize" into open source communities—Microsoft programmers certainly don't—and in fact open source communities still represent the exception rather than the rule. Neither the low transaction costs of a network nor the so-called law of large numbers can solve this problem by itself. There is something more than just the motivations and interactions of individuals, something else in the social structure that is autonomous and needs to be uncovered on its own terms to understand open source.

The second and probably more important reason I shy away from the heading "self-organization" has to do with normative peculiarities of the discourse that it prompts. Self-organization often evokes the cheerful feeling of a "state of nature" narrative, a story about the way things would naturally evolve if the "meddling" hands of corporations and lawyers and governments and bureaucracies would just stay away. Of course those non-self-organized organizations have their own narratives, which portray the state of nature as a chaotic mess. But the whole premise is faulty. To pose two state of nature narratives against each other creates a battle of assumptions, a tournament of null hypotheses, which is not productive. The underlying presumption—that there *is* in fact a state of nature without human agency—is even more damning to the discourse when we are talking about something like software code or knowledge production more generally. The coexistence of very different production processes and community structures in software simply illustrates the general point that *there is no state of nature on the Internet.* Knowledge does not want to be "free" (or for that matter, owned) more than it wants to be anything else—instead, it is people and institutions that want knowledge and the property rights around it to be structured in a particular way (Brand 1995). There is a simple but profound practical agenda here: we are going to be creating lots of new things in this technological space, and we can organize the creation of those things however we want.

This is social constructivism at its core, pitting different social constructivist narratives against each other (instead of against an imaginary state of nature). And this represents an opportunity since a technologically created space like the Internet makes power easier, not harder, to visualize. Lawrence Lessig's (2000) gloomy perspective notwithstanding, I believe that "code" is in fact *more transparent* than lots of other "architectural" features that shape traditional political spaces. Put simply, the implications of what we build for power relations are easier to analyze when it rests in software code, than when it is buried in layers of tradition, language, historical practices, and culture. We don't really need a Michel Foucault for the Internet in the same way as we needed him for other social settings because the architecture of the Internet is ultimately more visible. To hide behind the notion of self-organization short circuits that very important discussion.

So what then are generalizable principles of organization in open source? The answer to this question is embedded in a fundamentally different notion of property rights with which the open source movement is experimenting. The core notion of property in a modern market economy is the right to exclude, according to terms that are specified by the owner.[28] (In some formulations of international relations and economic "history," the sovereign state's core function is to secure those rights, and the medieval-to-modern transformation rests on that move) (Ruggie 1993; North 1990). Open source simply inverts this foundational notion of property rights, so that ownership now becomes the right to distribute, not exclude, and to do so with the only significant constraint being that the owner cannot constrain the freedom of others to do as they wish with the product. This has enabled a production process that is analogous to the end-to-end architecture of the Internet. It builds a technological and social commons that drives participation to it and appears to generate a level of distributed innovation that at least on the face of evidence from

[28] For more complicated versions of this idea, see Ostrom (1990).

market share can be more rapid and more efficient than innovation that is incentivized within traditional proprietary software firms.

This is not a unique phenomenon. Similar conceptions of property rights characterize other important parts of human life. Modern religious traditions are increasingly "owned" in the same way. Consider the question, what does a rabbi own? Clearly a rabbi doesn't own a tradition in the sense that he or she can exclude you from taking a piece of it and recombining it with pieces of other traditions, religious or otherwise, in ways that suit your own purposes. In practice people now modify, customize, recombine, and redistribute religious traditions in much the same way as they do open source software. This obviously changes the basis of the rabbi's power and leadership role in the community. Like Torvalds with Linux, leadership is certainly charismatic but it is also powerfully contingent on constant demonstrations of competence and judgment. If power is at least partly a story about assymetrical interdependence, then it is notable that leaders in these kinds of communities are at least as dependent on followers as the other way around.

Similar questions arise in the realm of copyright and the ways in which it is being reconfigured for the digital era. Leave aside property fundamentalists who believe that the right to exclude is a moral consequence of having created something. Copyright then is simply a social bargain made on familiar pragmatic grounds: legislate some excludability in order to incentivize people to create but as little excludability as necessary because it inefficiently limits distribution once the digital (and thus infinitely redistributable) good has been "made." Until recently the debates about the terms of the bargain really were debates on the margins. Now technology has placed the fundamental terms of the bargain up for grabs. Napster (as a business model, not a technology per se) was built on a completely different set of assumptions about incentivizing creativity with money and facilitating distribution. The courts cut off that experiment before the assumptions could be fully tested. There will be other experiments. The open source movement is one, and it is ultimately a much bigger experiment in the same kind of logic. The critical point is that the core assumptions behind the copyright bargain could be shown to be faulty in some very central facets of human creative and productive processes. If that happens, then the underlying notions of property rights could shift dramatically simply because there is very little to anchor them in place, except utilitarian calculations about what works.

A shift in property rights transforms some of the foundational principles of communities and cooperative relationships. Consider, for example, the mainstream literature on international regimes, which conceptualized international institutions principally as means to reduce transactions costs so that international relations could in some settings get closer to a

Coase-style equilibrium. (Ironically, the first generation of business literature about e-commerce and so on has followed a similar trajectory.) But of course low transaction costs are only one ingredient in the Coase theorem. The other is secure property rights. My argument here simply is that shifting foundations of property rights can and will likely destabilize the "Coasian" foundations of existing cooperative arrangements, and possibly in more radical ways than do changing transaction costs. It seems that the transition to a new set of "stable" property rights regimes (if that transition is ever really complete in a meaningful sense) would be the trickiest part to navigate—at least according to Coase. And it seems likely to me that we will be living through such a transition for at least the next decade.

Think then about some plausible consequences for international organization as one example of what this may mean in practice.[29] For much of the twentieth century, international organizations (IOs) could claim a special normative status because of their pluralist nature. By including representatives from all or nearly all states, and sometimes on a (nominally) equal basis, the policies that IOs promoted and the "truths" or bodies of "consensual knowledge" that they espoused acquired a distinctive legitimacy. The challenge for IOs is that the distinctiveness is going away. Pluralism at many different levels is being enabled and powered by the revolution in communications technologies, and more fundamentally by the alternative conceptions of property they engender, that is reducing the marginal cost of adding one more voice toward an asymptote of zero. International politics of course is not heading toward one big pluralist society where anyone can be part of any organization or community. But the default position is, indeed, changing—as the active choice becomes a matter of whom to *exclude* rather than whom to *include*. As more inclusive and pluralist organizations grow up in the space of international politics and economics, IOs themselves will come to look less special, and the legitimacy they have drawn from that special status will dissipate.

The same driving forces are chipping away at rational/legal authority manifested in large multifunctional bureaucracies, in a way that Max Weber might well have appreciated. It is difficult to see reinforcement of the bases of support for large bureaucracies as a way of organizing economic life in the future. Increasingly, a parallel phenomenon finds its way into political life as well. I think it is easy (particularly after the dot com bust, which ultimately was a financial market event, not a technological or social event) to underestimate the broad importance of organizational models that have created in the last fifteen years a massive technological revolution that now is on the verge of becoming a way-of-life revolution

[29] Some of these ideas I have drawn from a previous argument. See Weber (2000).

as well. Open source is just one example. Bureaucratic organizations did not achieve this. Flexible networks (often but not always fed by venture capital channels) did (Saxenian 1994).

The success of network organizing principles in economic settings is starting to work its way more deeply and broadly into other aspects of social and political life as well. Open source demonstrates the way in which it is possible for an increasingly large number of actors, both public and private, to enter the contest for control over the channels by which technical expertise (or claims of expertise) flow. Consider as another indicator of this trend new demands on, and ambitions by, corporations as well as NGOs to be political actors in a primary and self-conscious sense. The struggle for legitimacy in global politics is at least in part a story about who or what will be seen to solve problems. Large multinational firms understand this point well. Nike is now repositioning itself, quite deliberately, as a human rights organization. Royal Dutch Shell proposes to become an environmental organization; Monsanto, a global food and nutrition organization. Mutual funds with "green" or "social justice" criteria for investments now control around 10 percent of managed funds in America.

I think about this ongoing process as a kind of disaggregation of legitimacy. And I think this disaggregation and spreading of claims on legitimacy around to a variety of unexpected and unlikely actors is set to continue. This works to the disadvantage of traditional IOs. The United Nations in particular and the World Bank as well have been trying to reach out to nongovernmental organizations and other nonstate actors over the past decade, to reaggregate under their own roofs some of the legitimacy that has leaked outside. But the structures, ideologies, and historical legacies of IOs are mostly a burden rather than an asset in that process. Legitimacy can stay disaggregated for a very long time. The community that builds Linux is as "real" as is Microsoft or any other proprietary software company. When they meet within markets, different organizational forms often misunderstand how to deal with each other.

Markets and battlefields are similar in that way, as the events of September 11, 2001, make clear. To conceptualize war as a bargaining problem between two discrete and similarly structured actors (the main differences were their respective power and preferences) was an analytic convenience for international relations scholarship at the end of the twentieth century, but it assumed far too much. The deeper difference lies in how organizations are structured and how that complicates setting the terms of their interaction.

This problem I refer to in shorthand as the problem of interface between networks and hierarchies. When he was secretary of state in the Nixon administration, Henry Kissinger famously asked, "When I call 'Europe' who answers the phone?" Behind this glib comment lies a profound theoretical issue and a very practical set of questions for governments, corporations, and other organizations. What Kissinger really was asking is this: How does a hierarchically structured government (the United States) deal effectively (communicate with, cooperate with, or compete with) a powerful institution that is quite differently structured? Take this one step of generalization further and the question becomes, what are the dynamics of a relationship between a hierarchy and a network? For several years now the United States has been fighting an undeclared war against terrorist organizations al-Qaeda and others. Now we have de facto declared war on them. But the declaration of war is a social convention that has grown up between national governments. One state can declare war on another and the other can declare war back. Each makes demands on each other; their ambassadors meet; they try to negotiate an end to violence; they bargain and fight and then sign a treaty. It is a gruesome repertoire, but both sides understand how the game is played and what their roles are going to be in that game.

Now in the wake of September 11 a national government, perhaps the largest and most hierarchical national government on the planet, declares war on a network. That network certainly has committed an act that we call war against the United States. But it makes no explicit demands on us. In fact it does not even announce itself as the attacker. How do you bargain with an enemy that hides and doesn't tell you what it wants? It is a comforting fiction that Osama Bin-Laden is the equivalent of a president or a king, but it is only a fiction. Loosely coupled, cell-like structures act sometimes in coordination and sometimes not. They have a life cycle measured in scores of years because they do not depend on the leadership of one person or a conventional structure of succession.

Networks really are different in profound ways. The hard part is managing the interface, where the network and the hierarchy meet—whether that be in a cooperative relationship, a competitive relationship, or—in the case at hand after September 11—an explicitly hostile and conflictual relationship. The U.S. government is going to have to figure out this problem as it goes along. Social scientists may have much to learn from observing, and from comparing this problem to what we know in other realms. In fact, some of the practical business implications of this seemingly simple question about networks and hierarchies have been front and center for the open source community—as is evidence that there are no clear answers. In 2001 Microsoft essentially declared war on open source software. Now remember that (for better or for worse) Microsoft is one of the most profoundly successful organizations on the planet when it comes to strategy. It has an extraordinarily well-honed system for managing its relationships with other corporations (too successful by some accounts). It is just as expert and nearly as successful in managing its relationships with governments—in many ways, the business equivalent of U.S. hegemony.

But Microsoft has absolutely no idea how to manage a relationship, even an intensely hostile one, with the open source software community. You can't buy it; you can't drive it out of business; you can't "hire away" the talent; and you can't really tie it up in the courts. The glib version of this point would be to say, "When Microsoft calls Linux (even to issue threats, for example) who answers the phone?" In fact the question makes no sense; it presupposes a structural configuration of an organization that is not true of many networks and certainly not of the Linux community. Microsoft could easily buy Red Hat or drive it out of business in some other way if it wanted to. This would be the equivalent of the U.S. occupying Afghanistan. It complicates the life of the network but by no means undermines it.

Of course relationships between networks and hierarchies are not necessarily hostile. Just as Microsoft is trying to figure out the dynamics of conflict with the open source community, IBM has made a major commitment to cooperation with that community. Royal Dutch/Shell (in the context of the Brent Spar affair) tried determinedly to manage a conflictual relationship with Greenpeace; just as determinedly, Shell is now trying to develop cooperative relationships with networked NGOs.

The general point is simply this. One of the key government policy and business strategy questions for the next decade is, how do hierarchically structured organizations (like large governments or corporations) develop and manage their relationships with network organizations? Put differently, there exists no strategy template for how to understand the interface and build relationships between hierarchies and networks. People and organizations are figuring this out as they go along, through trial and error. And the problem is not going away anytime soon. Institutional sociologists particularly have developed a powerful body of theory about isomorphism, detailing some of the pressures driving organizations that are connected to each other in highly dense relationships to change so that they look more like each other structurally (Dimaggio and Powell 1991). This body of theory will likely prove itself quite useful in the longer term. But in the medium term isomorphic pressures are just that—pressures, not outcomes. It is important to get away from the fiction that national governments and big companies are all going to become networked organizations in the foreseeable future. They won't. And terrorist organizations are not soon going to become hierarchical structures with clear lines of command, ambassadors, and physical capitals. The reality is more complicated: both will coexist and have to find ways to relate to each other.

I am certain that some of the most interesting processes in international politics and business over the next decade are going to take place at the interface between hierarchies and networks, rather than solely within either one. And if I am correct in my claim that the open source process represents a distinctive form of political economy, then the places where the "open source economy" meets the "traditional," "proprietary" economy will be places of great creativity and interest from the perspective of social, organizational, and economic thought.

References

Adar, Eytan, and Bernardo A. Huberman. 2000. Free Riding on Gnutella. *First Monday* 5(10).

After the Microsoft Verdict. *The Economist,* April 8, 2000.

Baird, Davis. 1997. Scientific Instrument Making, Epistemology, and the Conflict between Gift and Commodity Economies. *Philosophy and Technology* 2(3–4): 25–45.

Becker, Gary S., and Kevin M. Murphy. 1992. The Division of Labor, Coordination Costs, and Knowledge. *The Quarterly Journal of Economics* 107(4) (November): 1137–60.

Brand, Stewart. 1995. *How Buildings Learn: What Happens after They're Built.* New York: Penguin USA.

Brooks, Frederick P. 1975. *The Mythical Man-Month: Essays on Software Engineering.* Reading, MA: Addison Wesley.

DiMaggio, Paul J., and Walter W. Powell. 1991. The Iron Cage Revisited: Institutional Isomorphism and Collective Rationality in Organzational Fields. In *The New Institutionalism in Organizational Analysis,* edited by Paul J. DiMaggio and Walter W. Powell, 63–82. Chicago: University of Chicago Press.

Ellickson, Robert C. 1991. *Order without Law: How Neighbors Settle Disputes.* Cambridge: Harvard University Press.

Free Software Foundation. 1991. GNU General Public License, vol. 2.0. Boston: Free Software Foundation. Available at http://www.gnu.org/copyleft/gpl.html.

Gancarz, Mike. 1995. *The Unix Philosophy.* Boston: Butterworth-Heinemann.

Ghosh, Rishab. 1998. Cooking Pot Markets: An Economic Model for the Trade in Free Goods and Services on the Internet. *First Monday* 3(3), March 2, 1998.

Ghosh, Rishab, and Vipul Ved Prakash. 2000. The Orbiten Free Software Survey. *First Monday* 5(7).

Kuwabara, Ko. 2000. Linux: A Bazaar at the Edge of Chaos. *First Monday* 5(3).

Lerner, Josh, and Jean Tirole. 2000. The Simple Economics of Open Source. *NBER,* February 25, 2000.

Lessig, Lawrence. 2000. *Code and Other Laws of Cyberspace.* New York: Basic Books.

Levy, Steven. 1984. Hackers: Heroes of the Computer Revolution. Garden City, NY: Anchor Press/Doubleday.

McGowan, David. 2003. Copyleft and the Theory of the Firm. *Illinois Law Review.*

Merges, Robert P. 1997. The End of Friction? Property Rights and Contract in the "Newtonian" World of On-Line Commerce (Digital Content: New Products and New Business Models). *Berkeley Technology Law Journal* 12(1): 115–36.

Moglen, Eben. 1999. Anarchism Triumphant: Free Software and the Death of Copyright. *First Monday* 4(8).

North, Douglas C. 1990. *Institutions, Institutional Change, and Economic Performance.* New York: Cambridge University Press.

Ostrom, Elinor. 1990. *Governing the Commons: The Evolution of Institutions for Collective Action.* Cambridge: Cambridge University Press.

Perens, Bruce. Debian Social Contract. Available at http://www.debian.org.

Raymond, Eric S. 1998. Homesteading the Noosphere. *First Monday* 3(10).

———. 1999. The Cathedral and the Bazaar. In *The Cathedral and the Bazaar: Musings on Linux and Open Source by an Accidental Revolutionary.* Sebastopol, CA: O'Reilly Publishing.

Ruggie, John. 1993. Finding Our Feet in Territoriality: Problematizing Modernity in International Relations. *International Organization* 47: 107–30.

Saxenian, Annalee. 1994. *Regional Advantage: Culture and Competition in Silicon Valley and Route 128.* Cambridge: Harvard University Press.

Smith, Marc A., and Peter Kollock, eds. 1999. *Communities in Cyberspace.* London: Routledge.

Stallman, Richard. 1999. The GNU Operating System and the Free Software Movement. In *Open Sources: Voices from the Open Source Revolution,* edited by Chris DiBona, Sam Ockman, and Mark Stone. Sebastopol, CA: O'Reilly Press.

Torvalds, Linus. 1998. What Motivates Free Software Developers. Interview in *First Monday* 3(2) March 1998.

———. 1999a. The Linux Edge. In *Open Sources: Voices from the Open Source Revolution,* edited by Chris DiBona, Sam Ockman, and Mark Stone. Sebastopol, CA: O'Reilly Press.

———. 1999b. Linux History. Available at http://www.li.org/li/linuxhistory.html.

Torvalds, Linus, and Andrew Tanenbaum. 1992. E-mail debate reprinted in *Open Sources: Voices from the Open Source Revolution,* edited by Chris DiBona, Sam Ockman, and Mark Stone. Sebastopol, CA: O'Reilly Press.

Tuomi, Ilka. 2000. Learning from Linux: Internet, Innovation, and the New Economy. Ms., April 15, 2000.

Vixie, Paul. 1999. Software Engineering. In *Open Sources: Voices from the Open Source Revolution,* edited by Chris DiBona, Sam Ockman, and Mark Stone. Sebastopol, CA: O'Reilly Press.

Weber, Steven. 2000. International Organizations and the Pursuit of Justice in the World Economy. *Ethics and International Affairs* 14 (Winter).

Womack, James P., Daniel T. Jones, and Daniel Roos. 1990. *The Machine That Changed the World.* New York: Rawson Associates.

DESIGNS AND INSTITUTIONS

Designing Information Resources for Transboundary Conflict Early Warning Networks

HAYWARD R. ALKER

THIS CHAPTER REVIEWS and reflects on design considerations related to the development of partly computerized information resources intended to be useful for early warners of impending, potentially violent, intergroup conflicts.[1] The early warners in question—most typically those associated with the London-based Nongovernmental Organizations (NGOs) International Alert (IA) and the Forum on Early Warning and Early Response (FEWER)—were, and are, committed to transboundary cooperation in the prevention, or amelioration, of violent intergroup conflicts. To this end, the design and prototype development of a set of information systems was undertaken by the Conflict Early Warning Systems (CEWS) research project[2] of the International Social Science Council, a Paris-headquartered confederation of global, regional, and national social science associations and agencies. From its inception, CEWS's job was seen as developing small, networkable, extensible information systems that could be helpful for partly decentralized, modestly resourced networks of

[1] I wish to thank Kumar Rupesinghe, Lincoln Bloomfield, Elise Boulding, Karl Deutsch, Ernst Haas, Dwain Mefford, Marvin Minsky, Thomas Schmalberger, Herbert Simon, and Stephen Toulmin for their especially important contributions to my understanding of this topic, without holding them in any way responsible for what I have chosen here to say. The editors of this volume, Robert Latham and Saskia Sassen, have also had a material impact of the improvement of my text.

[2] The work in question—covering a time period from 1992 through 2000, and reported on in Alker, Gurr, and Rupesinghe, eds. (2001)—was principally funded by the Carnegie Corporation of New York, in grants to the School of International Relations, including its Center for International Studies, at the University of Southern California (USC). The author of the present chapter was the principal investigator of several Carnegie grants and co-principal investigator of a related grant from the Annenberg Center for Communications at USC. Throughout its formal existence (December 1992–May 1999), the CEWS project was coordinated by Alker, with Kumar Rupesinghe as his co-coordinator. For much of this period, Rupesinghe was secretary-general of IA and/or chair of the FEWER supervisory committee. For more details, see the preface and the first two chapters of Alker, Gurr, and Rupesinghe (2001).

future early warners in regional and/or global collective security organizations and related NGOs, such as IA or FEWER.

CEWS was a novel, practically oriented attempt to merge organizational information processing research paradigms with newer, network-oriented information technologies. In that merger, it turns out that historically and hermeneutically oriented philosophies of inquiry associated with thinkers such as Jürgen Habermas have important resonances with the practical concerns of information systems designers working in computational traditions familiar to followers and reformulators of the work of Herbert Simon.

Such resonances can potentially shield against the tendency to apply early warning capacities to *automated* systems that reinforce "closed worlds" politics (Edwards 1996). That is, the CEWS project could be viewed as computationally linked, *open systems* design research seeking to enhance the conflict-relevant, historical information-handling capacity of *human-centered* complex adaptive systems, and to mitigate the closed world tendencies of North American automated early warning systems deployed in advanced domains of military surveillance, ballistic missile "shields," and "smart weaponry."

Also at stake in this chapter—and central to the concerns of this volume—is the question of how knowledge networks can be designed that are inherently dynamic, interactive, and potentially transformative. One way to do so, as I will show, is to create knowledge spaces for rewritable and contestable interpretations and histories of conflicts.

I will begin with a brief sketch of the historical and intellectual context out of which CEWS emerged as a viable pilot project.[3] This will be followed by an exploration of the heuristic logics that guided the design process, a consideration of the conceptual underpinnings of open system design, and an examination of the relevance of historical and hermeneutical philosophies. Finally, I will review how the pilot was carried out and then conclude with an assessment of its outcomes.

The Contradictory Context of Recent International Conflict Management/Resolution Practices

Soon after the end of the cold war, Secretary General Boutros-Ghali attempted to reinvigorate the UN collective security system by offering an

[3] This discussion briefly reviews, from my own individual perspective, contacts with various governmental and nongovernmental conflict management and specialists in the UN- and U.S.-related arenas, as well as the more scholarly discussion of epistemological and paradigm differences in ibid., chap. 1.

ambitious *Agenda for Peace* (Boutros-Ghali 1992). He called for increased attention to a variety of threats to the domestic and international security and integrity of the UN's member states. Not blocked by incessant cold war vetoes, Security Council decisions in the post–cold war period have often been able to launch useful and influential UN presences in troubled situations. UN "peace-building" and "state-building" missions have increased. But problems remain: Western, liberal states have ideological premises underlying much of what they propose or prefer as well. Those states identifying their worldviews with religious or non-Western civilizational identities are reluctant to concede to incremental, skeptical, secular, practical, or scientific rationality an autonomous basis for cross-civilizational truths. Parochial loyalties are strong everywhere. And levels of economic development, nationhood, and statehood differ widely. At best, pragmatic compromises or common-sense responses to particularly acute security disasters have held sway, creating a considerable, but unevenly legitimated, repertoire of potentially usable precedents and lessons concerning future conflict involvements.

Moreover, there is the power-related tension between an interdependent, globalizing world, conceived in supranational collective security or unipolar hegemonic terms, and the core logic of what has been called the Westphalian international system of legally equal, sovereign states. In an era of increasing globalization, "sovereign" states are very unequally divided. In the security realm, the most obvious division is between those with a Security Council veto and those without one. Among the veto powers, the United States now exercises a special role, given its superior, globally applicable, militarized, information-shaped, force projection technologies, its soft power, and its unparalleled logistic support capabilities. Both old and new states are being challenged by resurgent ethnic and religious identities; the European practice of affiliating conflicted local loyalties to a supranational community formation process has inspired many but not yet found many effective imitators.

At play here is the fundamental question of the basic units of world society: are they peoples, nations, states, or civilizations? If the UN Charter refers fundamentally to "the peoples" of the United Nations, almost all of its legitimated actions depend on the will of states. Yet the post–cold war world has been awash in conflicts between ethnic minorities and the states they reside in. Transnational actors and internationally supported ethnic and religious movements and forces rely on global connections for financial and media support. Many of them challenge the Weberian concept of modern statehood's association with the monopoly of legitimate coercion. Conflict prevention, management, and transformation practices are unequally associated with these often contradictory ways of seeing the world and attempting to handle its about-to-be violent conflicts. The UN

Charter specifically prohibits interference in the "domestic affairs" of states; but in a period where almost all violent intergroup conflicts can be described as "internal," that legal standard is being constantly rewritten and interpreted differently.[4]

With the possible exception of some high-tech military personnel, among seasoned practitioners of conflict management and prevention, skepticism concerning "quantitative" or "scientific" research approaches and their so-called imputations of causality is rampant. The statistical studies of many international relations scholars of war and peace have long been seen by conflict management or resolution practitioners to be just too abstract, too quantitative, too general, too ahistorical and decontextualized, too dehumanized to be of great relevance to their urgent, specific concerns. One heard this view often during the 1980s and early 1990s in conversations with bureaucrats or NGO personnel educated in traditional disciplines like history and law, especially in those from less developed countries, whose professional teachers were traditionally educated as well. It was almost as if the "traditional-modern" dimension of the last five centuries of world historical development—with its different priorities concerning different kinds of rationality[5]—was being replayed in their minds; at least that is what the unreflective scientist might think. Among academics, the related, 1960's debate between "scientific" and "classical" approaches to international relations (see my relevant review in the final chapter of Alker 1996) was endlessly being replayed as well.

In this context, conflict management research programs, like Ernst Haas's (1968, 1993) exceptional series of empirical studies of UN-related collective security regimes, fit more closely with the aspirations of UN secretary generals working within the limits of the Veto Powers' support, which is so essential for their incumbency and a legal order placing a high priority on noninterference in the domestic affairs of sovereign nation-states. They make good sense to believers in the complex interdependence of the late cold war world.[6] And they are realistic about the difficulties of

[4] Alker, Gurr and Rupesinghe (2001: 7) ask rhetorically: "What . . . has happened to Kant's [Enlightenment-motivated, anticipated future] world of war-weary republics?" They cite three data sources from the 1990s, arguing that "Less than 3, 4, or 10 percent of such [intergroup] violence is described as now being of the 'conventional' interstate variety."

[5] See Toulmin (1990) and Hodgson (1993) for different, but related, versions of the transformations associated with the culturally dominant forms of rationality in the modern period.

[6] Keohane and Nye (1975, 2000) have suggested this influential ideal type of international relations as a contrast with political realism's map of the world. Complex interdependence is characterized by multiple channels of interstate communication, nonunitary governments and transnational actors connecting societies, multi-issue agendas without clear hierarchies among them, and the unlikelihood of the use of military force in such relationships.

genuine conflict resolution when the agents of conflict intervention are foreign ones, often tempted to intervene in intrastate conflicts for their own partisan purposes. Lessening the frequency and level of violence, or limiting the spread of a conflict within a particular region, involving only a limited number of states, are worthy goals from this ameliorative, conflict management perspective.

By way of contrast, peace researchers interested in conflict transformation (Rupesinghe (1995) are more revolutionary in wanting to get to the bottom of conflicts, wherever they may be. In seeking conflict transformations, which may not be permanent conflict resolutions, they often see a vital role for nongovernmental actors. Especially in weak states, NGOs can sometimes intervene in domestic affairs in ways that would be very difficult for intergovernmental organizations (IGOs) to do.[7] Often their guiding vision sees transnational NGOs as members of a possibly emerging, significantly more peaceful, post-Westphalian, global civil society. Conflict early warning research could be said to evidence all three related orientations: a concern of status quo great powers anticipating threats to their world's continued existence and status (traditional international relations), the aspirations of reformers within or outside existing international institutions oriented toward the amelioration of violent conflicts (the Haas approach), and a focus on key capability enhancements leading eventually to the transformation of international systems from "the state of war," to the condition of integration, defined in terms of the expectation of peaceful changes and dispute settlements among nations, states, and peoples (peace research).

Designing a Cross-paradigm Early Warning Information System Prototype

As possibilities for change emerged as the cold war wound down, International Alert and later FEWER developed regionally oriented early warning teams to monitor and call attention to developments in ongoing conflict situations.[8] From its inception, CEWS's job was seen as develop-

[7] This aspect of what might be called "extended complex interdependence" is worth thinking more about theoretically in a world of weak, penetrated, or "failed" states, now seen by many in the United States, at least, as potential havens for globally oriented terrorists.

[8] International Alert's involvement in several years of the Sierra Leone situation became one of several comparable cases discussed by S. N. Anderlini, E. Garcia, and K. Rupesinghe in (Alker, Gurr, and Rupesinghe (2001: chap. 8). Information on three of FEWER's ongoing projects in Africa and the former Soviet Union was available at http://www.fewer.org in early 2002.

ing potentially portable information resources that could be helpful for decentralized networks of future early warners in such organizations. Based on the realization that a research team of academics could of course not have access to the confidential and strategically important information that practitioners need and sometimes have,[9] and that there would be lots of organizational information systems that were not of direct scholarly interest, the CEWS strategy was a narrowly focused one: to develop a way of encoding shareable information about previous conflicts that would be practically helpful vis-à-vis potential turning points of new or ongoing conflicts. Security-sensitive information would have to be encoded by practitioners not able or willing to share such information, hence academic inputs would of necessity be partial, at best a kind of prototypical development. A modular, portable, extensible, prototype information system—attractive to practitioners with different styles of handling conflicts and scholars with different approaches—became the specialized goal of the CEWS project. However, how to retool this prototype and, with sufficient resources, put such an open system in place has to date remained an unfulfilled challenge.

A two-stage CEWS project design emerged: first scholars interested in conflict anticipation and resolution from different disciplinary research paradigms—both quantitative and qualitative, conflict management and conflict transformation—and different world regions were asked to prepare intervention-suggestive chronologies or narratives of intergroup conflicts within multiphase historical frameworks of their own design. Then an inclusionary framework was to be induced, if possible, within which these chronologies or narratives would be recoded, highlighting both possible ameliorative interventions and contested historical perspectives concerning the supposedly "objective" coding of conflict phases. If the original producers of the narratives and chronologies could be persuaded that the new framework did not do serious violence to their original frameworks of conflict representation and analysis, then this would be an important, if preliminary, test for the wider implementation of a later generation of far more sophisticated information systems.[10]

I would argue that there is something more at stake in the development of the CEWS design for network-oriented information systems than an effectively operating, shared knowledge space. A resonant mixture of what I shall call "conversationally oriented" or "humanistic" philosophies of

[9] The strategic use and protection of information is a growing thematic focus within game-theoretically oriented international research. For conflict early warners, Stephen J. Stedman's work (e.g., Stedman 1997) is particularly relevant in this regard.

[10] A detailed account of the CEWS project and its collective accomplishments—such as they are—is told in (Alker, Gurr, and Rupesinghe (2001). It will not be repeated here.

sociopolitical inquiry (Alker 1988, 1996) were combined with recent generations of information and communications technologies (ICTs) to shape the human and mechanical aspects of the development of that project.[11] I believe there are what physicists might call generative "resonances," biologists might designate as "co-evolutions," and Hegelians could characterize as "internal relations" between these humanistic philosophies and technical open systems design efforts, a linkage I attribute to the constructive human impulses "within," or engendering, these communication-oriented technologies. If we think of digitally defined formations or technologies as constructed, relatively-enduring materializations of human intentions, purposes/functions, and plans,[12] the most radical impact of the newer ICTs may well be the re-idealization—with designed, reformulated, and unintended variations of these intentions, communicative purposes/functions, and plans—of such philosophically inspiring practical craftsmanship at the level of intergroup interactions.[13]

I have found that causal law-seeking, naturalistic philosophies of inquiry have a hard time resonating with and responding constructively to the detailed, specific, case-by-case concerns of locally or regionally oriented early warning/early response organizations; those from the more systematically oriented humanistic traditions of inquiry do not. Quite unlike those whose exposure to modern information technologies has reinforced a naturalistic orientation to social scientific practice, I have found hermeneutically accessible, humanly constructed and redesigned worlds within these information/communication architectures, worlds that humanistic philosophers, scholars, and engineers have long anticipated. New meanings have been given to older, precedentially oriented forms of bureaucratic rationality; virtual spaces for globally sharing summarized, prevention-relevant conflict histories, summaries that respect the contested historicities of major conflict protagonists, have been created.

[11] Although Alker (1996) subscribes to the nature-respecting, humanistic ideal of social scientific research that is both scientific and normatively oriented, or even artistic, in its motivation, I had forgotten Lasswell's much earlier but similar technical definition and advocacy of the "adoption of both [what they called] the manipulative and contemplative standpoints of inquiry [which they] designate[d as] the *principle of configurative analysis*" (Lasswell and Kaplan 1950: xiii).

[12] This "internal relations" ontological perspective is developed in my "Can the End of Power Politics Be Part of the Concepts with Which Its Story is Told?" (Alker 1996: chap. 5), a conference paper originally given in 1977; it thus antedates the similar inspirations of Bruno Latour noted by Bach and Stark in the present volume. See also Alker (1986).

[13] As cited in Alker (1996: 401), Hintikka suggests a contemporary version of Aristotelian practical reason to involve "reason in so far as it is occupied with human action, human doing and making, and with the results of such action . . . 'Maker's knowledge' [here includes] 'doer's knowledge,' for no distinction between *poiesis* [production] and *praxis* [practice] is intended."

The Emergence of an Open/Adaptive/Complex Systems Design Research Perspective

"Information engineering" offers a particularly fruitful way to get at the theoretical links between humanistic social science and the information system design logics relevant to the CEWS project. Just as Aristotle distinguished the useful knowledge of specialized professions from the eternal truths of the pure sciences, human-oriented information engineering has a rather similar normative, applied, practical character. More specifically, within human information engineering there is a specific tradition of architectural "design research" associated with what used to be called "open (biological) systems," those capable of negantropic, structure-modifying exchanges with their environments. Now these might more accurately be called "complex adaptive systems." Although humanly designed practices and institutions may only weakly approximate the awesomeness of Nature's "creatures" or "designs," as an architecturally inclined social scientist one has fewer quasi-religious inhibitions in treating human products as modifiable combinations of natural foundations, accidental factors, and human design-oriented activities. At most these products may be seen as "quasi-natural"; in reality they are artificial. And, in my view, such conceptions can help the design-oriented social scientists to think about ways of making contributions to the improvement of human international conflict management and resolution practices.[14]

Ontologically or phenomenologically speaking, the subject matter of Herbert Simon's (1969) "sciences of the artificial" include humanly synthesized artificial things, which may imitate the appearances of natural things without having all their real world features. He argues that

- artificial things may be characterized in terms of functions, goals, evolution-like adaptations, and, more revolutionary, partly designed transformations;
- like LISP programs, they may be described as *revisable* program objects; and

[14] It should be noted, however, that I rarely or never emphasized information engineering, Simonesque Artificial Intelligence, "design research on complex adaptive systems," or "conversational ontologies" in the activities of the CEWS Steering Committee. We had enough different disciplinary and cultural boundaries to cross as it was! Rather, for a group in which peace research was the closest disciplinary communality, I distributed and repeatedly referred to Alker (1996: chap. 10; originally published in 1988) on "Emancipatory Empiricism." The procedurally oriented computational modeling idea emphasized there is that of LISP—encodable and revisable data stories.

I don't think such behavior was deceptive, since there is considerable discussion of complex adaptive systems theory in the Aristotle chapter of that book, which was easily available, and a bibliographical discussion of the contributions of pioneering peace researchers like Kenneth Boulding, Karl Deutsch, Harold Guetzkow, and Anatol Rapoport would have easily retrieved many references to open systems theory.

- such program objects might also be treated as practical imperatives waiting to be actualized by practical reasoning.[15]

Herbert Simon's (1969) concept of the "sciences of the artificial" and the "architecture of [nature's] complexity" is preferable to the Popperian notion of "piecemeal social engineering" as the basis for understanding the design logics of conflict early warning systems. Not only is Simon more epistemologically, theoretically, and technologically innovative, but his approach has inspired a range of relevant humanly oriented computer science (I shall refer to this approach as "[architectural] design research," often dropping the construction-oriented first word of this phrase for the sake of brevity).

Humanly oriented computer scientists, who are inspired by Simon, his colleagues, and revisionist successors, are very much aware of the human communication networks, of the flow of human thoughts, criticisms, suggestions, and program proposals that constitute the public sphere and *res publica* of social life. Thinking about coordinative and constitutive human arrangements as artificially evolved means that, like organisms they resemble, they are open systems, that is, they evolve, persist (and perhaps reproduce themselves) through nonequilibrium exchanges of information and resources with their environments. Usable energy is extracted from those environments, and degraded waste products are also discarded into them. Persistence, change, and renewal are thus problematical questions, not to be taken for granted. (I shall return to this thematic below.)

This view is related to CEWS development if we recognize that *international system designs* or proposed *revisions in institutional architectures* may be thought of as *interfaces* between system-internal human environments (those of national citizens or bureaucratic office holders) and system-external natural or artificial environments. One should, therefore, focus on the needs, intentions, plans, and purposes of actors in connecting meaning and utility to the design of such interfaces. The perspective of improving the content of relevant shareable memories, and of improving accessible ways of interrogating the relevant past on a case-by-case basis is common among historians and experienced conflict management/transformation practitioners. It is not the perspective of social scientists trained to look for lawlike, timeless statistical generalizations in large aggregations of summary versions of somewhat similar cases.

When the relevant past experience is an improvement-oriented record

[15] Since Aristotle, practical reasoning has been distinguished by its orientation toward action. Simon's writings on administrative behavior, human problem solving and bounded rationality, Stephen Toulmin's many writings on practical argumentation, von Wright's philosophical explorations of practical understanding, and Martha Nussbaum's suggestive classicism are major sources of the several discussions of the topic in Alker (1996).

of context-specific past performances by humanly implemented *quasi-regimes*—international relations terminology for multilevel coordinative systems of rules and procedures of uneven extent, legitimacy, effectiveness, and institutionalization—the standard data bases on international conflict available through consortia like the Inter-University Consortium for Political and Social Research were not likely to contain such information. Nor were the aggregative statistical formalisms used by social scientists for summary approximations of large data-bases in general algebraic terms going to be the most helpful to those concerned with a single, troubling case. Statistical information summaries are not precedentially organized; hence the value for information systems designs of previous explorations of precedentially organized, narratively represented, procedurally suggestive case descriptions.

Simon-inspired design research is very sensitive to discussions of formal schemata for representing knowledge inside a bureaucratic or computational system because design-oriented information engineers know that different representations, like different programming languages, have very different emphases and utilities. Their focus on feasibility/desirability/implementation/orchestration questions are familiar to policy analysts but rarely the primary concern of causal modelers looking for general laws in the eternal pages of Nature.

A second or third generation of design-oriented open systems engineering focuses on what are now called "complex adaptive systems." Moreover, their representations of adaptive human systems are much more detailed and suggestive than those of the earlier generation. And I find a deep convergence between the quasi-evolutionary approaches to the reproduction and transformation of social systems emphasized by Parsons, Habermas, and other social systems theorists, and the representationally and architecturally suggestive variant of Complex Adaptive Systems theory that Axelrod and Cohen have recently proposed for "managing" or "harnessing" complex organizational practices.[16]

Axelrod and Cohen's framework of analysis can be summarized as follows:

[A population [physical
of **Various types** of] and conceptual]
Agents, using **Artifacts**, according to **strategies**, in **Spaces**, lead to **interaction patterns** [and
resulting events] within **systems**. (1a)

[16] Axelrod and Cohen (1999: esp. 152–60). In the text and equations of the next several paragraphs, all the boldface terms are taken from the cited pages of the Axelrod-Cohen book. Bracketed phrases are usually, but nonuniquely, my additions. Some of the key terms are rearranged in my presentation, which in the interests of readability mixes paraphrases and direct quotations without further acknowledgment.

Performance measures
[or **success criteria,** defined on events] are used by **agents/designers** in **selection processes,** resulting in
agent/strategy
changed **frequencies,** through processes of **copying and recombination** [within] **systems.** (1b)

Notice how schema (1b) focuses on reproduction, as well as on the quasi-evolutionary, or Lamarkian, transformations. Why these occur is linked to performance-based selection processes that account for more or fewer copies of a variety of agents or strategies in a later period. Notice also how reproduction processes can happen in both physical and conceptual spaces.

Also, quite relevant were the linguistically informed, narratively oriented modeling efforts emerging in related disciplines. By the late 1960s and early 1970s, thanks to Robert Abelson and Roger Schank, Allen Newell and Herbert Simon, Noam Chomsky, John Searle, Rom Harré, Jürgen Habermas, Marvin Minsky, and European structuralists, there emerged a whole new dramaturgical world of analyses of linguistically mediated conduct, and an associated, constructivist mathematics. Actual human *rewrites*[17] of underlying goals, plans, or scripts *generated* meaningful *texts* made up of *word strings,* through *speech acts* or *text acts, narratives, plot structures,* and *scripted understandings* performed on *sociohistorically situated stages.*[18] In the mutable performances of institutional

[17] Generalizing Chomsky's original insights, I have argued that a fundamental representational/ontological feature of second generational cybernetic models, which accounts for, or constitutes, much of their generative and interpretive power is the ability to rewrite, record, and recall intermediate, unobservable revisions of ordered, sentencelike, phrase structures, perhaps in a context-sensitive fashion. The fundamental constructive mechanisms of Newell and Simon's theory of General Problem Solvers, von Neumann's brilliant final work on self-reproducing automata, Abelson and Schank's powerful models of belief system dynamics, the post-Chomskyan fields of generative semantics and text linguistics, and Miller and Chomsky's paradigm-defining discussion of intentional action *all* depend on *rewrite* mechanisms that are constitutive rather than causal (Alker 1988, 2000) Thus Chomsky's famous formal hierarchy of the generative/interpretive power of languages, grammars, and automata can be used to show how human capabilities, when indefinitely specified, reflect *context sensitive rewrite rules,* transformational capabilities that are two levels of infinity higher than those of *all* causal models' production relationships.

[18] See especially Abelson (1973), Schank and Abelson (1977), and Alker (1975), (1996: chap. 5, 8). This partly italicized mapping sentence could be schematized like schema (1a,b) above, adding hermeneutical realism and complexity in the same way that (1b) adds richness to (1a). But within the present context, focused on information technologies, I limit this development to this mention of my fascination with potentially scientific computational/formal hermeneutics, and the following comment. When I tried to convey this new-for-me, verifiable mathematics of infinite meaning productions/understandings/constitutions with the last, Abelsonian example in my paper "Polimetrics: Its Descriptive Foundations" (1975), the face-to-face reactions I got were suggestions from rational choice theorists that "politimetrics" would have been a better spelling for the "metric" aspects of political science, and the

role occupants, one looked for, or tried to reconstruct, not the eternal causal laws of physics, chemistry, and neurophysiology of the embodied natural world we inhabit, but the changeable, creative, *generative* rule systems, the personally, socially, institutionally, historically situated *grammars* of *possible* speeches or linguistically informed actions. Nonspoken human actions did not escape similar grammatical complexities: what distinguishes all human actions from mere behavioral reflexes—winks from blinks, in Geertz's memorable example—are linguistically encoded *meanings*. As Habermas (1971) has argued in the defense of the autonomy of the hermeneutic knowledge interest, and Harré and Secord (1972) made equally clear from a dramaturgical, ethnomethodological perspective, meaningful understandings can be seen as products, conveyed through grammatically enabled *skilled performances* deserving of careful compositional investigations, not the "mere descriptions" of unreflective, positivistic analytics. Although speech production and interpretation were different processes, the complexity of human grammatical capabilities argued for the likelihood that most of the mechanisms involved in each were the same. Habermas's theory of communicative action (1970, 1979, 1984, 1987) was much more sensitive to the constitutive role of substantively persuasive speech acts, precedential-normative-historical story telling, and identity transformations in the possible development of pluralistically, sociologically integrated—that is, peaceful—international relations.

Trying to make sense of history in terms of the *practical grammars* that made it possible, and sometimes included possibilities for more conflict-lessening alternative paths not taken, meant doing something akin to what Habermas called *reconstructive research* rather than conventional *empirical-analytical science*. Moreover, Habermas's (1971) insistence on the multiplicity of human knowledge interests was critical to recognizing the existence of different knowledge trajectories. Besides the positive knowledge interest in prediction and control on the basis of general causal or developmental laws, Habermas emphasizes the importance of both hermeneutic and emancipatory knowledge interests. The former may be described as focused on interpretive understandings, but with the important addition that hermeneutics arises out of practical reasoning, the practical concern with developing a shared, rational basis for collective action. Here is where the differences between modern and traditional forms of rationality—highlighted in the discussion above of differences among conflict-oriented practitioners and scholars—make a methodologically

concession from the political statisticians that not all descriptions were simple. It was as if the grammar-reconstructing scientific work of linguistics did not exist, or was irrelevant to politics and political science.

relevant appearance, supporting the traditionalist position. Equally important are his and Bhaskar's articulation of the focus of emancipatory knowledge interests in terms of the promotion of self-directed transformations and the ending of unnecessary repressive relations, from "unwanted and unneeded to wanted and needed source[s] of determination."[19]

CEWS's Design: Decentralized, Regional Networks of ICT-Resourced Conflict Early Warners

Peace researcher Kumar Rupesinghe's vision of decentralized early warning networks for dealing with the mostly "internal" conflicts of the post–cold war era was a kind of institutional design research (e.g., Rupesinghe and Kuroda 1992; Rupesinghe 1995; Rupesinghe with Anderlini 1998). With support from Scandinavian and other like-minded governments and foundations, he encouraged the setting up of transnational networks combining, connecting, and attempting to empower regionally and locally situated conflict analysts and peacemakers. The umbrella organizations like International Alert and FEWER were affiliated with the United Nations and other intergovernmental organizations as NGOs, and in frequent contact with national bureaucracies and secretariats. At least provisionally compatible with a "complex interdependence" conception of international political order, Rupesinghe's network conception also resonated, in my view, with the interventionist, design-oriented perspective on complex adaptive systems summarized above.

What could social scientists contribute? Here is where an especially international comparison of successes and failures at preventively oriented conflict management and resolution seemed a reasonably familiar possibility. Social scientists could produce these case studies and compare the frameworks for digesting them and inferring practical lessons from them. If we were to use the power of newer ICT technologies to help resource the actors in these networks, we would have first to develop a hermeneutically informed framework for storing case histories, precedentially recalling and deriving forward-looking suggestions for violence-reducing conflict interventions. So what I have above described as "grounded, practically interpreted, alternative allowing, conflict narratives" would be informationally made available to "frontline" peacemakers in the "open/adaptive/complex systems" that Rupesinghe and his peacemaker as-

[19] This Bhaskar quotation and the longer paraphrase of Habermas's and Bhaskar's views is taken from my discussion of "Emancipatory Empiricism" (1996). Almost the same text first appeared in Wallenstein (1988), shortly before the beginning of the CEWS project.

sociates were designing, opening the way for "retrievable institutional memories."

As the CEWS project developed, the relevance of the Habermas-Bhaskar world of critical, emancipatory inquiry became quite important. Most of the regionally recruited practitioner-scholars did not see the relevance of quantitative research methods in the cases of their special concern in the conflict prevention domain; several others were pioneers in more quantitatively oriented "event data" research. Revising Alex George's "structured, focus comparisons" approach to lesson-drawing comparative case studies, the 1995 London meeting of the CEWS steering committee agreed on a first stage of the CEWS project: generating narrative or chronological case studies of conflict prevention successes or failures on which to build its future work.[20] Allowing each investigator to use different versions of the conflict life-cycle idea (as cross-culturally presented by Johan Galtung, among others), the original conflict accounts were to be "*preventively focused, life-cycle structured, trajectory comparisons*" (Alker, Gurr, and Rupesinghe 2001: 39, italics in the original).

In 1997, once the CEWS project had been funded, a second meeting of case authors, analysts, and CEWS steering committee members took place. The different narratives and chronologies, plus the beginnings of intended interpretations, reflected the authors' different approaches to conflict trajectory representation and analysis. Discussed were the possibility and difficulties of developing a synthetic sequential phase coding schema (Alker 1988).

Especially important was the notion that we think of structured narrative accounts as "LISP encodable data stories" (Alker 1988).[21] Here the unusual ontological property of LISP *programs*—that they could also be treated as *data* modifiable by other interpretive procedures—was a direct source of the operational idea of revisable or annotatable, action-suggesting precedents that governed the development of the CEWS web site and the LISP-implemented CEWS Explorer. Emancipation as a peace research knowledge interest meant attempting to move conflict trajectories away from steps where higher violence levels were likely to follow.

In the report on the CEWS project, the most pedagogically useful example is that of the Guatemalan conflict and peace process as retold by a

[20] Somewhat revised, these are now on the CEWS web site: http://www.usc.edu/dept/LAS/ir/cis/cews/index.html.

[21] Lesson 10, p. 350, reads: "Think of conflict and cooperation case descriptions as LISP encodeable data stories. These descriptions are then executable programs, situation specific practical accomplishments, procedural enactments that constitute the cases, analogous to, but possibly different from the practical actions constituting the observed realities they refer to" (italics omitted). James Bennett of Syracuse University is responsible for the "data stories" notion.

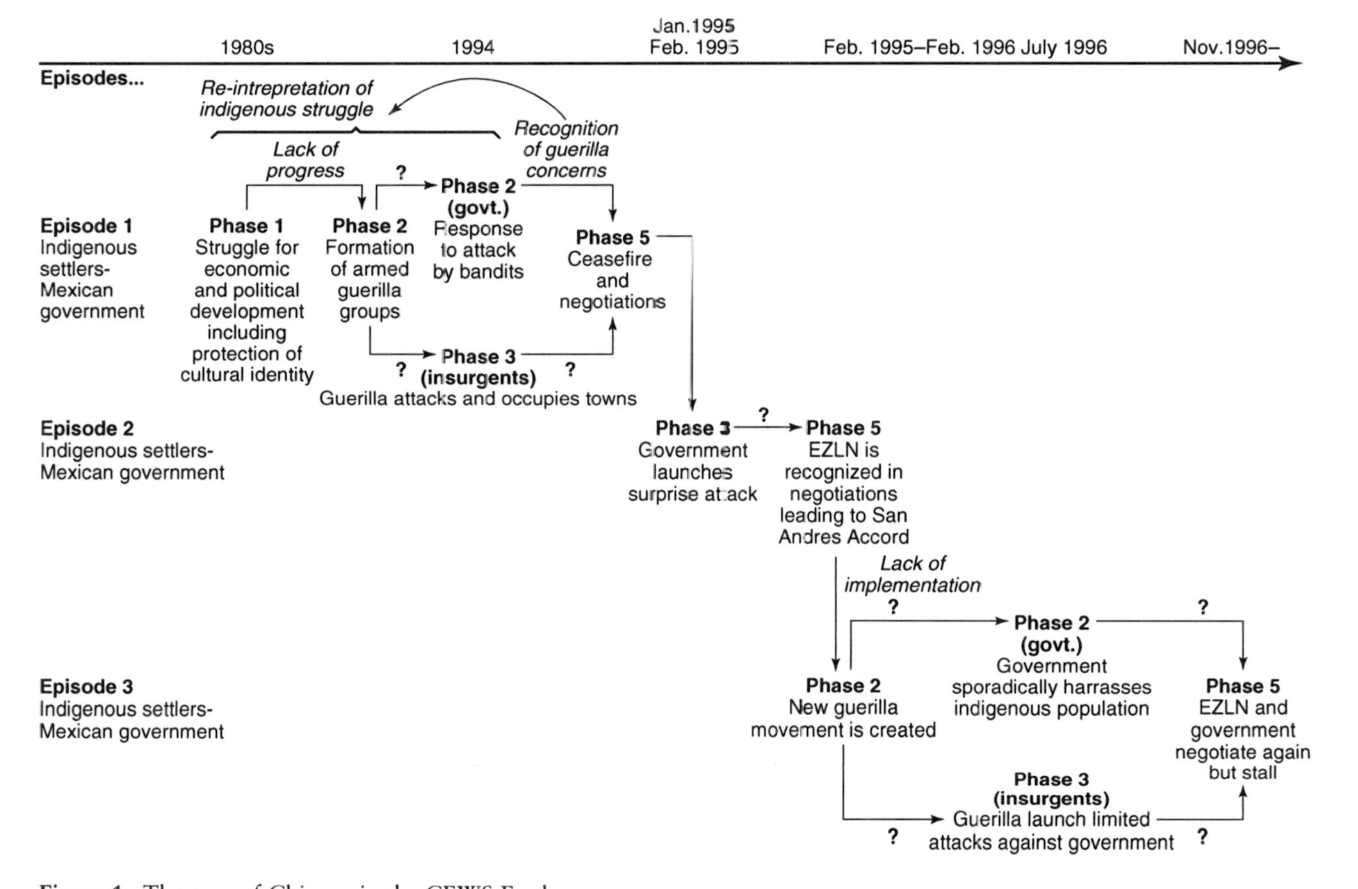

Figure 1. The case of Chiapas in the CEWS Explorer

knowledgeable observer of that process, Luis Albert Padilla, and reanalyzed by Thomas Schmalberger and myself (Alker, Gurr, and Rupesinghe 2001: chaps. 3, 11). Downloaded from the CEWS web site, figure 1 shows a different but similar example of the CEWS graphic schematization, the unfinished case of Chiapas. The figure was generated by CEWS researchers, based on a narrative (provided by Rudolfo Stavenhagen) that is also available on the web site. From a precedential perspective, the inconclusive state of the Chiapas conflict is not a paradigm likely to be consciously copied elsewhere; rather, one might think, as Stavenhagen does, of the more comprehensive peace processes in Guatemala and El Salvador as possibly useful models—revised to omit some, it is hoped, avoidable parts—for application to the Chiapas case. The recent efforts of Mexican President Vincente Fox to unpack a stalemated situation could be further investigated in this light.

What figure 1 is meant to convey is the contextually grounded, practically interpreted, alternative allowing, narrative-like schematization of conflict trajectories developed by the CEWS project. This suggestive, flexible schematization would probably not have occurred without the immersion of its principal developer—Thomas Schmalberger—in the Sylvan-Majeski tradition (most systematically explicated publicly in Sylvan and Majeski 1998) of sociohistorically grounded, computer-supported, constitutive analysis, which has often used LISP-encoded representations. Also necessary was the focus on alternative, LISP-encodable "data stories," as previously discussed.

The foundational, grammatical idea for figures like figure 1, visible in earlier work in the precedent logics tradition, is of a historical-based grammar/flowchart of actual and counterfactual conflict trajectory possibilities. The actual history of a particular case—as schematically summarized in the figure—breaks its analytical, historical reconstruction into episodes and phases. Presupposed is a sequential pattern of possible movement from phase 1 (a dispute) to phase 2 (a crisis) to phase 3 (limited violence) to phase 4 (massive violence) to phase 5 (abatement) to phase 6 (conflict settlement); our conflict grammars allow conflict-exacerbating reversals, treating them as generating new episodes. Conflict management practices under intergovernmental auspices often move disputes only to an abatement phase; much fuller involvement from elements of domestic societies, and perhaps NGOs, is required to bring about an effective phase 6, a settlement that IGOs like the United Nations can help to ratify, solemnize, and modestly support.

Several additional special features of the CEWS representational format are illustrated in the figure. First notice that the historical sequencing of phase descriptions can be contested—for example, on many occasions governments, in the interest of buttressing their precarious legitimacy, like to downplay the existence of a genuine civil war led by organized rebels claiming that they can provide an alternative, more legitimate government for a country. But whereas quantitative, behavioral coding practices try to sidestep or settle such "partisan" disagreements in the name of objectivity and objective, less interpretation-laden assessments, the CEWS representational format highlights with the use of bifurcated paths/trajectories, and treats as valuable data, such differences when they speak of genuinely different assessments of possible historical developments.

The CEWS web site's prototypical representational strategy for computerized case-history storage, retrieval, and analysis follows the Sylvan-Majeski tradition. It takes a Kripkean essentialist approach to phase identification: dispute, crisis, limited violence, massive violence, abatement, and settlement phases are constitutively defined in terms of oppositions, the level of violence, and sequential expectations about violence levels. Each phase also can be characterized in terms of sometimes manipulable, transition-relevant, contingent characteristics, such as the undertaking of reforms during a dispute phase, the suppression of opponents during a crisis phase, or the making of concessions in an abatement phase. Conflict trajectories composed of different phase sequences are new entities. Alternative possibilities for different phases, or movement to new episodes, are latent within an analytically framed, but historically derived, three-phase-sequence-based grammar of sequential possibilities. Possibly hopeful precedents can be searched for: one can ask what other cases in the extensible database have the same phase sequence dyads, and see what trajectory triads they produced. Searches in such a virtual world of historical possibilities can suggest historically plausible, possibly less violent, alternative third-phase sequential possibilities.

How are alternative pathways represented and explored? A more complex characterization of historical possibilities allows the past-looking history also to be redefined, as in conflict transformations—indicated in the figure and elsewhere by backward arrows and big horizontal braces. Peace, "truth," and "reconciliation" commissions often attempt to revise national "myths" in terms of such rewritings of intergroup histories. More modest versions of that kind of process involve a government's entering into negotiations with guerillas whose position and concerns were not previously, officially recognized—the labeled brace in figure 1. But what about reactionary governments that balk at partial reinterpretations of the parties and their past interactions in a peace process they reject? According to present coding conventions, such conflict-exacerbating past revisionism would merit a brace as well, if it where shared by the principal parties. Under more inclusive coding rules concerning the rewriting of past histories, a bigger, wider brace could be rewritten under the whole first episode, trumping the smaller brace above it, etc. Historical politics like this needs to be represented, as we have done, in terms of the *contested historicities*—the time-ordered self-understandings of continuing human groups, parties, or societies.

Two more ways of treating trajectories, their determinants and alternatives, should be mentioned. What CEWS did was somewhat different from more conventional statistical approaches;[22] its alternative orientation was linked to its judgmentally oriented, intervention-sensitive, narrative counterfactuals approach. Narrative/chronology constructors were asked to indicate points of intervention where paths might have been redirected in less violent directions. Historical actions accounting for these and other phase shifts where also to be noted, when it was possible to identify them. Italic labels between phase and between episode paths respond to this crucial analytical suggestion, and the inadequacy of the narrative/chronology's account of such transitions is indicated by question marks or the absence of labels. In Stavenhagen's Chiapas narrative account, for example, we were unable to find sufficient information as to why and how in early 1995 the government broke off negotiations and initiated episode 2, starting in a low-violence stage 3 with surprise attacks, nor how and why these eventually led to a new abatement phase in which a new guerilla agent, the EZLN, was recognized, and a new, never implemented, San Andres Accord was tentatively reached. As is often the case in historical research, here further situation-specific investigation is pointed to and needed, including additional possible intervention points suggested by other, knowledgeable researchers seeking neglected peace possibilities.

The second way alternative trajectories were tracked and partially explained in a constitutive fashion was through the further theoretical/empirical specification, and historical identification, of contingent phase subtypes (Alker, Gurr, and Rupesinge 2001: table 12.1, 364ff.). For example,

[22] The CEWS project did a preliminary empirical validity test of early warning indicators, taking a look at the anticipatory power of indicators derived from a large quantitative-empirical study, sponsored by the U.S. government, of "Failed States" (Esty et al. 1999; see also King and Zeng 2001); Alex Schmid compared the prognostic power of these state failure indicators with alternative early warning indicators suggested and used by a Dutch human rights early warning organization he directed, and found the former to be slightly more accurate. There is no fundamental difficulty in incorporating such findings into the CEWS approach, although the context specificity of such statistically oriented findings needs further investigation by those not willing to sidestep "as if random" errors. Computationally, and visually, one could overlay the results of such studies, appropriately reconfigured, on figures like figure 1. It should be further noted that Esty et al. took a big step forward methodologically in applying nonlinear, neural net estimation/construction computational algorithms to their sequential data, an approach also applied, more systematically, by King and Zeng, with suggestive, richer results.

phase 1 in episode 1 of the Chiapas crisis, as represented in figure 1, is not business as usual in that it has associated with it the expectation of a possible subsequent crisis phase. On the basis of Stavenhagen's account, one might reasonably judge this to be codified as a phase subtype 1a, a dispute subphase associated with the "separation [of a protesting group] from its opponent," indicated by the mention of indigenous groups' demands for the recognition of their cultural identity, but taking place within the existing political system. The sensational announcements of the existence and program of the Zapatista/EZLN guerillas also involves a subsequent phase 2, which is distinguished as phase subtype 2a, characteristically involving the formation of separate armed groups not willing to work within the system.

This kind of more detailed and differentiated set of distinctions could also be annotatively added to figure 1. Such phase subtype discriminations are used extensively by Schmalberger's CEWS Explorer LISP/SCHEME software (Alker, Gurr, and Rupesinghe 2001: chap. 12) in suggesting possible alternative conflict trajcctories on the basis of historically similar precedent cases.

Conclusion

Diplomats in the UN system have regularly talked about being driven by present crises, about the difficulty of being able to develop, train, fund, and position capabilities—like rapid-reaction peacekeeping forces—generally available for unpredicted emergencies. Their capacities for command and control, and for logistical supply and multiunit force coordination, are minimal. The lead nation of an ad hoc peacekeeping or peace-building force takes primary responsibility in these areas, perhaps with help from the United States or another great power. The UN's High Commissioner for Refugees has developed considerable capabilities for coordinating responses to humanitarian crises but has shied away from strictly early warning functions as too political.

Obviously, the United Nations is largely dependent on the will of its members, and the will of its members has been very reluctant to give it significant, autonomous capabilities in a world where knowledge can be power. Compare the funding of UN management efforts concerning its peacekeeping missions with the costs of the contributions, both lethal and constructive, of individual national efforts. Ratios of 1:10 or 1:100 seem appropriate here. Even in design-oriented research pointed toward improving early warning forecasting capabilities, the high-quality "failed states" project had a level of U.S. governmental funding surely ten or more times the level of CEWS's financial support or of the computerized information processing budgets of the now defunct UN Secretariat Office for Research and Collection of Information. The United Nations had a small office that used some of the same methodologies of the failed states project; when I visited it in the late 1990s, it has only two or three professional-level technical staff.

National governments of wealthy and powerful states, on the other hand, have many more resources for such purposes, but the failures of central United States decision makers to predict the breakup of the Soviet Union, the testing of nuclear weapons by India and Pakistan, or the attacks of September 11, 2001, show how far the world is from an effective, widely accepted, and widely implemented collective capacity to anticipate major security threats to its major states, let alone its more numerous groupings of nations and peoples.

Groups like FEWER have only modest support; they are better financed than most small individual nations' general early warning efforts, and able to do pretty decent work in the specific local regional areas they are focusing on. In particular, FEWER uses Internet, telephone, and fax networks of trained observers for this purpose in regions where it is active, such as the Great Lakes region of East/Central Africa, and in parts of the former Soviet Union. Although I have not made a detailed evaluative study of their anticipatory achievements, the continued existence of International Alert and FEWER in particular is a sign that at least their sponsors—governments and foundations—deem their contributions worthwhile. This is a hopeful sign.

Even though CEWS-type historical information systems have so far not been deployed in the UN-centered conflict-prevention domain, cruder data-based practices have begun to be selectively implemented, by FEWER and the Canadian government, for example. What these early warners say, however, is that they need much more work on learning how to catalyze and sequentially compose effective early responses to situations of potential violence that they are aware of. CEWS has made only a small beginning in this regard: at best, separate analyses of efficacious strategies could be wedded to its graphical conflict trajectories, perhaps as annotations or commentaries placed on top of underlying trajectory figures. Implementations of second-generation versions of CEWS's data-making and reanalysis practices could, I believe, significantly enhance the conflict-handling capacities of both intergovernmental and NGO members of a complexly interdependent world society; perhaps they could also be used to make more selective and effective the calls of "wolf" that so many governments do not want to hear until disaster has already struck.

Excepting the increased availability and use of virtually instantaneous world news coverage, and associated technologies of digesting such high-volume news sources,[23] the newer technologies of information and communication have had more impacts on the *design and development of prototype systems* for early warning concerning intergroup conflict than on the effectiveness of cross-boundary early warning *practices*. Good, detailed, sharply patterned data on the effectiveness of specific, violence-anticipating and preventing interventions by individual, NGO or IGO organizations is hard to come by; successful preventive diplomacy, by its nature, deals with specific detail, and is often nearly invisible. To call attention to such successes can embarrass those who have made important concessions, hence it is not good peacemaking to single them out. The building of better, relevant, usable but discrete institutional memories is a hard, but no less important, task.

The CEWS project was a modest effort to make a UN-oriented change in such practices, seeking to empower peacemakers affiliated with NGOs trying to improve upon or supplement the UN's efforts at preventive diplomacy. The main contribution of the CEWS project to early warning practices has been the making available of an extensible, prototype early warning information system. It encodes and retrieves extensible, historically focused, computationally enhanced, collective memories, expressed in potentially suggestive, annotable, summary visual and narrative representations.[24] Many more cases, more analyses of interventions, and eventual integration with empirically oriented studies seeking to discover effective intervention mixes are all needed for the successful development of second-generation CEWS-type systems. Although IA and FEWER are ongoing organizations, the CEWS modular, networkable, information systems design has yet to be implemented.[25] If and when it is, strategically sensitive versions of such systems will take into account the possible responses of conflict management "spoilers" interested more in their own gains than in the peaceful resolution of a conflict.

What differences, more generally, have conflict-oriented early warning

[23] See Davis and Gurr (1998) for a fairly systematic survey of different violence-oriented early warning practices within IGOs, NGOs, and the American government. CEWS early decided not to focus on media-based, short-term warning events data approaches, for which businesses and governments struggle to keep ahead of CNN reporting by a few minutes or a few days. Several authors associated with the project looked at longer term, computerized uses and analyses of such data, as well as other data forms.

[24] Compare Manuel Castells (1998: 462): "The timelessness of multi-media's hypertext is a decisive feature of our culture. . . . History is first organized according to the availability of visual seconds of frames to be pieced together, or split apart, according to specific discourses." I disagree with Castell's description of postmodernity in terms of the suspension of recursive rewrite procedures.

[25] Thus the jury is still out whether digital formations built out of co-evolving technologies and NGO networks are likely to occur in the intergroup conflict early warning domain, as proposed by Bach and Stark in this volume.

ICTs made in a world of international relations, a world repeatedly evidencing the features of extended, complex interdependence? Volumes can be, and are being, written about the revolution in military affairs associated with the development of domestically and internationally oriented semi-automated surveillance systems, and the design and development of "smart" weapons (e.g., Latham 2003). The United States spends many billions of dollars on signals intelligence annually. The reorganization of U.S. military strategy and force deployments sought by Secretary of Defense Rumsfeld has many features associated with network-oriented re-understandings of locally effective fighting forces, antiterrorist strategies, as well as nonstate terrorist organizations.

More ominous have been the successes and failures of the reconnaissance and targeting technologies used by American forces to awesome effect in their only partly UN-sanctioned efforts in Kosovo, Kuwait, and Afghanistan, and the even more technologically advanced ballistic missile early warning systems whose development was a moving force behind George W. Bush's administrative commitment to withdrawal from the ABM treaty with the Soviet Union (and its successor Russia). The minutes or hours in which such warnings must be responded to, combined with the level of violence associated with such increasingly automated signals and responses, raise issues whether dangerously "closed worlds" are being built which keep decision makers away from crucially relevant human features of the issues and conflicts in question.[26]

I wish here briefly to note some design features of the CEWS effort that are specifically tied to preventing potential deployers of a fully implemented CEWS-type ICT-resourced network from falling into such traps. First, the reliance on local peacemakers as ongoing input-givers to early warning NGOs provides just the kind of intelligence that the "smart" bombers of the Chinese Embassy in Belgrade did not have; this kind of information can be kept confidential, if necessary. Second, the system is extensible in two important ways: first, both new cases and contesting reinterpretations of existing case analyses can be contributed to the system. A minimal but enlargeable provision for overlays representing different perspectives on important essential or contingent features of past conflicts exists on the web site. The hermeneutic style of coding and recoding conflict trajectories, allowing for grounded differences in telling conflict histories in any initial representation, are all ways of sensitizing CEWS users to the importance of different perspectives, including those identified with different parties to a particular, ongoing conflict. The full CEWS network

[26] For relevant discussions, see Edwards (1996), Der Derian (2001), and the related, extensive discussion of the roles of Norbert Weiner, Jay Forrester, Herbert Simon, and John von Neumann in Mirowski (2002).

design—to which Rupesinghe, IA, and FEWER were important contributors, which the CEWS project assumed and did little to contribute to—was and is a humanistic research methodology designed to capitalize on fundamental differences in perspectives, not computationally to paper over such differences.

Moreover, I want to suggest that even though complex adaptive systems theory has many connections with the institutional development and research funding of the American military establishment, the technologies that I have tried to input to the computational part of such a broader CEWS-type network should be seen as usable for other than military purposes as well. The user should keep to their open, extensible, mixed human-machine character, and their multiple narrative encodings. These are a real improvement over merely quantitative representations!

Additionally, the complex adaptive systems framework that Axelrod and Cohen (1999) so brilliantly deploy—in a book that was at least partly funded by the U.S. Defense Department—is highly suggested regarding next steps in the conflict early warning effort. I can suggest just some of the kinds of questions that Axelrod and Cohen link to the framework of schema (1a,b) above, turning them into the area of further research on the selection, mutation, and transformation of early warning practices. For example, what interventions, by which governmental agencies, NGOs or foundations, account for an increased variety of forecasting or intervention strategies? Why are these agents themselves increasing or decreasing in size or power? In the competition for conflict manager of the year, or the decade, why has the United Nations gone up and down in the number of cases it has handled? What competences or incompetences play a role in the selection process for conflict management practices? How about UNHCR, the Red Cross, NATO, or the Organization of African Unity and their descendants? Within the most powerful actors, who or what is helping to reproduce, that is, help select, which patterns of neglect toward festering problems? Which processes of copying and recombining tactics of intervention result in the overmilitarization of some states' intervention dispositions, as compared with those processes increasing the frequency for some states of playing peacekeeping and developmental assistance roles? How can we design a selection process that identifies and reward improved early warning and early response behaviors?

Axelrod and Cohen use their framework to suggest (re-)design possibilities;[27] similarly, I suggest the following. Arrange organization-specific conflict-monitoring routines in a way that nicely mixes exploration of new

[27] See especially their last, summary chapter, from which many of the phrases in this paragraph are taken. Of course, I should be held responsible for my selections from, and reinterpretations of, such material.

modes of conflict diminution with the exploitation of proven approaches. Given their relative virtues, build networks of reciprocal interaction that foster trust and cooperation among the different kinds of conflict monitors. Assess strategies for conflict transformation in terms of how their consequences might be spread. Promote effective neighborhoods in which would-be cooperators in conflict containment can more easily recognize the role each other can play. Figure out what else besides the Nobel Peace Prize can be used to support the growth and spread of peacemaking activity. And in a world where small successes can be quiet stepping stones to a better future, look for fine-grained success measures, such as promotion criteria and associated selection processes within early warning and response organizations that can usefully stand in for bigger, longer-range goals like the gradual transformation of world society. Isn't that a journey others ought to continue?

References

Abelson, R. P. 1973. The Structure of Belief Systems. In *Computer Models of Thought and Language,* edited by R. C. Schank and K. Colby. San Francisco: W. H. Freeman.

Adler, E., and M. Barnett 1998. *Security Communities.* Cambridge: Cambridge University Press.

Alker Jr., H. R. 1974. Are There Structural Models of Voluntaristic Social Action? *Quality and Quantity* 8:199–246.

———. 1975. Polimetrics: Its Descriptive Foundations. In *Handbook of Political Science,* vol. 7, edited by F. Greenstein and N. Polsby, 139–210. Reading, MA: Addison-Wesley.

———. 1977. A Methodology for Design Research On Interdependence Alternatives. *International Organization* 31:29–63.

———. 1986, 1988. Bit Flows, Rewrites, Social Talk: Towards More Adequate Informational Ontologies. Published in the Proceedings of the Centennial Conference of Todai University, *Information and its Functions;* reprinted in *Between Rationality and Cognition,* ed. M. Campanella. Turin: Meynier, 1988.

———. 1996. *Rediscoveries and Reformulations: Humanistic Methodologies for International Studies.* New York: Cambridge University Press.

———. 2000. Learning from Wendt. *Review of International Studies* 26:141–50.

Alker Jr., H. R., L. P. Bloomfield, and N. Choucri, 1974. *Analyzing Global Interdependence.* Cambridge: Center for International Studies, MIT.

Alker, Jr., H. R., and C. Christensen. 1972. From Causal Modeling to Artificial Intelligence: The Evolution of a UN Peace-making Simulation. In *Experimentation and Simulation in Political Science,* edited by J. A. Laponce and P. Smoker. Toronto: University of Toronto Press.

Alker, Jr., H. R., and W. J. Greenberg. 1971. The UN Charter: Alternative Pasts and Alternative Futures. In *The United Nations: Problems and Prospects,* edited by Edwin H. Fedder. St. Louis: Center for International Studies, University of Missouri.

Alker, H. R., T. R. Gurr, and K. Rupesinghe, eds. 2001. *Journeys through Conflict: Narratives and Lessons*. Lanham, MD: Rowman & Littlefield.

Axelrod, R. 1984. *The Evolution of Cooperation*. New York: Basic Books.

Axelrod, R., and M. Cohen. 1999. *Harnessing Complexity: Organizational Implications of a Scientific Frontier*. New York: Free Press.

Axelrod, R., and R. O. Keohane. 1985. Achieving Cooperation under Anarchy: Strategies and Institutions. *World Politics* 38(1): 226–54.

Boutros-Ghali, B. 1992. *An Agenda for Peace: Preventive Diplomacy, Peacemaking and Peacekeeping*. New York: United Nations.

Burckhardt, J. 1990 (1860). *The Civilization of the Renaissance in Italy*. London: Penguin.

Castells, M. 1998. *The Information Age, Economy, Society and Culture*. Vol. 3: *End of Millenium*. Malden, MA: Blackwell Publishing.

Collingwood, R. G. 1994 (1946). *The Idea of History*, edited by Jan van der Dussen. Rev. ed. Oxford: Oxford University Press,.

Davis, J., and T. R. Gurr, eds. 1998. *Preventive Measures: Building Risk Assessment and Crisis Early Warning Systems*. Lanham, MD: Rowman and Littlefield.

Der Derian, J. 2001. *Virtuous War: Mapping the Military-Industrial-Media-Entertainment Complex*. Boulder: Westview Press.

Deutsch, K. W., et al. 1957. *Political Community in the North Atlantic Area*. Princeton: Princeton University Press.

Edwards, P. 1996. *The Closed World*. Cambridge: MIT Press.

Esty, D. C., et al. 1998. The State Failure Project: Early Warning Research for U.S. Foreign Policy Planning. In *Preventive Measures*, edited by J. Davis and T. R. Gurr. Lanham, MD: Rowman and Littlefield.

Fogel, L. J., A. J. Owens, and J. J. Walsh. 1966. *Artificial Intelligence through Simulated Evolution*. New York: Wiley & Sons.

Greenstein, F., and N. Polsby, eds. 1975. *Handbook of Political Science*. 7 vols. Reading, MA: Addison-Wesley.

Haas, E. B. 1968. *Collective Security and the Future International System*. Denver: University of Denver Monographs.

———. 1993. Collective Conflict Management: Evidence for a New World Order? In *Collective Security in a Changing World*, ed. T. G. Weiss. Boulder: Lynne Rienner.

———. 1996. *Nationalism, Liberalism, and Progress*. Ithaca: Cornell University Press.

Habermas, J. 1970. Science and Technology as Ideology. *Toward a Rational Society*. Boston: Beacon Press.

———. 1971. *Knowledge and Human Interests*. Boston: Beacon Press.

———. 1979. *Communication and the Evolution of Society*. Boston: Beacon Press.

———. 1984, 1987. *Theory of Communicative Action*. 2 vols. Boston: Beacon Press.

Harré, R., and P. Secord. 1972. *The Explanation of Social Behavior*. Totowa, NJ: Rowman and Littlefield.

Hodgson, M.G.S. 1993. *Rethinking World History.* Cambridge: Cambridge University Press.
Jervis, R. 1985. From Balance to Concert. *World Politics* 38(1): 58–79.
Keohane, R. O. 1984. *After Hegemony.* Princeton: Princeton University Press.
———. ed. 1986. *Neorealism and Its Critics.* New York: Columbia University Press.
Keohane, R. O. and J. S. Nye, Jr. 1974. Transgovernmental Relations and International Organizations. *World Politics* 27:39–62.
———. 1975. Integration and Interdependence. In *Handbook of Political Science,* edited by F. Greenstein and N. Polsby. Vol. 7. Reading, MA: Addison-Wesley.
———. 1977, 1989, 2000. *Power and Interdependence.* Boston: Little, Brown.
King, G., and L. Zeng. 2001. Improving Forecasts of State Failure. *World Politics* 53(4): 623–58.
Lasswell, H. D., and A. Kaplan. 1950. *Power and Society: A Framework for Political Inquiry.* New Haven: Yale University Press.
Latham, R., ed. 2003. *Bombs and Bandwith.* New York: The New Press.
Lindblom, C. E. 1965. *The Intelligence of Democracy: Decision Making through Mutual Adjustment.* New York: Free Press.
Mirowski, P. 2002. *Machine Dreams: Economics Becomes a Cyborg Science.* Cambridge: Cambridge University Press.
Onuf, N. 1989. *World of Our Making: Rules and Rule in Social Theory and International Relations.* Columbia: University of Southern Carolina Press.
Oye, K. A. 1985. Explaining Cooperation under Anarchy: Hypotheses and Strategies. *World Politics* 38(1): 1–24.
Rupesinghe, K., ed. 1995. *Conflict Transformation.* Basingstoke, UK: Macmillan Press.
Rupesinghe, K., with S. N. Anderlini. 1998. *Civil Wars, Civil Peace; Introduction to Conflict Resolution.* London: Pluto Press.
Rupesinghe, K., and M. Kuroda, eds. 1992. *Early Warning and Conflict Resolution.* London: Macmillan Press.
Schank, R. C., and R. P. Abelson. 1977. *Scripts, Plans, Goals and Understanding: An Inquiry into Human Knowledge Structure.* Hillsdale, NJ: Lawrence Earlbaum Associates.
Simon, H. A. 1957. *Models of Man.* New York: Wiley & Sons.
———. 1969. *The Sciences of the Artificial.* Cambridge: MIT Press.
Skinner Q. 1978. *The Foundations of Modern Political Thought.* Vol. 1: *The Renaissance.* Cambridge: Cambridge University Press.
Snidal, D. 1985. The Game Theory of International Politics. *World Politics* 38(1): 25–57.
Stedman, S. J. 1997. Spoiler Problems in Peace Processes. In *Synergy in Early Warning,* edited by S. Schmeidl and H. Adelman, 333–374. Toronto: Centre for International and Security Studies, York University.
Sylvan, D., and S. Majeski. 1998. A Methodology for the Study of Historical Counterfactuals. *International Studies Quarterly* 42(1): 79–108.
Toulmin, S. 1990. *Cosmopolis: The Hidden Agenda of Modernity.* Chicago: Chicago University Press.

———. 2001. *Return to Reason*. Cambridge: Harvard University Press.
Wallensteen, P. 1988. *Peace Research: Achievements and Challenges*. Boulder: Westview.
Wendt, A. 1992. Anarchy Is What You Make of It: The Social Construction of Power Politics. *International Organization* 46 (Spring): 391–425.

Discourse Architecture and Very Large-scale Conversation

WARREN SACK

HISTORICALLY, NEW spaces for public discussion have been invented every few centuries (the agora, plaza, town square, town hall, café, newspaper). The introduction of electrical and electronic technologies in the twentieth century accelerated the rate of change in public spaces to a pace measured in decades (film, radio, television). Now with the increasing ubiquity of computer networks, new spaces for public discussion and exchange are invented, introduced, and updated on an almost continual basis (e-mail, newsgroups, Internet Relay Chat (IRC), weblogs, instant messaging, Napster, Gnutella).

This exponential increase in the rate of change has reached the escape velocity of the disciplines and professions normally accorded the responsibility to design, build and analyze public spaces. No longer is it only architects, civil engineers, and urban planners who design spaces for public discussion. Symptomatic of this transformation is a proliferation of new architectures of computers and networks that are not designed by traditional architects, for example, computer architectures, network architectures, and information architectures. Conversely, traditional architecture has become increasingly involved in efforts to extend its methodologies to cover computer networks by rendering them as "cyberspaces."

The gaps between discourse, code, and architecture have now been bridged to the extent that it is crucial for us to understand issues such as the legal ramifications of network architectures on free speech.[1] Today public spaces for discussion include bits as well as bricks and boards. This convergence of language and architecture has frequently produced an assemblage that fails like the Tower of Babel. Discourse specialists (linguists, sociologists, legal scholars, political scientists) have not often enjoyed the reputation of great designers of spaces and architectures. On the other hand, artists, designers, engineers, and architects—renowned for their abilities to envision and execute the configuration and mixing of spaces

[1] Cf. Lessig (1999).

and materials—have often been typified as inept in the skills of writing and speaking. But we are now at a point in time when the future of the public space depends upon the ability to mix discourse and architecture in a new area of endeavor called *discourse architecture*.

Network architecture is the computer science of connecting machines to machines. Information architecture is primarily practiced by librarians and database and web designers to connect people to machines by making it easy for people to find information on networked machines. Discourse architecture is the practice of designing environments to connect people to people through networked computers. Or, more specifically, discourse architecture is the practice of designing networked environments to support conversation, discussion, and exchange among people.

Prior work in this area includes that of the original Discourse Architecture Laboratory, a research group at Apple Computer.[2] Closely related is a large variety of work in Computer-Supported Cooperative Work (CSCW), Computer-Human Interaction (CHI),[3] and Computer-Mediated Communication (CMC). Most recently a number of research groups have emerged to focus on what has been called *social computing* and *social informatics*. Groups of this sort now exist at a number of industry research labs, universities and nonprofit organizations. Unlike many scholars who work in CSCW and CMC, researchers in the area of social computing have identified earlier work in architecture and urban design as useful and interesting for the design of networked spaces. Discourse architecture is an area of social computing in which environments for discussion are of primary importance.[4]

The practice of discourse architecture entails two kinds of work: one concerns the extension and use of methods from art and design; the second employs and further develops ideas from the humanities and social sciences. First, as a practice of design, discourse architecture concerns the design and implementation of new computer network technologies for discourse; that is, the means to shape the conversation that takes place within a given system. Just as physical architecture facilitates certain activities and inhibits others (compare, for instance, the exchanges sup-

[2] Founding members of this research group at Apple included Dave Curbow, Paul Dourish, Tom Erickson, Jed Harris, and Austin Henderson, with consulting help from Niklas Damiras, Sha Xin Wei, Brian Cantwell Smith, and Helga Wild. See http://pliant.org for more information about this group.

[3] See especially Erickson, Herring, and Sack (2002) and Munro, Hook, and Benyon (1999).

[4] The following definition of discourse architecture is a direct outgrowth of the writing I have done together with Susan Herring and Thomas Erickson. The following paragraphs should be compared with our co-authored work on the subject (Erickson, Herring, and Sack 2002). However, Herring and Erickson may not agree with the version that appears here.

ported by amphitheaters versus those supported by cafés), so do system architectures facilitate certain types of conversations. For example, media architectures like television broadcasting facilitate one-to-many exchanges but do not directly support a democratic, many-to-many exchange between people. In contrast, the Usenet newsgroup network protocol, for instance, does support many-to-many exchanges. Prior work exists in the fields of architecture, urban design, and the arts.

Second, the criteria for evaluating any given discourse architecture depends upon some means to critique the form, character, content, and extent of the supported discourse. Thus, discourse architecture is concerned with the structure of conversation itself; that is, with how the utterances of a conversation interrelate and build upon one another. Discourse architects are interested in analytical techniques for identifying conversational structure and explicating the forces that shape it. Relatively little research has been done to understand how network architectures influence existing patterns of discourse or facilitate new patterns. Furthermore, the work that has been done is spread across a wide array of humanities and social science disciplines such as linguistics, literature, theater, philosophy, anthropology, communications, computer science, information science, political science, psychology, rhetoric, and sociology and draws on diverse theories and methods. Consequently, the practice of discourse architecture entails the extension, synthesis, and production of new knowledge appropriate to disciplines of the social sciences, arts, and humanities.

This chapter is an introduction to discourse architecture. It is an introduction by example. First a new area of discourse is identified; an area that will be referred to as *very large-scale conversation* (VLSC). It is usually conducted on the Internet through the exchange of e-mail. VLSC facilitates many-to-many exchanges among citizens across international borders. I argue that VLSC poses a fundamental challenge to existing social science methodologies because it constitutes a different scale of conversational interaction, a scale that has not been previously addressed by social science. I propose a computationally enabled means to understand and theorize VLSC and illustrate this proposal with a prototype piece of software, the Conversation Map. Finally, I argue that the Conversation Map is not just a tool but also a *technology of the self*, a means of self-reflection.

Very Large-scale Conversation

On the Internet there are now very large-scale conversations in which hundreds, even thousands, of people exchange messages across interna-

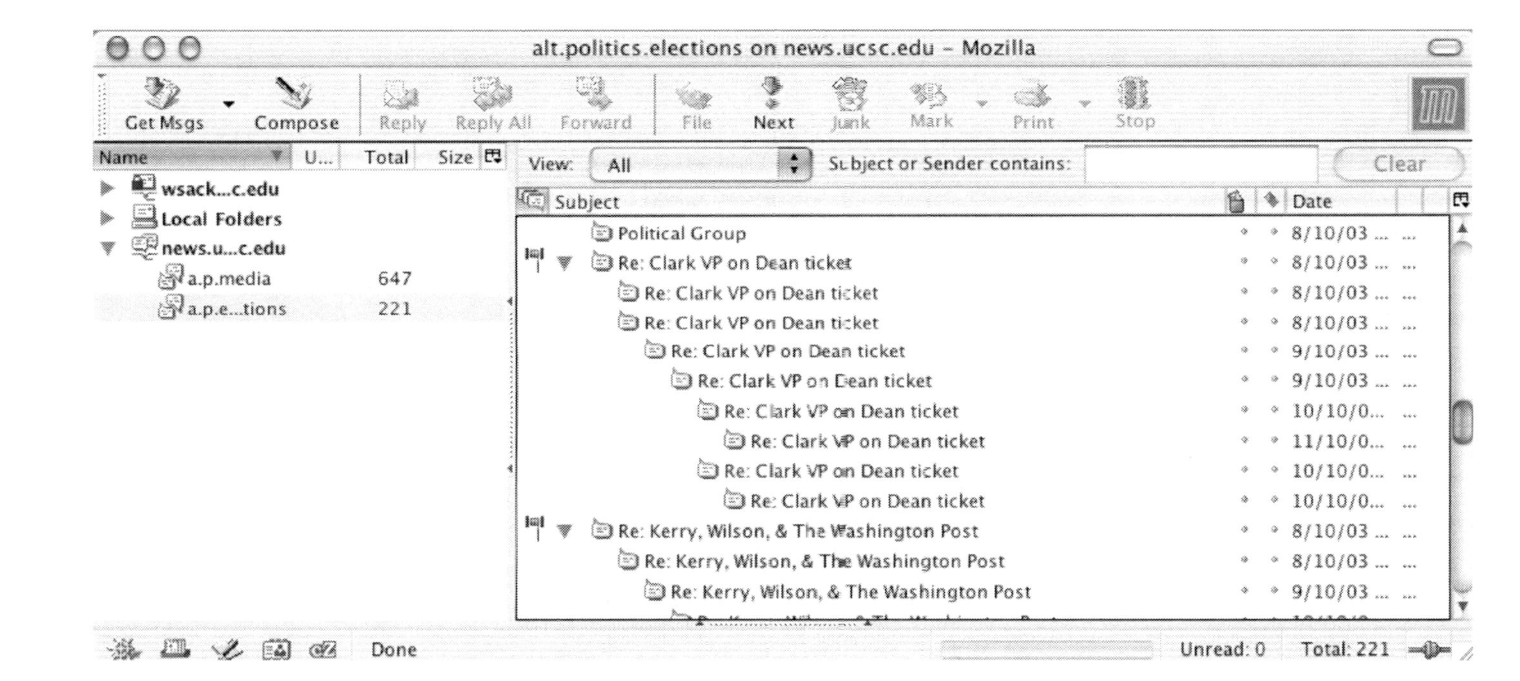

Figure 1. Mozilla News—a typical, contemporary view of VLSC

tional borders in daily, many-to-many communications. VLSC is an emergent communication medium that engenders new social and linguistic connections among people. It poses fundamental challenges to the analytical tools and descriptive methodologies of social science previously developed to understand conversations of a much smaller scale.

VLSC is both a well-known phenomenon and, simultaneously, something as yet largely unexamined by designers and social scientists. On the one hand, VLSC is well known in the form of busy Usenet newsgroups and large, electronic mailing lists and weblogs.[5] For participants and observers alike, VLSC manifests itself as huge lists of messages in a conventional e-mail reader like RN, Eudora, or Netscape Messenger.

On the other hand, VLSC is largely unexamined. What does it mean to have a conversation that involves hundreds or thousands of people? Existing theories of conversation and discourse do not cover this scale of conversation. Moreover, very little design work for VLSC has been done. For example, why is VLSC usually represented as a long list of e-mail messages? Isn't something better possible? In fact, with a better theory of VLSC, better software for navigating VLSCs can be designed.

Detailed, micro analyses of face-to-face conversation usually involve a very different kind of work and produce a very different type of research result—that is, a very different type of knowledge—than do macroscale analyses of discourses involving thousands or millions of people. This micro/macro divide is a recurrent one in many of the social sciences and has been widely discussed in, for example, economics and sociology. Bridging this divide for the analysis of VLSCs is necessary because, on the one hand, the phenomenon under examination is macroscale by definition, but, on the other hand, one of the most ethically important motivations for analyzing VLSCs is to give participants a means to find their way and locate their position in a VLSC. Consequently, standard social scientific methods of dealing with macroscale phenomena by working with norms and averages are unworkable because they risk effacing the contributions of particular individuals.

I will argue that a bridge can be found between micro- and macroscale analyses of online conversations. This bridge is the lexicon or what might be called the "thesaurus" of a group conversation. On the micro scale, contributions to a conversation are judged to be coherent and cohesive partially according to whether or not they are taken to be "on topic" by the participants. Knowledge of deviation or convergence with a given topic is based on knowledge of a lexicon; that is, according to the rela-

[5] Readers unfamiliar with these forms of online exchange might be interested in exploring http://www.google.com/grphp, an index of tens of thousands of Usenet newsgroups; http://groups.yahoo.com/, a selection of tens of thousands of web-based discussion groups; and/or http://blogdex.net/, an index of thousands of weblogs.

tionships between and the definitions of words. But, over the course of the lifetime of a group, new (e.g., slang) words are coined, some words gain new meanings, and others lose their currency, their connotations, or the controversy that surrounds them. Thus, conversation both depends upon and changes the lexicon or "thesaurus" of a group.

This conceptualization of VLSC—as the substrate and catalyst of community—is concordant with a large amount of work in sociolinguistics and the sociology of knowledge. Roughly speaking, what characterizes many of these sociolinguistic and sociological approaches to conversation and discourse is this: through the production and reproduction of a way of speaking and/or writing about certain pivotal subjects, a group is formed and distinguishes itself from other groups. Thus, chemists in the eighteenth century distinguished themselves from alchemists by developing a new discourse that we now recognize as the science of chemistry. Rather than talking about water as an essential element, chemists talk of the combination of hydrogen and oxygen. So, a new way of speaking and writing simultaneously produces a new group (e.g., chemists) and unravels or divides itself from a preexisting group (e.g., alchemists).

A way of speaking and writing (re)produces limits and possibilities for the way a subject can be spoken and/or written about and, simultaneously, (re)produces a social structure (e.g., a group or community). This way of thinking about the process and product of verbal interaction is well known in, for instance, conversation analysis.[6] This way of describing the product or production of written and conversational forms has been termed "a discourse" by various European "continental" theorists:

> [C]ontinental discourse theorists such as Foucault, Lyotard, Donzelot, Pêcheux, and De Certeau tend to use the term "discourse" to refer to relatively well-bounded areas of social knowledge. So, at any given historical conjuncture, it is only possible to write, speak, or think about a given social object (madness, for example) in specific ways and not in others. "A discourse" would then be whatever constrains—but also enables—writing, speaking, and thinking within such specific historical limits. Thus while a discourse can be thought of as linguistic in one sense, it also has to be treated in terms of the conditions of possibility of knowing a specific social object. (McHoul 1994: 944–45)

[6] Harvey Sacks, the inventor of conversation analysis, provides several examples of the coconstruction of a lexicon and a social group: "Now if we can take it that to some extent 'hotrodders' is a category that is by and large employed by kids to characterize themselves, and whose use, to some considerable extent, they enforce, and whose properties they enforce, and obviously it's, at least to some extent, a category that rebellious persons can use, then at least one of the initial questions we might ask is: Why should it be the case that at least some people who go about doing kinds of rebellion, do it by formulating themselves as a particular type? That is, why do they set up a type? Why don't they try to make themselves observable as 'individuals,' so to speak?" (1992: 1:172).

From this continental perspective it is therefore possible to talk about, for instance, "the discourse of chemistry." This usage of the term "discourse" (i.e., the use of the term discourse preceded by a definite or indefinite article like "the" or "a") is sometimes at odds with or appears more or less incomprehensible to practitioners of other sorts of Anglo-American forms of discourse analysis.[7]

In her book that compares and contrasts six different Anglo-American approaches to discourse analysis (speech act theory, pragmatics, ethnomethodology, interactional sociolinguistics, ethnography of communication, and variation theory), Deborah Schiffrin (1994: 339) states: "Discourse has often been viewed in two different ways: a structure, i.e., a unit of language that is larger than the sentence; and the realization of functions, i.e. as the use of language for social, expressive, and referential purposes." In other words, from an Anglo-American perspective, "discourse" is a name for a sequence of sentences (a structure) or a certain kind of language use (a function). But from a European, continental perspective, "discourse" is the result of either language use or the background conditions or context for a given sequence of sentences. Borrowing the trope of "figure/ground" from art history, one might say the difference between scholarly approaches to discourse analysis arises from the use of the term discourse to describe *figure* versus its use to describe *ground.* Or, alternatively, the conflict involves the use of discourse as a name for *text* versus its use as a name for *context.*

Rather than sort out this knotty conflation and conflict of terminology, I will try to find a way around it. From a continental perspective, one might talk about how a VLSC produces or reproduces a given or new discourse. From an Anglo-American perspective, one might say that a VLSC is a discourse. Instead, I will simply state that a VLSC produces, reproduces, and relies on a set of social and semantic relationships. In the language of mathematics, one might say that there exists a mutually recursive relationship between a VLSC and a set of social and semantic networks. Or, one might say, the coherence of a VLSC depends upon social and semantic background knowledge, but this background knowledge is also, at least partly, a product of the VLSC.

Three Dimensions of Conversational Commonsense

For conversations of a smaller scale (i.e., smaller than VLSC), it is possible to see when the background knowledge of a conversation is being

[7] Some of the practical implications of these incommensurable differences between Anglo-American and European approaches to discourse are described in Pennycock (1994).

abused or flouted. Commonsense, conversational background knowledge can be described in a variety of ways: as a set of common associations and common terms, as a set of social and semantic networks, or—as will be elaborated below—as a set of metafunctions named *interpersonal, textual,* and *ideational* by Michael Halliday (1994: 179).

Divergences or differences of routine, conversational background knowledge can produce misunderstandings and conflict, but they can also produce comedy. Consider the following one-liner from comedian Stephen Wright: "I was driving down the highway one morning and I saw a billboard advertising a restaurant that said 'Breakfast any time' so I stopped and ordered French toast in the Renaissance." The social coherence of a group underwrites conversation and depends upon a number of things. Semantics is just one of these things, but Wright's joke illustrates how the production of common terms—a shared semantics—is important to conversation.

If the terms of conversation are followed, but the conventional turn-taking "rules" are not, another sort of nonsense is produced. Lewis Carroll (1960: 97) illustrates the "rules" of riddles when he has the characters of *Wonderland* violate them:

> "Have you guessed the riddle yet?" the Hatter said, turning to Alice again.
> "No, I give it up," Alice replied: "what's the answer?"
> "I haven't the slightest idea," said the Hatter.
> "Nor I," said the March Hare.
> Alice sighed wearily. "I think you might do something better with the time," she said,
> "than wasting it in asking riddles that have no answers."

The common terms and rules of conversation are tightly coupled in the production of the cohesion of a conversation. When the cohesion is deliberately undone, the conversation is unhinged, as this snippet from Eugene Ionesco's (1958) absurdist play *The Bald Soprano* illustrates. Suddenly, in this dialogue sequence, all of the people being discussed are named Bobby Watson:

> MRS. SMITH: But who would take care of the children? You know very well that they have a boy and a girl. What are their names?
>
> MR. SMITH: Bobby and Bobby like their parents. Bobby Watson's uncle, old Bobby Watson, is a rich man and very fond of the boy. He might very well pay for Bobby's education.
>
> MRS. SMITH: That would be proper. And Bobby Watson's aunt, old Bobby Watson, might very well, in her turn, pay for the education of Bobby Watson, Bobby Watson's daughter. That way Bobby, Bobby Watson's mother, could remarry. Has she anyone in mind?

MR. SMITH: Yes, a cousin of Bobby Watson's.
MRS. SMITH: Who? Bobby Watson?
MR. SMITH: Which Bobby Watson do you mean?
MRS. SMITH: Why, Bobby Watson, the son of old Bobby Watson, the late Bobby Watson's other uncle.

Obviously, writers and comics know and bend the common terms and rules of conversation in order to produce these sorts of effects. Using insights of this sort, scholars like Roman Jakobsen (1985) have been able to explain the linguistic workings of avant-garde artistic literature, but the wittiness of more common performances also often depends upon an explicit understanding of how conversation engenders social cohesion and/or how the norms can be manipulated to reveal or break the underpinnings of social cohesion.[8] It is equally as obvious that anyone who finds

[8] In linguistics there exists a principle called Ziff's Law (Ziff 1960). Ziff's Law is the observation that any arbitrary string can be interpreted as a proper name. This is often mentioned as a serious difficulty for the construction of computer programs to parse natural language texts. However, it is also the main observation underlying Bud Abbott and Lou Costello's famous "Who's on first?" skit first performed on the *Kate Smith Radio Hour* in 1936:

COSTELLO: Look Abbott, if you're the coach, you must know all the players.
ABBOTT: I certainly do.
COSTELLO: Well you know I've not met the guys. So you'll have to tell me their names, and then I'll know who's playing on the team.
ABBOTT: Oh, I'll tell you their names, but you know it seems to me they give these ball players now-a-days very peculiar names.
COSTELLO: You mean funny names?
ABBOTT: Strange names, pet names . . . like Dizzy Dean . . .
COSTELLO: His brother Daffy
ABBOTT: Daffy Dean . . .
COSTELLO: And their French cousin.
ABBOTT: French?
COSTELLO: Goofé
ABBOTT: Goofé Dean. Well, let's see, we have on the bags, Who's on first, What's on second, I Don't Know is on third . . .
COSTELLO: That's what I want to find out.
ABBOTT: I say Who's on first, What's on second, I Don't Know's on third.
COSTELLO: Are you the manager?
ABBOTT: Yes.
COSTELLO: You gonna be the coach too?
ABBOTT: Yes.
COSTELLO: And you don't know the fellows' names.
ABBOTT: Well I should.
COSTELLO: Well then who's on first?
ABBOTT: Yes.
COSTELLO: I mean the fellow's name.
ABBOTT: Who.

these manipulations funny or absurd has a set of well-developed intuitions about the rules and terms of conversation: the *commonsense knowledge of conversation.*

Each of the comedic examples above illustrates a different metafunction of language. According to Michael Halliday (1994: 179), language has at least three meta-functions: (1) *ideational:* language can represent ideas; (2) *interpersonal:* language functions as a medium of exchange between people; and, (3) *textual:* language functions to organize, structure, and hold itself together; this function allows the various devices of cohesion, including citation, ellipsis, and anaphoric reference, to be used. The Steven Wright joke shows what can happen when the ideational metafunction breaks down. The selection from *Alice in Wonderland* illustrates the breakdown of the interpersonal metafunction. And in Ionesco's dialogue, the textual metafunction is thwarted by a breakdown of lexical cohesion. The point of these examples is simply to give examples of what might be considered the three different dimensions of commonsense knowledge about conversations that must be in place for a conversation—and so, transitively, a group of interlocutors—to hold together.

When one or all of these sorts of conversational background knowledge fall apart, the result can be funny.[9] But by citing only the absurd and the comedic, it is difficult to picture what can be lost if the terms or rules of conversation are questioned or broken. While these questions and breaks can be funny, they can also arouse anger or mistrust.

Harold Garfinkel (1967: 43–44) asked his students to document this, the breakdown of common terms assumed in conversation; that is, to document the breakdown of the *ideational* metafunction. In the course of everyday conversation, Garfinkel's students questioned the assumed, common terms. The results were graphic. In the following accounts, Garfinkel's students play the role of the so-called experimenter (E).

> The subject was telling the experimenter, a member of the subject's car pool, about having had a flat tire while going to work the previous day.
>
> (S) I had a flat tire.

COSTELLO: The guy on first.
ABBOTT: Who.
COSTELLO: The first baseman.
ABBOTT: Who.
COSTELLO: The guy playing . . .
ABBOTT: Who is on first!
COSTELLO: I'm asking you who's on first.
ABBOTT: That's the man's name.

[9] The violation of these sorts of commonsense knowledge can be seen as funny, as can the violation of a large variety of everyday expectations. See Freud (1960).

(E) What do you mean, you had a flat tire?

She appeared momentarily stunned. Then she answered in a hostile way: "What do you mean, 'What do you mean'? A flat tire is a flat tire. That is what I meant. Nothing special. What a crazy question!"

. . .

"On Friday night my husband and I were watching television. My husband remarked that he was tired. I asked, 'How are you tired? Physically, mentally, or just bored?'"

(S) I don't know, I guess physically, mainly.

(E) You mean that your muscles ache or your bones?

(S) I guess so. Don't be so technical.

(*After more watching*)

(S) All these old movies have the same kind of old iron bedstead in them.

(E) What do you mean? Do you mean all old movies, or some of them, or just ones you have seen?

(S) What's the matter with you? You know what I mean.

(E) I wish you would be more specific.

(S) You know what I mean! Drop dead!

. . .

The victim waved his hand cheerily.

(S) How are you?

(E) How am I in regard to what? My health, my finances, my school work, my peace of mind, my . . . ?

(S) (*Red in the face and suddenly out of control.*) Look! I was just trying to be polite. Frankly, I don't give a damn how you are.

These examples make the risks clear. By questioning the common terms of conversation, the students threaten the social contracts, or at least the smooth functioning, of various small groups of people: the car pool, the marriage, the friendship.

Questioning the common terms—the ideational metafunction of language—has risks. Analogously, there are risks to questioning the textual and interpersonal metafunctions. Using an ethnographic methodology, John Gumperz and his colleagues have documented how the textual and interpersonal metafunctions of language can break down in cross-cultural conversational situations. Consider the following utterances spoken by a Malaysian-born Indian immigrant in a London Adult Education class discussion about mortgages: "Mortgages. If you are to buy a house. Who can get and who cannot get. What assumptions we made, what? If you work. If you don't work, can you get a mortgage?" Gumperz et al. comment on this example: "the difficulties here are in following the connections that are being made, and consequently in understanding the intention of the final questions. [The] example starts with a string of noun phrases that appear to announce the intended topics. Is the final question intended to elicit a review of the assumptions made at another time, or is it the commencement of discussion of the topic of 'who can get and who cannot get' a mortgage?"

Of course, the "difficulties" that Gumperz et al. mention are their difficulties, not the difficulties of the speaker or her audience, who are also, largely, English-speaking Indian immigrants. Gumperz et al. show how the structures and resources of grammar, prosody, and intonation of Hindi, Urdu, Punjabi, Gujerati, and Marathi are employed to join together multiple sentences when speakers of these North Indian languages speak English. Thus, the difficulties in resolving the cohesion are mostly difficulties for Anglo-American English speakers, not English-speaking Indians and Pakastanis. Consequently, even in a situation where the language being spoken is English and everyone in the situation is perfectly fluent in English, cross-cultural ties cannot, at times, be created because the textual and interpersonal metafunctions are produced very differently by members of different cultural groups.

This can have grave repercussions in legal, medical, and employment situations. In such situations bilinguals are sometimes thought to not be telling the truth because their testimony seems to be self-contradictory when interpreted by monolinguals; or the bilingual does not receive the medical care they need because the doctor does not understand them; or the bilingual does not get the job because the monolingual thinks the bilingual is hard to understand. In other words, in such situations—unlike the example situations of Garfinkel—the social fabric of a group is not ripped; rather, the group or social relationship is never threaded together or is clipped off right from the start.

Obviously the ideational, interpersonal, and textual relations established through inter- and intracultural conversational interactions do not remain static. Some groups become closer knit over time. Others fall apart. Intercultural, multilingual interactions can produce creoles and new forms of intelligibility; or, unfortunately, such situations can deteriorate through repeated miscommunication, and so cross-cultural conversation can become more and more difficult. To understand these shifts it is necessary to understand how a series of conversational interactions add up and thereby influence the performance of the metafunctions of language. For instance, how can good first impressions make interactions thereafter easy? How can a set of misunderstandings lead to diminished rapport between people who have gotten along for years?

The Micro-Macro Divide

It is quite easy to roughly characterize the difficulties of visualizing VLSC as a substrate and catalyst for community. It is a "chicken and egg" problem. The communities of VLSC—and thus, also, the conversational common sense of the community—do not preexist the VLSC except in some very vague manner. The texture and ideas of online communities come through collective actions and individual interactions, but it is difficult to see how a multiplicity of such (inter)actions might add up to, for instance, a coherent conversation, or what in continental theories might be called a discourse. The difficulty is what social theorists often refer to as the *micro-macro problem:* how can a large number of individual interactions add up to a larger social or political force and, vice versa, how does a larger social force act on small-scale, even intimate interactions?

> Social theory has been in general terms concerned with different levels of analysis. In economic theory we are familiar with the idea of micro and macro economics to describe these different levels. Micro economics is concerned with the economic activity of individual economic units such as the household. Macro economics considers the behavior of the economy as a whole. Political science and sociology also work with such a distinction. In commonsense terms the micro level is the level of everyday interaction typically involving face-to-face negotiation between individuals. By contrast the macro level refers to the global structure of societies, and the analysis of major institutions such as the interface between the economy and politics; it also deals with large-scale collective action such as global social movements. The majority of social theorists recognize implicitly some form of this distinction, and various social theories have attempted to explicate the relationship between the micro and macro levels. (Turner 1996: 222).

Just as there is a micro-macro divide in economics and sociology research, there is also a micro-macro divide in discourse and conversation analysis work. The great majority of work done on conversation, by linguists and sociologists, consists of micro analyses of interactions between a small group of people. For example, work in conversation analysis often examines interactions between two or three people (e.g., Sacks 1992). Larger-scale work includes analyses of individual classrooms or small group interactions involving ten, twenty, or thirty people (e.g., Sinclair and Coulthard 1975). But large-scale work in examining interactions among hundreds or thousands of people, for instance, in online newsgroups or interchanges in scientific literatures, usually effaces so many of the rich language details that microanalyses take particular care with that these large-scale investigations are a completely different species of work. These studies are, in other words, macro analyses, and it is difficult to see whether or how they complement the work of micro analyses. Historically, the most expedient thing to do has been to choose either a micro- or a macroanalysis methodology and then ignore the results of the other. However, this is not an option for VLSC because it is large scale, thus macro, in size, but its rich details are what makes it a conversation rather than just, for example, an "information superhighway."

For instance, recent work by Steve Whittaker, Loren Terveen, Will Hill, and Lynn Cherny (1998) on the dynamics of massive interaction analyzes the headers (the *to:*, *from:*, *references:*, etc. forms) of several million e-mail messages to investigate online conversational dynamics, but they do this analysis by completely ignoring the contents of the messages. Arguably, this sort of methodology—like a lot of work in sociology on social networks (e.g., Wasserman and Galaskiewicz 1994) and cocitation analysis (e.g., Garfield 1979)—is an exploration of some of the interpersonal dimensions of the medium of VLSC, but it leaves untouched the textual and ideational relations established or broken by VLSCs. Since the production and reproduction of social groups through VLSC is a function of at least all three of these aspects of language (the interpersonal, the textual, and the ideational), a strictly social network–based examination (who is responding to whom) is not sufficient as a complement to detailed microanalysis work.

Conversely, much other large-scale work has been done on text corpora that reveals recurrent patterns of ideational and textual relations but ignores how a series of texts can produce or reinforce a social network, a set of interpersonal relationships. For example, corpus-based, computational linguistics work has developed technologies for automatically compiling rough-draft thesauri given a large archive of texts;[10] or, given an archive of tagged and bracketed texts, machines have been developed to automatically generate a grammar and a parser;[11] or, given a set of texts that mention many of the same people or places, some newly developed machines can now automatically hyperlink the texts so that entities in one text are automatically connected to mentions of the same entities in other texts.[12] Many of these same techniques have been taken up by sociologists of science working in the area of actor-network theory (e.g., Law and Hassard 1999). For example, Geneviéve Teil and Bruno Latour (1995) describe a machine that uses measurements of conditional probability and mutual information to automatically compile a rough-draft thesaurus from a corpus of scientific abstracts.[13]

The difficulty with visualizing the conditions and productions of VLSC is therefore the following. Even though it would be ideal to simply "scale-

[10] For contemporary work, see Grefenstette (1994); Hearst (1998); Harabagiu and Moldovan (1999). For the history of this field, see, for example, Soergel (1974).

[11] See, for example, Magerman (1994).

[12] See Bagga, Baldwin and Shelton (1999); Green (1997).

[13] Teil's and Latour's work is one of the latest outgrowths of a long line of such computerized text analysis work (on co-word analysis) conducted at the Centre de Sociologie de l'Innovation, Ecole des Mines de Paris. See also, for example Callon, Law and Rip (1986).

up" the methodologies of the micro analysis of conversations and discourse, such methodologies can no more be scaled-up than the rich insights into bird flight gathered by a keen-eyed ornithologist can be scaled-up to analyze the dynamics of jet airplane travel. This is because these microscale analyses require too much of the analyst. Often a micro analysis of a conversation demands that the analyst identify the intentions of the participants; this is nontrivial if not impossible to do for a discussion that involves thousands of people.

Moreover, previous attempts to create theoretical tools for the examination of large text corpora have often neglected one or another metafunction of language (e.g., the interpersonal, the textual, the ideational) that is clearly important for conversational interaction. These large-scale theories of language are not adequate as large-scale theories of conversation because they leave out too many details. Large-scale work tends to fall into either (a) a social network type of work, which usually leaves out a lot about the contents of the text or talk exchanged; or (b) a corpus-based linguistics style of work that tends to overlook too much of the interpersonal or social structure of the texts examined. If, however, these two different styles of macroanalysis could be combined, then a richer picture of the combined social and semantic (re)productions of VLSC could be painted.

Thesauri and Conversational Common Sense

If these different sorts of macroanalysis (social network–based and computational linguistics–based) are to be connected together, a linkage point must be found. By looking at the history of discourse analysis, a linkage point can be found: it is the thesaurus.

In the first essay in linguistics to mention discourse analysis, Zelig Harris (1952) provided a rough-draft version of this linkage point. Harris explained that the key to discourse analysis is to find corpus-specific equivalencies:

> Suppose our text contains the following four sentences: The trees turn here about the middle of autumn; The trees turn here about the end of October; The first frost comes after the middle of autumn; We start heating after the end of October. Then we may say that *the middle of autumn* and *the end of October* are equivalent because they occur in the same environment (The trees turn here about—), and that this equivalence is carried over into the later two sentences. On that basis, we may say further that *The first frost comes* and *We start heating* occur in equivalent environments. (1952: 6).

In the Anglo-American traditions of discourse analysis, no one has followed up on Harris' work.[14] However, the description provided by Harris on how to find "equivalencies" is a technically unnuanced description of the sort of work that some researchers in contemporary, corpus-based, computational linguistics have undertaken to automatically generate or extend thesauri. Harris's insight about what he called his "distributional analysis of discourse" was that regularities within a given discourse, rendered as "equivalencies," could be descriptive of the cultural specificities produced and reproduced within a given discourse. By blending the technology of contemporary corpus-based linguistics with Harris's insight, it is possible to use this insight as a pivot point through which different kinds of macro analysis connect together with the concerns of micro analysis of conversation.

Consider the following conversational exchange:

A: What sorts of fruit do you like?
B: Oh, apples and bananas.
A: What about strawberries? Do you buy them when they're in season?
B: No, I don't really like berries.

To find the lexical cohesion between the statements in this exchange, it is necessary to know that apples, bananas, strawberries, and berries in general are all kinds of fruit. Knowing this, it is possible to say that this short sequence concerns fruit, but it is also possible to say that A and B have had a verbal exchange concerning fruit. In short, thesaurus-like knowledge about fruit provides a means for more specifically describing interpersonal and textual relations of the conversation.

Within the micro analysis specialty of conversation analysis, scholars have noted the key role that thesaurus-like knowledge of categories plays in the construction of coherent sequences of dialogue. For instance, in elaborating his theory of categories, the inventor of conversation analysis, Harvey Sacks, provided the following definition and example:

> Sacks refers to activities which imply identities as category-bound activities (CBAs). His definition is as follows. *Category-bound activities:* "many activi-

[14] ". . . with Chomsky's appropriation of the notion of transformations as an intrasentential feature, and with the overwhelming dominance of linguistics by the transformational-generative movement which Chomsky came to lead, Harris' early attempt with longer stretches of texts was not followed up, and the models of discourse analysis described below [discourse analysis as influenced by Michael Halliday and conversation analysis as influenced by Harold Garfinkel] cannot be seen as direct developments of Harris's model (Malmkjaer 1991: 100–101). However, Michel Pêcheux and his colleagues and students in France did attempt to use Harris's (or at least Harris-like) insights to examine differences and similarities between specific discourses. See, for instance, Pêcheux (1995).

> ties are taken by Members to be done by some particular or several particular categories of Members where the categories are categories from membership categorization devices" (Sacks 1992: 249). CBAs explain why, if the story read "The X cried. The Y picked it up," we might have guessed that X was a baby and Y was a mommy. Crying, after all, is something that babies do and picking up (at least in the possibly sexist 1960s) is something that mothers did. (Silverman 1998: 83)

Similar observations about the key role of semantic and pragmatic associations for given terms in the construction of the coherence and cohesion of texts were realized within some work in computational linguistics (cf. Carbonell 1980). However, this computational linguistics work, like most other non–Harris-like computational work on discourse analysis, has been—for all practical purposes—a methodology of micro analysis of conversation and discourse.[15]

What has been left unexplored is the fact that there now exist empirical methods applicable to large-scale corpora that can provide a means for documenting the emergence of categories of terms, what Harris called equivalencies between terms. But, it is not the case that these new techniques from corpus-based linguistics can automatically bridge the theoretical chasms dividing micro from macro conversational analysis and social-network versus computational-linguistic macro analyses. One more theoretical insight is necessary: even as much as the textual and interpersonal relations are influenced by the ideational relations (i.e., the semantic links articulated in thesaurus-like compilations), the inverse is also true. In other words, the social and semantic aspects of VLSC are related in a mutually recursive manner: ideational → textual → interpersonal → ideational → . . . .

From the word usages (what Ferdinand de Saussure would call *parole* or what Noam Chomsky would call *performance*) in a corpus of texts, a set of equivalencies and thus a rough idea of semantic relations between terms can be derived with the procedures of corpus-based, computational linguistics. These equivalencies can be compiled as a kind of rough draft thesaurus. The categories and equivalencies in the thesaurus have, in turn, an influence on how cohesion (i.e., textual) and social (i.e., interpersonal) relations are labeled. By looking at which terms are important to a conversation (i.e., which terms label a large number of social and cohesion relations present in a corpus of multiauthored texts, (such as an archive

[15] Most "discourse analysis systems" that have been built in the fields of artificial intelligence and computational linguistics have been very elaborate productions constructed to illustrate the analysis of interchanges that can be transcribed into one or two pages of text (e.g., Allen et al. 1996).

of email messages), one can get a feel for which parts of the rough-draft thesaurus are important. The ways in which these highlighted elements of the rough-draft thesaurus are "spoken about" by members of the conversation provide a means for characterizing the conversation as a whole. Thus, for instance, a conversation that associates water with hydrogen and oxygen might be characteristic about a conversation of chemistry rather than a conversation about alchemy.

As conversations and so cultures and common sense evolve, so do the thesauri that can be derived from them. This is true too of more official, hand-compiled reference works.[16] Very large-scale conversation is an eclectic domain because, as it is currently practiced on the Internet, participants can come from a wide diversity of cultural backgrounds and so what is or is not commonsensical cannot be enumerated beforehand. An understanding of VLSC requires a perspective that allows one to see, for instance over the course of a long-term conversation, how common sense is produced, reproduced, extended, and changed by a group of potentially culturally diverse participants. The political philosopher Antonio Gramsci gives us just such a picture of common sense: "Every social stratum has its own 'common sense' and its own 'good sense,' which are basically the most widespread conception of life and of men. Every philosophical current leaves behind a sedimentation of 'common sense': this is the document of its historical effectiveness. Common sense is not something rigid and immobile, but is continually transforming itself, enriching itself with scientific ideas and with philosophical opinions which have entered ordinary life. . . . Common sense creates the folklore of the future, that is as a relatively rigid phase of popular knowledge at a given place and time" (Gramsci 1971: 326, as cited in Hall 1982: 73).

From this perspective, common sense is accumulated and transformed through the process and productions of science, philosophy, and other powerful conversations, discourses, and practices. This is a perspective that has been useful for understanding the workings of older media (newspapers, television, film) and could be of use to understand and design for new forms of mediation like VLSC.[17]

[16] The literary theorist Roland Barthes speaks of the contents of reference books, like thesauri, as "cultural codes" central to the process of reading. "The **cultural** codes, which are extremely numerous and heterogeneous, to a very large degree subsume all the other categories. They speak the familiar 'truths' of the existing cultural order, repeat what has 'always been already read, seen, done experienced.' . . . Barthes underscores the discursive basis of the 'reality' to which cultural codes refer by equating it with 'the set of seven or eight handbooks accessible to a diligent student in the classical bourgeois educational system'" (Silverman 1983: 241, 274).

[17] According to Stuart Hall (1982), Anglo-American media studies of the early twentieth century saw the media (newspaper, television, etc.) as producers of content that "re-

Maps of Very Large-scale Conversation

Discourse architecture entails two kinds of work: (1) the design and implementation of new computer network technologies for discourse, that is, the means to shape the conversation that takes place within a given system; and (2) the production and employment of analytical techniques for identifying conversational structure and explicating the forces that shape it. This chapter has, so far, discussed only the second kind of work.

I have argued—following Michael Halliday—that conversational common sense has at least three crucial dimensions: the interpersonal, the textual, and the ideational. Using examples from art, comedy, and sociology, I have illustrated how the breakdown of conversational common sense can have effects both comedic and/or dire for social cohesion. I have hinted how new and old thinking about thesauri gives one insight into the constitution of conversational common sense. Finally, I have asserted—with Antonio Gramsci and Stuart Hall—that common sense is an "accretion" dynamically produced and transformed by the groups that it links: conversational common sense is defined in a mutually recursive relationship with the social group that invents and reproduces it.

To illustrate the other aspect of discourse architecture—the design and implementation of new computer network technologies for discourse—I present a system, the Conversation Map, designed to visualize the three dimensions of VLSC common sense and its emergence and transformation. The Conversation Map system accepts a corpus of hundreds or thousands of e-mail messages and analyzes those messages using a set of computational linguistics and sociology techniques to generate a summary of the messages that includes (1) who is reciprocally replying to or quoting from whom—i.e., the interpersonal dimension of the conversation; (2) the themes of discussion that are important to the conversation embodied in the messages—i.e., the textual aspect of the conversation; and (3) what can be understood as some of the emergent definitions or metaphors of the discussion that are apparent if, in a certain sense, all of the participants' language inscribed in the text—i.e., the content—of the e-mail messages is analyzed and summed together. This last aspect is performed through the automatic calculation of a rough-draft thesaurus from the

flected" the "common sense" of the larger public. The media was said to objectively write down and distribute the consensus, or *sensus communus,* that was produced by the public independent of the media. Hall argues that, later, media studies came to recognize the media's role in producing, rather than simply reflecting, community values and common sense. By being the only "voice" that could reach across the nation and even across the world, the electronic and print media framed public discourse, and thus public "common sense," simply through the editorial choice of which stories should be broadcast and which should be left untold.

written content of the e-mail messages. In short, the Conversation Map is designed to make the three dimensions of VLSC common sense visible: the interpersonal, the textual and the ideational. More specific and technical descriptions of the Conversation Map system can be found elsewhere (Sack 2000, 2002). Here we will examine a few examples of the maps that were automatically generated by the system.

One unprecedented activity that the Internet has made possible is the debate about international politics by ordinary citizens in different countries on a daily basis in a public "space" where people do not necessarily know one another before the debate begins. Such debates occur regularly in weblogs, listservs, and Usenet newsgroups. The following examples are all drawn from public, Usenet newsgroup discussions, and they all illustrate ways of understanding these new horizontal, transnational relations conducted by "citizen diplomats."

The first map (fig. 2) was generated from several hundred messages posted to the Usenet newsgroup soc.culture.palestine during the period August 1–7, 2001. The upper left corner of the map displays a social network. Nodes in the network are message authors. A link between nodes indicates that two authors have mutually responded to and/or quoted from one another. Note the visible evidence of a relatively tight-knit group: there is one large cluster of authors with only a couple of author pairs floating off to the side. The upper center menu lists a series of dis-

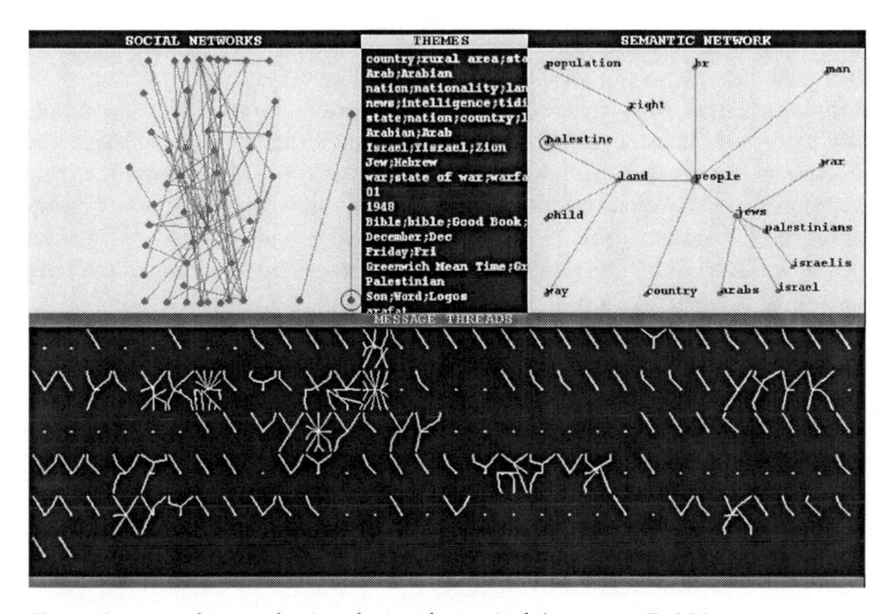

Figure 2. soc.culture.palestine during the period August 1–7, 2001

cussion themes that the Conversation Map has counted in the bodies of serially posted messages (message threads). Note that the terms "Arab and Arabian" are counted as frequent themes of discussion. The upper right corner displays the output of the automatic thesaurus computations: terms that are linked are calculated to be potentially similar terms within the VLSC. Here one can see that "Israelis" and "Palestinians" are counted as possibly similar terms, as are "Jews" and "Arabs" and "lands" and "peoples."

Each of these three analyses (social network, themes, and semantic network) are cross-linked with one another. This is built into the software so that clicking on one panel highlights terms in the others. Thus, for instance, clicking on a theme highlights that part of the social network in which the participants have discussed the theme. The Conversation Map illustrates how the interpersonal, textual, and ideational aspects of a VLSC are interrelated and furthermore shows how a generated thesaurus reveals some of the crucial equivalences under discussion by the group: How are Palestinians the same as/different from Israelis? What makes Jews like Arabs? To what extent does the definition of a people depend upon a definition of land or country?

The next Conversation Map was generated from the same news group, soc.culture.palestine, but messages were collected a few days later; messages analyzed for this map include those posted between August 4 and 11, 2001. In other words, there are some overlaps with the messages analyzed for the map in figure 2. In figure 3, one can see that the terms "Arab" and "Arabian" have ceased to be central themes of discussion. The social network has increased in size and fractured apart into several nonoverlapping clusters, thus indicating that there is not just one conversation taking place but rather several in the same space, that is, the same newsgroup. Note also the possible equivalence drawn between entities common to conventional international relations—the posited equivalences between "government," "nation," and "state." By comparing figure 2s and 3, one can see how the VLSC of the newsgroup soc.culture.palestine changed over the course of a week and a half.

Figure 4 is a Conversation Map generated from over one thousand messages posted to the Usenet newsgroup soc.culture.afghanistan in the time period September 24–28, 2001. Note the highly fractured social networks. Note also the extremely generic semantic network including only abstract terms like "state, "country," "government," and "people." Unlike Israel and Palestine, Afghanistan had, at the time, only one Internet service provider. Writers knowledgeable about the specifics of Afghanistan are, consequently, relatively rare online. It is highly unlikely that anyone logged in from Afghanistan to post their side of the story to the newsgroup. In other words, it may be possible that some of these newsgroups

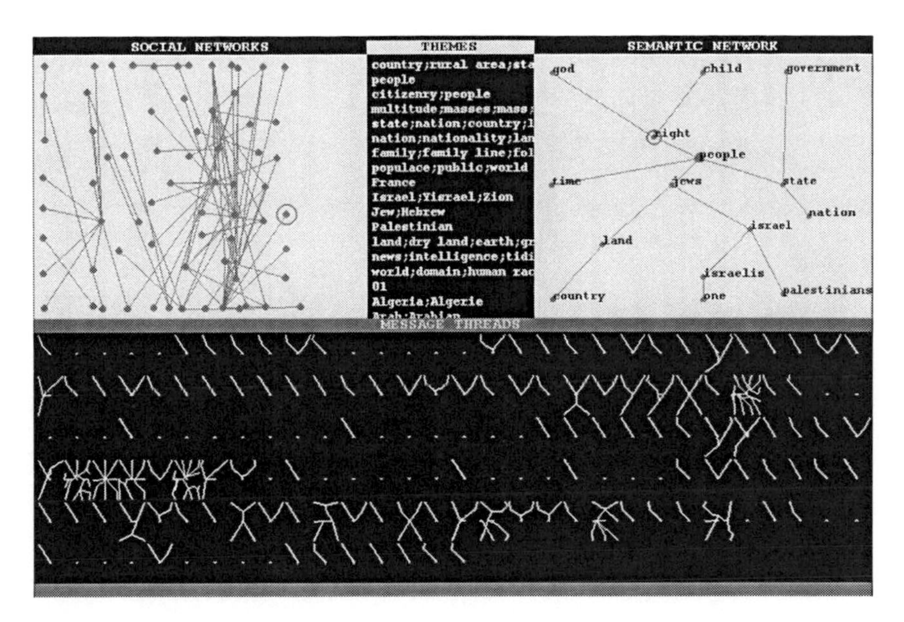

Figure 3. soc.culture.palestine during the period August 4–11, 2001

do support a truly new kind of cosmopolitan, citizen-centered diplomacy. But what happens in these cyberspaces is also inflected by what is happening offline. For U.S.-Afghani negotiations to take place in Usenet, the necessary infrastructure for the Internet would have to exist in Afghanistan.

The conversation maps in figures 2, 3, and 4 graphically summarize three collective productions achieved by groups of hundreds of people over the course of several days of online conversation:

1. The social networks shown in the maps give some indication of the interpersonal relations of the groups: they indicate how often and with whom members of the groups are reciprocating messages.

2. The lists of calculated discussion themes are created from a computational analysis of the words that are quoted and repeatedly taken up in sequences of messages. Discussion themes are listed according to the number of participants who have exchanged messages about the theme. Thus, the menus in the figures above can be understood as representations of the intertextual structure of hundreds or thousands of e-mail messages and also as representations of a group's current focus: the themes listed at the top of the list are themes addressed in the messages of many of the participants.

3. Finally, by parsing the contents of all of the messages and recording which words, specifically which nouns, are used in a manner similar to which other

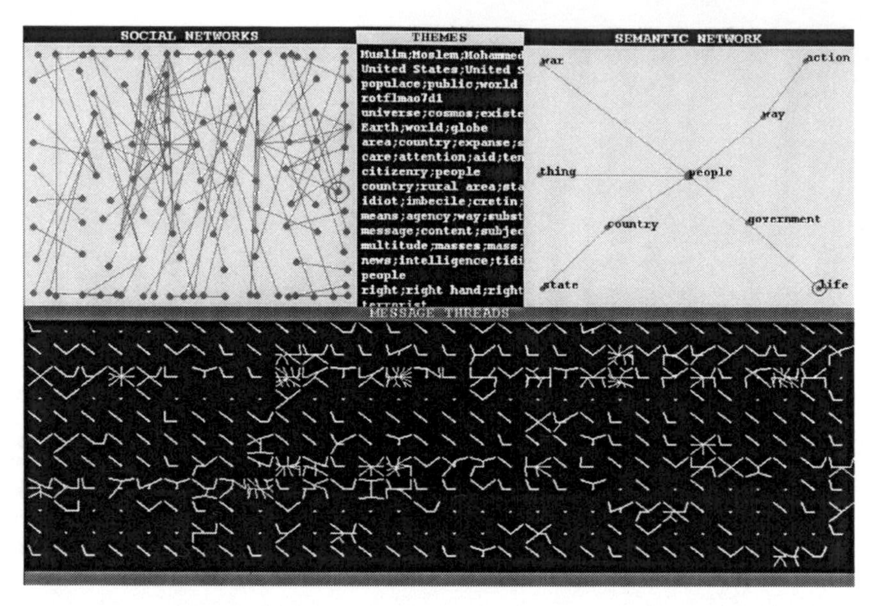

Figure 4. soc.culture.afghanistan during the period September 24–28, 2001

nouns (i.e., which nouns are written about like which other nouns), a rough-draft thesaurus is computed for the group and partially displayed as a semantic network. Two nouns are plotted next to one another in the semantic network if, for instance, many participants have used the same set of adjectives to describe them, have associated the same set of verbs with them, and collocated them in messages alongside a similar set of other nouns. Thus, for example, if in the text of many participants' messages in a given discussion group two words, like "building" and "argument," are both repeatedly described as having foundations (as being solid, strong, shaky, or weak; as collapsing, falling down, or standing up), then they may very well appear next to one another in the semantic network computed by the Conversation Map for that discussion group. This semantic network can be read as symptomatic of the emerging synonyms or metaphors that participants are in the process of collectively creating or defining.

Note, however, that the procedure I have outlined above—keeping track of and comparing the lexical contexts of each noun used in the discussion—does result in a set of many semantic networks. For discussions of the size mapped in the figures above, this results in an analysis of several thousand nouns. Although all of these semantic networks for the thousands of nouns are computed by the Conversation Map, only one semantic network is displayed: the semantic network that contains nouns frequently used as themes of discussion, and thus the nouns that are arguably the current, collective focus of the group. Interestingly, in all three maps, the noun "country" is within the groups' collective foci.

The maps constitute *representations* of the VLSCs, but they also can be used as *interfaces* into the message archives. Another aspect of the Conversation Map that is not discussed in this chapter is the fact that the maps it outputs are executable as Java applets on the Web. They are therefore in principle accessible to most of the discussion participants since they can be viewed as web pages. By clicking on the various aspects of the maps, one can see how each piece of the map is cross-indexed with other parts of the map. Thus, for instance, clicking on a discussion theme will highlight that portion of the social network that has exchanged messages about the selected theme. (See the main web site for the Conversation Map to learn more about how the maps function as interfaces as well as representations: http://www.sims.berkeley.edu/~sack/cm.)

Deliberative Democracy and the New Public Sphere

Is there a politics of very large-scale conversation? There are many such politics, but large, many-to-many exchanges between citizens are of special interest as new forms of deliberative democracy.

> The most exciting and potentially revolutionary political application of a computer conferencing system is the facilitation of the direct participation and voting of citizens on important state or national issues. . . . Perhaps the first operational use of computer conferencing systems to facilitate "participatory democracy" will be J. W. Huston's Constitutional Convention project in Hawai'i. Funded by grants from local and mainland foundations, it is being designed to establish 21 community centers throughout the state to allow public participation in the 1978 Hawai'i Constitutional Convention. (Hiltz and Turoff 1994: 195, 197)

For at least a quarter century, many have been excited about the possibilities of computer networks as a means of facilitating democratic participation. Reviewing the area in the mid-1990s, sociologist Manuel Castells (1997: 350) noted that local democracies appeared to be flourishing around the world and that "When electronic means are added to expand participation and consultation by citizens, new technologies contribute to enhanced participation in local government." Collections, such as Tsagarousianou, Tambini and Bryan's *Cyberdemocracy* (1998) documented how these experiments in local, online democracy were progressing in Amsterdam, Athens, Berlin, Bologna, Manchester, Santa Monica, and elsewhere.

At the national level, there has been less interest in citizen-to-citizen communication and more emphasis on delivering government services and documents (tax documents, legal codes) to citizens via the Internet. This understanding of the Internet as a one-way publishing and distribution network rather than as a many-to-many medium is due to a variety of entrenched economic and political interests. It is interesting to compare these political and technical efforts to produce the Internet, especially the Web, as a one-way broadcasting medium with centralized control with similar efforts that have been suffered over the course of the last century, such as the re-creation of radio as a one-way medium (e.g., Neuman, McKnight, and Solomon 1997). In fact, it appears to be the case that—as municipal web sites become more and more common—even local governments seem most intent on supporting a one-way, "services" model of information delivery rather than many-to-many deliberative discussion.

Interesting and powerful exceptions do exist. For example, Stephen Coleman and his colleagues at the Hansard Society have initiated online, public forums to elicit public opinion and encourage democratic deliberation on issues of national importance and communicated these results of online deliberation to the U.K. Parliament.[18]

Many of the political scientists and communications experts now exploring the area of online, democratic deliberation have been deeply influenced by philosopher Jürgen Habermas's (1991) conception of the "public sphere" and its transformations over the past three centuries.[19] Habermas' contention was that the public sphere constituted a set of norms and forums (the newspaper, the café, etc.) that allowed bourgeois society to meet and, through rational debate and deliberation, find consensus. His diagnosis was dark: rational-critical debate largely disappeared in the twentieth century as citizens became consumers and so consumption—rather than conversation—dominated the forums of the public sphere. However, contemporary work in "community informatics" has proceeded with the hope that computer networks can provide a basis for a renewed public sphere where deliberative democracy can be supported (e.g., Schuler and Day 2000).

Some of the more practical work necessary for the goal of supporting a new, online public sphere is well defined. Exemplary organizations, like the Seattle Community Network (http://www.scn.org), provide community members with e-mail accounts; host web sites, online discussion forums, and public calendars; provide help and computer training; and fa-

[18] See Coleman (2004). See also the Hansard Society's website, http://www.hansard-society.org.uk/eDemocracy.htm

[19] See also more recent commentary in Calhoun (1992).

cilitate low-cost or free distribution of computers and other necessary hardware. Some might wonder why this list of technical foundations for a new public sphere does not also include, for instance, multiway video conferencing and/or streaming audio servers for all participants. In principle this would be possible and would allow many citizens to, essentially, run their own television and/or radio stations. But, the practical essence is that most of this technology is too expensive and/or too complicated to support for large groups of people. So community networking of today is especially dependent upon textual exchanges such as e-mail, newsgroups, and weblogs.

Once the technical infrastructure has been put into place, the work that remains is not so well defined: How can online, deliberative discussions be engendered and facilitated? Activists and technologists attempting to support new forms of online democracy have had to turn to philosophy and political theory to help define the crucial issues: What is democracy? What is the "public sphere"? What constitutes deliberative discussion? These seemingly abstract questions have become pressing concerns for community networks. I contend that few, if any, have a working definition of deliberative discussion when the discussion involves asynchronous, online exchanges among hundreds or thousands of people. Even at the local, civic level, online exchanges of this sort quickly reach the size of very large-scale conversations.

Attempts to produce working definitions of new, electronic forms of the public sphere and of large-scale, deliberative discussion can be found in the literatures of the arts, humanities, and social sciences. One body of work is critical insofar as it points out the weaknesses of a Habermasian ideal of the public sphere and its goal of consensus through rational discussion. Habermas's original focus on the *bourgeois* public sphere was scrutinized, and it has been pointed out that participants in the stated ideal were limited, for instance, by class (e.g., Negt and Kluge 1993), by gender (e.g., Fraser 1992), or by activity—specifically the democratic potential of rational discussion has been questioned by Jean-Francois Lyotard (1984) and others. These critiques have yielded alternative ideals, and alternative ideals for online exchanges have been articulated.[20]

Another set of work takes as given a specific set of ideals of democratic discourse and then attempts to measure online exchanges against these ideals. This type of work can be problematic if either (1) the stated ideals of democratic discussion recapitulate the weaknesses already scrutinized and critiqued in Habermas's original work; or (2) the stated claims of the authors exceed the possible reach of their empirical work. The sec-

[20] For example, Jodi Dean (2001) points out the limitations of the "public sphere" ideal and prefers the term "civil society."

ond flaw is caused by a misunderstanding of the scale of online discussion. There are, for instance, many researchers who have closely read a few hundred e-mail messages and now claim to have an evaluation of Usenet as a possible, new form of the public sphere. These sorts of claims are mistaken. To understand the enormity of Usenet, it helps to know that as of late 2003, Google (http://www.google.com) had an archive of over eight hundred million messages exchanged on Usenet newsgroups. At best, with an analysis of a few hundred messages, one might claim to have some insights into a fleeting moment of one newsgroup within Usenet. It is with this specific problematic in mind that I propose a theory and technology of very large-scale conversation.

The best of this literature is worth close examination because it points toward many interesting new possibilities. In his book *Democracy in the Digital Age,* Anthony G. Wilhelm examines five hundred messages (fifty messages apiece from six Usenet newsgroups and four AOL discussion groups) and then makes sweeping claims like these: "If a democratic discussion is to be defined at least in part by the quality of the conversation, then the newsgroups analyzed in this study are not very deliberative" (2000: 98).[21] Examination of fifty messages is unlikely to provide enough evidence to warrant such an evaluation. For example, if one downloads even just the past week's worth of messages from the six Usenet newsgroups examined by Wilhelm, one gets the following message counts: alt.politics.elections (220 messages posted in the past week); alt.politics.libertarian (1,081); alt.politics.media (647); alt.politics.org.cia (104); alt.politics.reform (62); alt.politics.white-power (199).[22] Thus, the number of messages per group chosen by Wilhelm does not cover even a given week's worth of messages exchanged on the smallest of the groups.

What is most interesting about Wilhelm's study is his effort to define a set of necessary conditions for deliberative discussion. He then attempts to operationalize those criteria to determine if online discussions are deliberative. Wilhelm's criteria of deliberative discussion are stated as a series of research questions:

> 1. **Reciprocation:** . . . To what extent do participants of virtual groups solely provide ideas and information versus seeking information from other forum members? . . . [Do] reciprocal acts occur . . . in which participants . . . articulate their interests through talking, sharing ideas, and negotiating differences[?] (88)
>
> 2. **Interactivity:** . . . To what extent do participants of political groups ex-

[21] For an earlier version of Wilhelm's work in which he arrives at even more sweeping conclusions, see Wilhelm (1999: 154–78).

[22] Messages from these groups were downloaded on October 12, 2003, from the Usenet news server news.ucsc.edu.

change opinions as well as incorporate and respond to others' viewpoints? (88–89)

3. Heterogeneity: . . . To what extent is there in-group homogeneity of political opinion on Usenet newsgroups? . . . In this case, homogeneity is defined as the extent to which individual messages adhere to a certain political affiliation. (89).

4. Rationality: . . . To what extent are substantive, practical questions debated rationally in contradistinction to *ad hominem* argumentation not susceptible to criticism and grounding? (90)

Insofar as Wilhelm is one of many researchers inspired by Habermas, his criteria are representative of a larger literature on the issue of online, deliberative democracy. Moreover, his suggestions for operationalizing these criteria into a means of empirically investigating these questions suffer from the same problem as other empirical work in the literature: the work does not scale to the size necessary to address the questions posed about public, online discussions like those of Usenet newsgroups. Wilhelm attempts to address these questions with a form of content analysis[23] that requires a panel of judges to comb through and categorize the messages: a very laborious process that would cost a fortune to apply to thousands of messages. Other tested methodologies—for example, survey methods—suffer from the same problem.[24]

To adequately address the questions posed by Wilhelm regarding Usenet newsgroups and other VLSC, it is necessary to have a theory and methodology of VLSC that is at least partially embodied in a piece of software. This genre of "theoretical software" has been, historically, well known in social science but is recently less common than it used to be.[25]

The Conversation Map simultaneously embodies and articulates a theory of VLSC and allows one to begin to address the questions posed by Wilhelm. Consider, for example, a Conversation Map of one week's worth of messages posted to one of the Usenet newsgroups studied by Wilhelm, alt.politics.elections. The map shown in figure 5 was calculated from over a thousand messages posted to the group in the week before the 2000 U.S. presidential election. Illustrated below are examples of how

[23] For a definition of content analysis, see Krippendorf (1980).

[24] See, for example, the survey methods employed by Cappella, Price, and Nir (2002: 73–93). Although these researchers were able to survey a large number of people, their methodology entailed the creation of a set of small (approximately 20 to 30 people), moderated, nonpublic discussion groups that ran periodically for one hour. Thus, it is unclear whether their results have anything to say about VLSC: large, online, ongoing, public, unmoderated discussions involving hundreds or thousands of people at once.

[25] See, for example, Robert Abelson's early, computational/theoretical analyses of ideology, belief, and opinion that were embodied in working simulations; e.g., his "Goldwater Machine" (Abelson 1973).

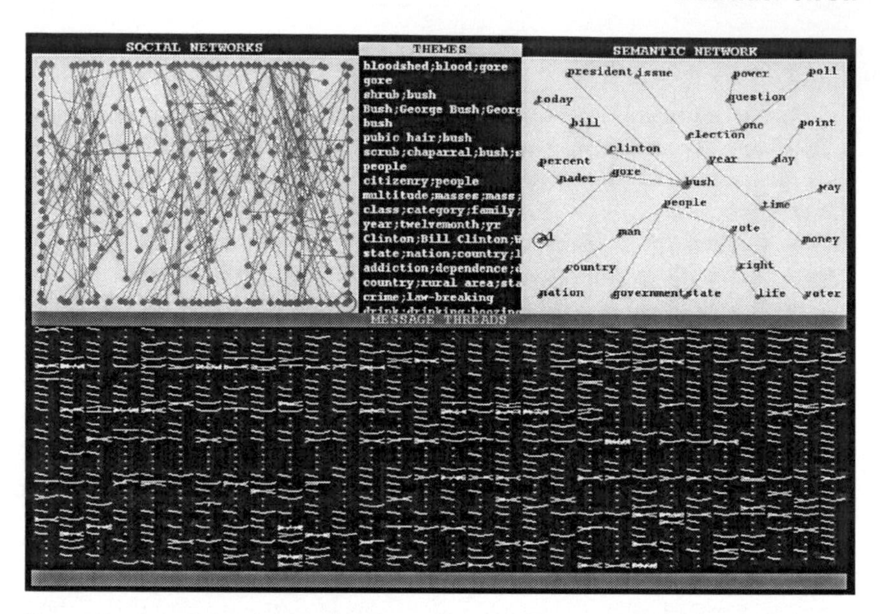

Figure 5. alt.politics.election for the week prior to the 2000 U.S. election

one might begin to address Wilhelm's questions with maps calculated by the Conversation Map system.

1. Reciprocation: Are participants reciprocating with one another; that is, are they responding to and/or quoting from the messages of other participants? The social networks provide a partial representation with which one can explore this question. In the map shown in figure 5, the answer to this question is not a simple "yes" or "no." Here it is possible to see a great number of social networks (recall that two participants are connected in the network if they have replied to each other and/or quoted from one another). Compare the social network shown here with the one shown in figure 2, where practically all the visible participants are integrated into a single network.

2. Interactivity: The Conversation Map graphically shows two ways in which participants are (or are not) incorporating or responding to others' postings.

2a. The calculated "themes of discussion" indicate which topics were repeatedly addressed in sequences of messages exchanged. These sequences are normally termed "message threads." The lower half of every Conversation Map is a graphical representation of all of message threads analyzed. By clicking on a given theme in the menu of themes, one can see which threads address the selected theme. The ovals highlighting the lower half of the Conversation Map shown in figure 6 indicate all of the messages threads where then–Vice President Gore was a theme of discussion.

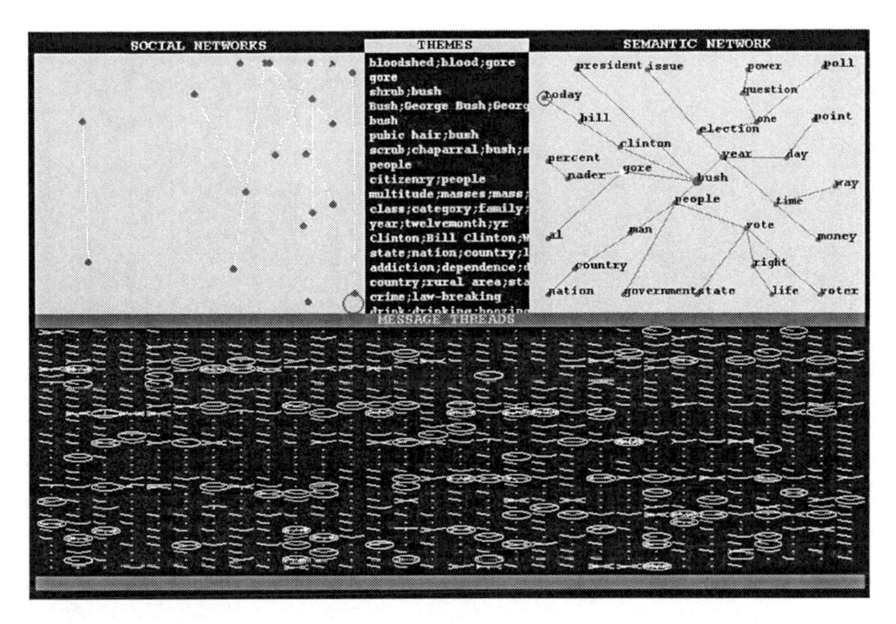

Figure 6. The same conversation map as shown in figure 5 with the discussion theme "Gore" selected

2b. The portion of the social network shown in figure 6 contains those pairs of discussants who have exchanged messages on the theme of Vice President Gore.

Well-focused, interactive newsgroups tend to have several themes of discussion that are repeatedly addressed in multiple message threads and cover substantial portions of the social network. In a sense, these two graphics show the extent to which a group of discussants stays on topic.

3. Heterogeneity: When "summed together," does the language of the messages exchanged link together a diverse semantic field, or is it a relatively homogeneous set of comments? Examination of the calculated semantic networks (through double-clicking on terms and sets of terms in the semantic network) reveals the diversity of terms employed to describe the themes of discussion. For example, after selecting both "Bush" and "Gore" in the semantic network, one can double click to demand a list of the terms used in the text of messages to describe Bush, the terms used to describe Gore, and the terms that were applied to both Gore and Bush (see fig. 7). In the interface, the different lists of terms are distinguished by color.

Symptomatic of the hetrogeneity of opinion for a given topic is the diversity of lexical terms associated with the topic. For example, from the (partial) list of verbs displayed in figure 7, one can see that Bush was the

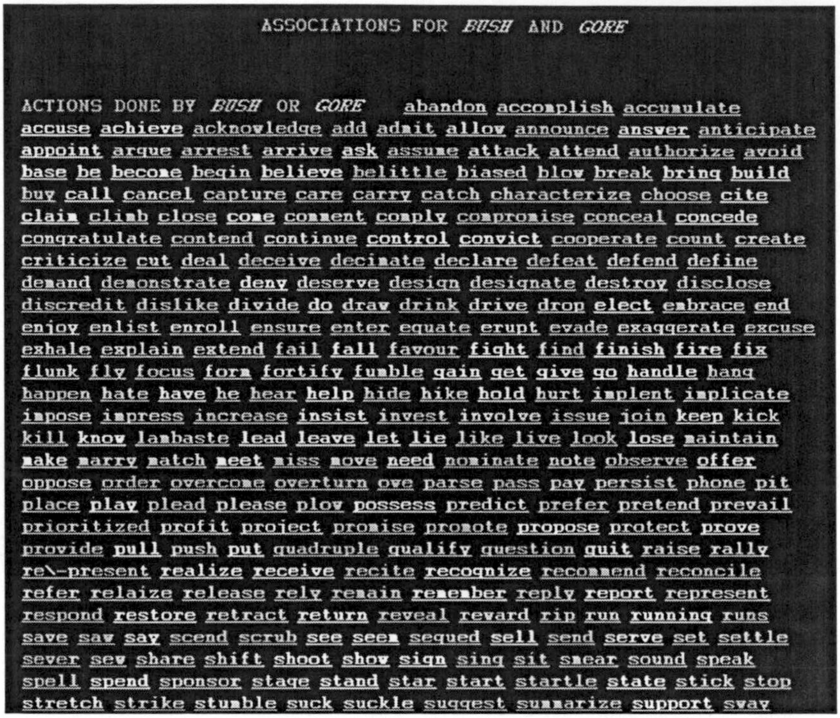

ASSOCIATIONS FOR *BUSH* AND *GORE*

ACTIONS DONE BY *BUSH* OR *GORE* abandon accomplish accumulate accuse achieve acknowledge add admit allow announce answer anticipate appoint argue arrest arrive ask assume attack attend authorize avoid base be become begin believe belittle biased blow break bring build buy call cancel capture care carry catch characterize choose cite claim climb close come comment comply compromise conceal concede congratulate contend continue control convict cooperate count create criticize cut deal deceive decimate declare defeat defend define demand demonstrate deny deserve design designate destroy disclose discredit dislike divide do draw drink drive drop elect embrace end enjoy enlist enroll ensure enter equate erupt evade exaggerate excuse exhale explain extend fail fall favour fight find finish fire fix flunk fly focus form fortify fumble gain get give go handle hang happen hate have he hear help hide hike hold hurt implent implicate impose impress increase insist invest involve issue join keep kick kill know lambaste lead leave let lie like live look lose maintain make marry match meet miss move need nominate note observe offer oppose order overcome overturn owe parse pass pay persist phone pit place play plead please plow possess predict prefer pretend prevail prioritized profit project promise promote propose protect prove provide pull push put quadruple qualify question quit raise rally re\-present realize receive recite recognize recommend reconcile refer relaize release rely remain remember reply report represent respond restore retract return reveal reward rip run running runs save saw say scend scrub see seem sequed sell send serve set settle sever sew share shift shoot show sign sing sit smear sound speak spell spend sponsor stage stand star start startle state stick stop stretch strike stumble suck suckle suggest summarize support sway

Figure 7. A partial list of the terms associated with Bush and/or Gore

subject of the following verbs in messages posted to the group: Bush acknowledges, adds, admits, announces, authorizes, avoids, belittles, compromises, conceals, drinks, promises, reveals, suggests, sways, etc. Clicking on any one of these terms causes the Conversation Map to produce a hyperlinked list of sentences in which the term appears. Figure 8 shows such a hyperlinked sentence generated by clicking on the verb "drink." Clicking on a sentence allows one to examine the message in which the sentence appears.

4. Rationality: Although Wilhelm is interested in evaluating whether or not a group is debating questions "rationally" (according to criteria of knowledge, truth, and conditions of validity outlined by Habermas) he—like many others interested in these criteria (e.g., Cappella, Price, and Nit 2002) has had to evaluate "rationality" according to the structure of the arguments advanced and the number of reasons included to substantiate an argument. In short, what is empirically decidable is not the rationality of argumentation, but rather the rhetorical structure of the online exchanges. This should make sense because rational arguments tend to be well structured. Some of this information is avail-

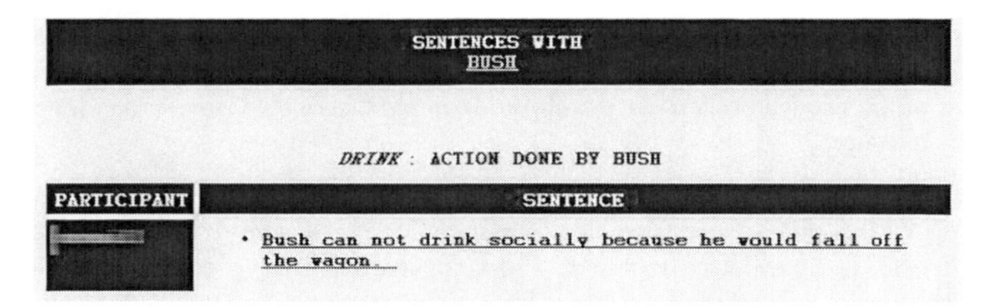

Figure 8. A sentence from the newsgroup associating a term (Bush) with a verb

able through (a) a close examination of the characteristic message thread structures of a given group, some is visible in (b) the quoting and citation patterns manifest in the messages, and some could be calculated automatically given (c) a procedure for rhetorical structure parsing (e.g., Marcu 1997). The Conversation Map incorporates a means for computing (a) and (b) and displays those results in thread structures seen in the lower half of each Conversation Map (which can be further magnified and explored by clicking on each thread, as has been done for one thread in figure 9), and in the social network and the messages themselves (in which quotes are identified automatically and hyperlinked to the message(s) of origin). A partial implementation of (c) is imple-

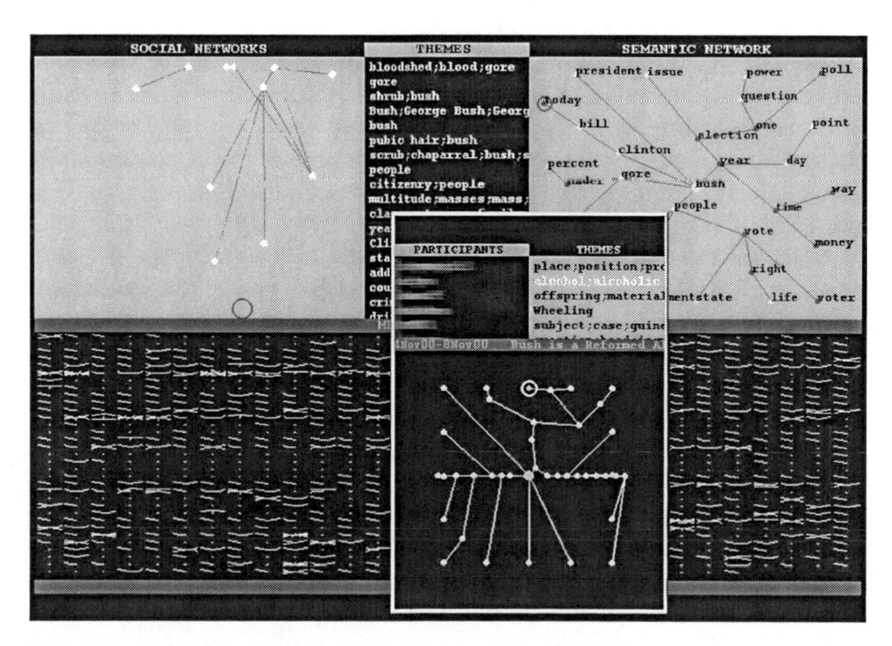

Figure 9. Close examination of the structure of a thread

mented in the Conversation Map: discourse markers indicative of structured argumentation ("because," "therefore," etc.) are tagged; however, the results of this tagging process are not currently visible within the Conversation Map interface.

Examination of the pattern displayed in the lower half of a Conversation Map gives some idea of characteristic length and structure of the threads in an online discussion. The Conversation Map plots the threads in a radial, spider web pattern: the initial message is located in the middle of the window, responses to the message are plotted in a circle around the middle and connected to the middle, responses to response are plotted in a ring slightly larger than that, and so forth. If each message had exactly the same number of responses, then the thread would look like a spider web. The more common case is a thread with many asymmetries, like the one shown in figure 9: messages differ widely according to their number of responses. Note that threads containing only one message (and no responses) appear as a dot; threads with one post and one response appear as a line. Figure 9 reveals that many of the posts to the group are unrequited and/or garnered only one response. Compare this to the alternations of simple and elaborately structured threads that one can see in figures 2, 3, and 4.

While the Conversation Map does not answer all of the questions posed by Wilhelm concerning the deliberative quality of discussion in a newsgroup, it does provide a means for exploring those questions because it incorporates a theory of VLSC into a computerized analysis procedure and a graphical interface. Consequently, it is possible to see how variable and quickly changing a discussion group can be with respect to any of these criteria. For instance, figure 10 shows a map of the same discussion group (alt.politics.elections) for the week immediately following the U.S. presidential election of 2000.

Note, among other things, how one can see in figure 10 that the conversation has shifted away from a conversation about the candidates (Bush, Gore, Nader) and is now a discussion about the technicalities of the election: vote, count, ballot, election, and so on are the central themes of discussion. If nothing else, this map illustrates how quickly a discussion can change, and therefore how careful one should be about generalizing from a one-time analysis of a newsgroup.

These Conversation Map images also illustrate the utility of the theoretical topography used to visually investigate online discussions and which I have argued for throughout this chapter. The topography of discussion proposed here is one of links, associations, graphs, and networks. The topography of a Habermasian democratic exchange is one of idealized spaces, territories, and logics—like the so-called public sphere and

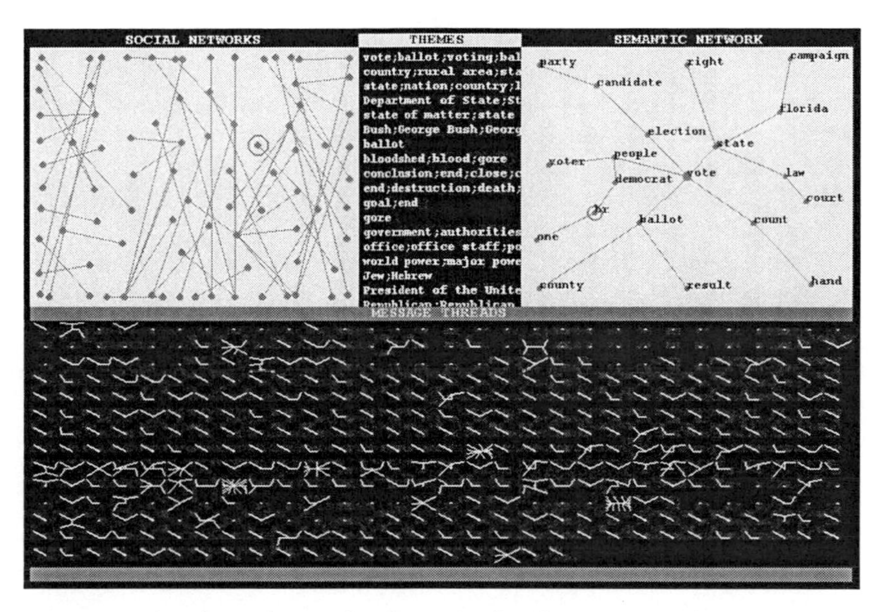

Figure 10. alt.politics.election for the week after the 2000 U.S. election

rationality. As the following quote shows, this difference in theoretical vocabularies—that based on idealized spaces, territories, areas, and logics versus that based on links, associations, and networks—is a centuries-old dispute. In his comments on sixteenth-century political theorist Guillaume de la Perrière (La Perrière 1567), Michel Foucault (1991: 93) argues that

> what government has to do with is not territory but rather a sort of complex composed of men and things. The things with which in this sense government is to be concerned are in fact men, but men in their relations, their links, their imbrication with those other things which are wealth, resources, means of subsistence, the territory with its specific qualities, climate, irrigation, fertility, etc.; men in their relation to that other kind of things, accidents and misfortunes such as famine, epidemics, death, etc.

While Foucault's words help to explain the longer genealogy of political analysis that the Conversation Map takes part in (and which also serves to differentiate it from the Habermas-inspired work of Wilhelm and others), they also point out the many inadequacies in links and networks displayed by the Conversation Map. The links visible in the Conversation Map are only those between people, between words, and between words and people. The larger set of possible relations between people and things is not visible. This larger set crucial to any expanded understanding of democracy and governance is termed a "Parliament of Things" by Bruno Latour (1993: 144): "we do not have to create this Parliament out of whole cloth, . . . [w]e simply have to ratify what we have always done. . . . Half of our politics is constructed in science and technology. The other half of Nature is constructed in societies. Let us patch the two back together, and the political task can begin again."

These insights from Foucault and Latour provide a direction for future development of the Conversation Map and a practical means for theorizing and analyzing online, deliberative, democratic, very large-scale conversations.

Conclusions: Technologies of the Self and Design Ethics for VLSC

The majority of this chapter has been devoted to what might be called an epistemological inquiry into VLSC. I have attempted to show what kinds of knowledge are necessary to understand these large-scale, online discussions. I have articulated a theory of VLSC and compared it to related work in political science, sociology, linguistics, and philosophy. And I have incorporated this theory into a technology, the Conversation Map, intended to graph the shapes and forms of online discussions. Yet, what kind of technology is the Conversation Map and other work comparable to it?

Michel Foucault distinguished techniques and *technologies of the self* from *technologies of power.*[26] For instance, a practice that makes good sense to do for oneself—for example, seeing a doctor on a regular basis and keeping detailed records of one's health—can shift from being a technology of the self to becoming a technology of power if a third party—like an insurance company—is allowed to collect and analyze health

[26] "My objective for more than twenty-five years has been to sketch out a history of the different ways in our culture that humans develop knowledge about themselves: economics, biology, psychiatry, medicine, and penology. The main point is not to accept this knowledge at face value but to analyze these so-called techniques that human beings use to understand themselves.

As a context, we must understand that there are four major types of these "technologies," each a matrix of practical reason: (1) technologies of production, which permit us to produce, transform, or manipulate things; (2) technologies of sign systems, which permit us to use signs, meanings, symbols, or signification; (3) technologies of power, which determine the conduct of individuals and submit them to certain ends or domination, an objectivizing of the subject; (4) technologies of the self, which permit individuals to effect by their own means, or with the help of others, a certain number of operations on their own bodies and souls, thoughts, conduct, and way of being, so as to transform themselves in order to attain a certain state of happiness, purity, wisdom, perfection, or immorality. (Foucault 1997: 224–25)

records. However, a technology of the self can be designed in such a way that makes it more resistant against such a transformation. Thus, it is a good idea to encrypt medical records stored in computer databases and design the database system so that any third parties must request the permission of the patient to get the "key" to records (e.g., Rind et al. 1997: 138–41).

The imperatives of design that shape technologies like the Conversation Map are not strictly epistemological in form. The designer of such technologies and techniques must instrumentalize a set of ethical considerations that make them either, more clearly, a technology of power or a technology of the self. Especially tools of democracy should be designed to be technologies of the self.

There is a long history of the use of media as technologies of the self, as reflective and communicative media for the construction of social, psychological, economic, and political self-governing people and peoples. Diaries have been used for millennia by particular people as a medium for self-reflection, for writing down and shaping the person's image of self. The diary is a medium that functions as a technology of the self where "self" is understood to be the self of one person. The oral story telling practices of folktales function in an analogous manner for the formation and description of a slightly larger self, a self of a small group of people. Oral story telling of folktales is a means for articulating the values and identity of small, tightly knit clusters of people. The facilitation of the production of larger selves, of the selves of self-governing nations, for instance, requires a different kind of medium. Scholars have shown how the mass production capabilities of high-speed printing presses made possible the media of novels and newspapers that were essential to the formation of the modern nation-state (Anderson 1983). VLSCs do and can function as the substrate for new kinds of selves, new sorts of groups of people, that are as yet unnamed (Deleuze and Guattari 1988: 469–71). These new groups of people can be transnational or international in scope. The Conversation Map is intended to be a technology of the self for VLSC.

But the metrics and graphics computed by the Conversation Map could be mishandled as technologies of power. Consideration of the history of "social metrics" and their graphics is sobering, as Ian Hacking points out in his article "How Should We Do the History of Statistics?" (1991: 181): "Statistics has helped determine the form of laws about society and the character of social facts. . . . Moreover the collection of statistics has created, at the least, a great bureaucratic machinery. It may think of itself as providing only information, but it is itself part of the technology of power in a modern state." In short, statistics has been and is "*state*-istics"—a technology of that not necessarily democratically governed political form that we know today as the nation-state. As VLSC flows over and across the boundaries of today's nation-states, new, as-yet-unnamed political formations are created through text and talk. These new non-nation-state entities demand new forms of representation that exceed the statistics of the state. Participants within VLSC-based groups need not only representation but also orientation: maps, charts, interfaces, and instruments of navigation in order to locate political position and agency. If these instruments of navigation are to be for the people of democratic organization, then they cannot be—as statistics now is—a tool for only specialists and powerful "decision makers."

The Conversation Map has been designed to make it difficult to find and follow any given individual author. It has also been designed to output qualitative diagrams rather than quantitative summaries. In these ways the Conversation Map has been made "surveillance resistant." Furthermore, the "output" of the Conversation Map is a format that can be widely distributed on the Web (a Java applet that runs as an interface to the archive of messages) and thus is, in principle, accessible to any participant of an online discussion. The maps generated by the system are intended to be representative and evocative enough to allow them to function as a means of reflection for the group of people involved in the VLSC—that is, to function as a technology of the self where the self in question is the collective of discussants. But the maps are also intended to be abstract and vague enough to be difficult to use as a technology of power. It is a challenge to find a form of representation that works as a technology of the self but does not work as a technology of power. Others in the social sciences are attempting to find and/or develop such forms of representation, notably ethnographers who write for the benefit of their informants instead of, or in addition to, their fellow anthropologists (e.g., Clifford and Marcus 1986).

Obviously the calculations and interface components of the Conversation Map are far from being universally accessible, and so what is outlined above is simply the first step in a long search for democratic representations and interfaces for VLSC. This search for new forms of self-representation is intrinsic to the development of an ethics of discourse architecture, that is, design for the medium of VLSC. As Guattari states (1991: 2), "In addition to ecology, the question of ethics of media and the future direction of new communication technologies of artificial intelligence and command-and-control constitutes one of the two axes in which to rethink the idea of progress for today's planet."

References

Abelson, Robert P. 1973. The Structure of Belief Systems. In *Computer Models of Thought and Language,* edited by R. C. Schank and K. M. Colby. San Francisco: W. H. Freeman and Company.

Allen, James F. et al. 1996. A Robust System for Natural Spoken Dialogue. In the *Proceedings of the 34th Meeting of the Association for Computational Linguistics,* Santa Cruz, CA.

Anderson, Benedict. 1983. *Imagined Communities: Reflections on the Origin and Spread of Nationalism.* London: Verso.

Bagga, Amit, Breck Baldwin, and Sara Shelton, eds. 1999. *Coreference and Its Applications, Proceedings of the Workshop.* New Brunswick, NJ: Association for Computational Linguistics, June 22.

Calhoun, Craig, ed. 1992. *Habermas and the Public Sphere.* Cambridge: MIT Press.

Callon, Michel, John Law, and Arie Rip, eds. 1986. *Mapping the Dynamics of Science and Technology.* London: Macmillan.

Cappella, Joseph N., Vincent Price, and Lilach Nir. 2002. Argument Repertoire as a Reliable and Valid Measure of Opinion Quality: Electronic Dialogue during Campaign 2000. *Political Communication* 19:73–93.

Carbonell, Jaime G. 1980. Towards a Process Model of Human Personality Traits. *Artificial Intelligence* 15:49–74.

Carroll, Lewis. 1960. Alice's Adventures in Wonderland. In *The Annotated Alice.* New York: Clarkson N. Potter.

Castells, Manuel. 1997. *The Information Age: Economy, Society and Culture.* Vol. 2: *The Power of Identity.* Oxford: Blackwell.

Clifford, James, and George E. Marcus. 1986. *Writing Culture: The Poetics and Politics of Ethnography.* Berkeley: University of California Press.

Coleman, Stephen. 2004. Connecting Parliament to the Public via the Internet: Two Case Studies of Online Consultations. *Information Communication & Society* 7(1)(March):1–22.

Dean, Jodi. 2001. Cybersalons and Civil Society: Rethinking the Public Sphere in Transnational Technoculture. *Public Culture* 13(2): 243–65.

Deleuze, Gilles, and Félix Guattari. 1988. *A Thousand Plateaus: Capitalism and Schizophrenia,* translated by Brian Massumi. London: Athlone Press.

Erickson, Tom, Susan Herring, and Warren Sack. 2002. Workshop Description: Discourse Architectures: Designing and Visualizing Computer-Mediated Conversation. In *Proceedings of ACM CHI 2002.* Minneapolis, April 22.

Foucault, Michel. 1991. Governmentality. In *The Foucault Effect: Studies in Governmentality,* edited by Graham Burchell, Colin Gordon and Peter Miller. Chicago: University of Chicago Press.

———. 1997. Technologies of the Self. In *Ethics: Subjectivity and Truth; Essential Works of Foucault 1954–1984.* Vol. 1. ed. by Paul Rabinow. New York: The New Press.

Fraser, Nancy. 1992. Rethinking the Public Sphere: A Contribution to the Critique of Actually Existing Democracy. In *Habermas and the Public Sphere,* edited by Craig Calhoun, 109–42. Cambridge: MIT Press.

Freud, Sigmund. 1960. *Jokes and Their Relation to the Unconscious,* translated and edited by James Strachey. New York: Norton.

Garfield, Eugene. 1979. *Citation Indexing: Its Theory and Applications in Science, Technology and Humanities.* New York: John Wiley.

Garfinkel, Harold. 1967. *Studies in Ethnomethodology.* Cambridge: Blackwell.

Gramsci, Antonio. 1971. *Selections from the Prison Notebooks.* London: Lawrence and Wishart.

Green, Stephen. 1997. Automatically Generating Hypertext by Computing Semantic Similarity. Ph.D. diss., University of Toronto.

Grefenstette, Gregory. 1994. *Explorations in Automatic Thesaurus Discovery.* Boston: Kluwer Academic Publishers.

Guattari, Félix. 1991. Pour une éthique des médias. *Le Monde,* November 6: 2.

Gumperz, John J., Gurinder Aulakh, and Hannah Kaltman. 1982. Thematic Structure and Progression in Discourse. In *Language and Social Identity,* edited by John J. Gumperz. Cambridge: Cambridge University Press.

Habermas, Jurgen. 1991. *The Structural Transformation of the Public Sphere: An Inquiry into a Category of Bourgeois Society.* Cambridge: MIT Press.

Hacking, Ian. 1991. How Should We do the History of Statistics? In *The Foucault Effect: Studies in Governmentality,* edited by Graham Burchell, Colin Gordon and Peter Miller. Chicago: University of Chicago Press.

Hall, Stuart. 1982. The Rediscovery of "Ideology": Return of the Repressed in Media Studies. In *Culture, Society, and the Media,* edited by Michael Gurevitch et al. New York: Routledge.

Halliday, Michael A. K. 1994. *An Introduction to Functional Grammar.* 2d edition. London: Edward Arnold.

Harabagiu, Sanda, and Dan Moldovan. 1999. Enriching the WordNet Taxonomy with Contextual Knowledge Acquired from Text. In *Natural Language Processing and Knowledge Representation: Language for Knowledge and Knowledge for Language,* edited by S. Shapiro and L. Iwanska. Cambridge: AAAI/MIT Press.

Harris, Zelig. 1952. Discourse Analysis. *Language* 28:1–30, and 474–94.

Hearst, Marti. 1998. Automated Discovery of WordNet Relations. In *WordNet: An Electronic Lexical Database,* edited by Christiane Fellbaum. Cambridge: MIT Press.

Hiltz, Starr Roxanne, and Murray Turoff. 1994 (1978). *The Network Nation: Human Communication via Computer.* Rev. edition. Cambridge: MIT Press; first edition: New York: Addison-Wesley.

Ionesco, Eugene. 1958. *The Bald Soprano and Other Plays,* translated by Donald M. Allen. New York: Grove Press.

Jakobsen, Roman. 1985. *Verbal Art, Verbal Sign, Verbal Time,* edited by Krystyna Pomorska and Stephen Rudy. Oxford: Blackwell.

Krippendorf, Klaus. 1980. *Content Analysis: An Introduction to Its Methodology.* New York: Sage Publications.

La Perrière, Guillaume de. 1567. *Le miroir politiqve, contenant diverses manieres de govverner & policer les republiques, qui sont, & ont está par cy deuant: ocuure, non moins vtile que necessaire à tous monarches; rois, princes, seigneurs, magistrats & autres qui ont charge du gouuernement ou administration d'icelles.* Paris: Pur V. Norment, & I. Bruneau.

Latour, Bruno. 1993. *We Have Never Been Modern.* Cambridge: Harvard University Press.

Law, John, and John Hassard, eds. 1999. *Actor Network Theory and After.* Oxford: Blackwell/The Sociological Review.

Lessig, Lawrence. 1999. *Code and Other Laws of Cyberspace.* New York: Basic Books.

Lyotard, Jean-Francois. 1984. *The Postmodern Condition: A Report on Knowledge.* Minneapolis: University of Minnesota Press.

McHoul, A. 1994. Discourse. In *The Encyclopedia of Language and Linguistics,* edited by R. E. Asher et al., 944–45. New York: Pergamon Press.

Magerman, D. 1994. Natural Language Parsing as Statistical Pattern Matching, Ph.D. diss., Stanford University.

Malmkjaer, Kirsten. 1991. Discourse and Conversational Analysis. In *The Linguistics Encyclopedia,* edited by Kirsten Malmkjaer. New York: Routledge.

Marcu, Daniel. 1997. The Rhetorical Parsing, Summarization, and Generation of Natural Language Texts. Ph.D. diss., University of Toronto.

Munro, Alan, Kristina Hook and David Benyon, eds. 1999. *Social Navigation of Information Space.* New York: Springer Verlag.

Negt, Oskar, and Alexander Kluge. 1993. *Public Sphere and Experience: Toward an Analysis of the Bourgeois and Proletarian Public Sphere.* Minneapolis: University of Minnesota Press.

Neuman, W. Russell, Lee McKnight, and Richard Solomon. 1997. *The Gordian Knot: Political Gridlock on the Information Highway.* Cambridge: MIT Press.

Pêcheux, Michel. 1995. *Automatic Discourse Analysis,* edited by Tony Hak and Niels Helsloot, translated by David Macey. Amsterdam: Rodopi.

Pennycock, Alastair. 1994. Incommensurable Discourses. *Journal of Applied Linguistics* 15(2).

Rind, D. M., et al. 1997. Maintaining the Confidentiality of Medical Records Shared over the Internet and World Wide Web. *Annals in Internal Medicine* 127(2):138–41.

Sack, Warren. 2000. Conversation Map: An Interface for Very Large-Scale Conversations. *Journal of Management Information Systems* 17(3) (Winter): 73–92.

———. 2002. What Does a Very Large-Scale Conversation Look Like? *Leonardo: Journal of Electronic Art and Culture* 35(4) (August).

Sacks, Harvey. 1992. *Lectures on Conversation. Vols. 1 and 2,* edited by Gail Jefferson. Cambridge: Blackwell.

Schiffrin, Deborah. 1994. *Approaches to Discourse.* Cambridge: Blackwell.

Schuler, Doug, and Peter Day, eds. 2000. *Proceedings of DIAC 2000: Shaping the Network Society: The Future of the Public Sphere in Cyberspace.* Seattle: Computer Professionals for Social Responsibility.

Silverman, David. 1998. *Harvey Sacks: Social Science and Conversation Analysis.* New York: Oxford University Press.

Silverman, Kaja. 1983. *The Subject of Semiotics.* New York: Oxford University Press.

Sinclair, J. M., and M. Coulthard. 1975. *Towards an Analysis of Discourse: The English Used by Teachers and Pupils.* London: Oxford University Press.

Soergel, Dagobert. 1974. *Indexing Languages and Thesauri: Construction and Maintenance.* Los Angeles: Melville.

Teil, Geneviéve, and Bruno Latour. 1995. The Hume Machine: Can Association Networks Do More than Formal Rules? In *Stanford Humanities Review* 4(2): 47–65.

Tsagarousianou, Roza, Damian Tambini, and Cathy Bryan. 1998. *Cyberdemocracy: Technology, Cities and Civic Networks*. New York: Routledge.
Turner, Bryan S. 1996. The Micro-Macro Problem. In *The Blackwell Companion to Social Theory,* edited by Bryan S. Turner. Malden, MA: Blackwell.
Wasserman, Stanley, and Joseph Galaskiewicz, eds. 1994. *Advances in Social Network Analysis: Research in the Social and Behavioral Sciences*. Thousand Oaks, CA: Sage Publications.
Whittaker, Steve, et al. 1998. The Dynamics of Massive Interaction. In *Proceedings of the International Conference on Computer-Supported Cooperative Work,* 257–64. Seattle: Association for Computing Machinery.
Wilhelm, Anthony G. 1999. Virtual Sounding Boards: How Deliberative Is Online Political Discussion? In *Digital Democracy: Discourse and Decision Making in the Information Age,* edited by Barry N. Hague and Brian D. Loader. New York: Routledge.
———. 2000. *Democracy in the Digital Age: Challenges to Political Life in Cyberspace*. New York: Routledge.
Ziff, P. 1960. *Semantic Analysis*. Ithaca: Cornell University Press.

Transnational Communication and the European Demos

LARS-ERIK CEDERMAN AND PETER A. KRAUS

Digitization is triggering a global communication revolution that has the potential to constitute new social domains. Emerging at the intersection of technological and societal processes, digital formations are creating new political topologies and reconfiguring existing networks and organizations. The European Union (EU)is often thought of as such a border-transcending communicative space. Since the heady pioneering days of Jean Monnet, technological advances have inspired the architects of the European integration project. Today, information technology plays a prominent role in the debate about how to promote a closer union of Europe's peoples.

In this chapter, we focus on the question whether digitization could constitute a new political realm coinciding with the European Union. Can democracy take root at the European level without a "demos" encompassing the whole of the Union? How would such a popular unit have to be constituted in order to form a communicative space supporting democratic deliberations and decisions? More specifically, what are the prospects of a digital demos in Europe?

Arguing that democracy does not depend on the preexistence of an ethnic nation, Jürgen Habermas insists that the formation of a postnational polity at the European level is both desirable and possible. Inspired by Habermas's theory of communicative action, many analysts and practitioners agree that the EU constitutes a nascent public sphere offering new opportunities for democratic participation in areas that were previously beyond the reach of national governments. In their view, revolutionary advances in information technology, including the emergence of the Internet, are currently laying the foundations for a viable communicative structure spanning across Europe's national borders.

We would like to thank the anonymous reviewers, as well as Cathleen Kantner, Ruud Koopmans, Dieter Rucht, Tobias Theiler, and James Tully for their useful comments.

Other theorists, such as Dieter Grimm and Anthony D. Smith, beg to differ. They argue that despite the rapidly accelerating information revolution, democracy is culturally and institutionally rooted in the nation-state. Thus, to the extent that the latter is threatened by the integration process, the democratic credentials of the entire Union are under attack. Some of them are so alarmed that they even recommend a "renationalization" of EU legislation. In their eyes, technological fixes and obtrusive communication are likely to erode representative democracy at the national level.

Finding both positions too extreme, we introduce a third, "bounded-institutionalist" approach derived from cultural and sociological institutionalism. Without reifying the identities as ethnic essences, such a perspective rejects abstract, cosmopolitan attempts to sever political institutions from culture. This approach considers explicitly the way that the demos and democracy emerged as mutually constituting ingredients in a macrohistorical process that involved specific identity-forming mechanisms. These mechanisms generate a deeper level of socialization than that expressed by formal politico-legal arrangements, such as constitutions, elections, and party politics. While sympathizing with Habermas's emphasis on public debate, our institutional perspective turns the attention to the cultural institutions that sustain democratic governance by creating civic spaces.

Rather than favoring a predominantly ethnic definition of the demos, as many of the Euro-skeptics seem to do at least implicitly, we suggest that without an internally cohesive and externally bounded notion of polity membership, the new information technology will fall short of establishing the public sphere that Habermas calls for. What is worse, unbounded deliberation may even under some circumstances accentuate elitist tendencies, thus threatening to undermine democracy at both the national and European levels. To conceive of a truly mass-based communicative space at the European level, which remains notoriously weak at the present point, more attention needs to be paid to the formation and maintenance of cultural identities within the communicative infrastructure of the European Union.

The chapter is structured as follows. First, we define the key concepts, democracy and demos, within three ideal-typical frameworks. To add historical depth to these abstract models, the second section applies them to the development of democracy within the framework of the classical European nation-state. Exploring primarily the impact of modern information technology, section three takes the final step to today's European Union. A concluding section elaborates on the consequences for theory and policy.

Conceptualizing Democracy and the Demos

Drawing a stark distinction between domestic and international life, international relations theorists have traditionally considered democracy beyond the nation-state to be utopian at best. More recently, however, there has been a surge of interest in questions relating to democracy and legitimacy in international affairs. As the revolution in information technology makes the world increasingly interconnected, this quest has moved from being a purely academic one to becoming a practical project. Whereas international organizations have long been evaluated in terms of their efficiency and effectiveness, they are now routinely subjected to scrutiny with respect to their legitimacy (Held 1995).

Given the extraordinary density of transnational communications in Europe, it is hardly surprising that the legitimacy debate has been particularly intense in that part of the world (Stein 2001). Thanks to institutional reform, especially in the 1990s, the European Union has transformed itself from a cooperative regime with supranational overtones to a nascent polity. It is only natural, then, that the Union's democratic qualities have become hotly debated among academics and policymakers alike. As we have seen, the current sense of malaise is clearly reflected in the weakening trends of public support, as recorded in recent Eurobarometer surveys, and in the perennial ratification crises that followed the signing of major treaties.

Whereas the pioneering studies concentrated on institutional impediments to democratization in relatively technical terms (Kaiser 1971; Williams 1991), more recently the debate has come to focus on the societal infrastructure of democracy (Weiler 1999). It is precisely this agenda of polity building that is the center of our attention. To grasp the logic of the main positions in this debate, it is necessary to clarify the master concepts: demos and democracy.

The etymological meaning of democracy is "rule by the people." This formula raises the question of what the people stands for and what rule entails. The standard answer to the first question is usually all adult members of a polity, namely, the citizens. Collectively, this group is referred to as the "demos," the body of citizens (Dahl 1989: 109). Clearly, this popular unit has to be more than a random collection of individuals. For active citizenship to be meaningful, the members of the demos must be able to communicate with each other. Thus, to have any political meaning at all, the demos needs to be embedded in a "communicative space." Moreover, the citizens need to share a common identity such that they are willing to make at least some sacrifices on behalf of the collective. Traditionally, the nation has served as the prototypical example of a demos, as in

Benedict Anderson's (1991) celebrated formula of an "imagined community." Yet it would be premature to rule out other types of demoi that go well beyond the nation-state.

If we want to assess the democratic quality of a given regime, we also have to ask: *how* is political power channeled by institutions? Procedural attempts at defining democracy are likely to follow a more formal approach that takes into account the mediating role of institutions, even if efforts are usually made to arrive at as general a conceptualization as possible. Consider, for example, the following compact definition by Schmitter and Karl, who characterize modern representative democracy as "a regime or system of governance in which rulers are held accountable for their actions in the public realm by citizens, acting indirectly through the competition and the cooperation of their representatives" (quoted in Schmitter 2000: 3). In this definition, the "rule of the people" is limited to the public realm; it manifests itself in the principles of accountability, competition, and representation. Nonetheless, citizens remain the main point of reference for a due application of these principles. They "provide the most distinctive element in democratic regimes" (Schmitter 2000: 5). Thus, in this case, the demos is implicitly present in the definition.

The crux is that, quite often in history, juridical stipulations have not coincided with social and political realities (Weiler 2000). As we will argue, democracy did not develop primarily thanks to democratic motivations. The exercise of democratic sovereignty is contingent upon a collective identity that sustains the polity conceived of as sovereign. This identity, however, is itself hardly a product of democratic decision making; its roots go back to an unavoidably predemocratic past (Mann 1999).

It was Rousseau (1987 [1762]: 164) who offered one of the first clearcut descriptions of this dilemma in his *Social Contract:*

> For an emerging people to be capable of appreciating the sound maxims of politics and to follow the fundamental rules of statecraft, the effect would have to become the cause. The social spirit which ought to be the work of that institution, would have to preside over the institution itself. And men would be, prior to the advent of laws, what they ought to become by means of laws. Since, therefore, the legislator is incapable of using either force or reasoning, he must of necessity have recourse to an authority of a different order, which can compel without violence and persuade without convincing.

The paradox has remained one of the most significant blind spots in democratic theory and democratic practice. Its topicality became evident in several cases of the "Third Wave" (Huntington 1991) of democratization, in which the process of regime change led to the breakdown of established states. Rousseau's paradox appears to retain its force in those democracies where the identity of the people is a politically contested issue. Unsurprisingly, the dilemma also haunts the process of European polity-building (Dahl 1994).

Logically, there are three ways of responding to the challenge of Rousseau's paradox. One is to accept it and to argue that there has to be a predemocratic demos before democracy can be achieved. Another is to reject the whole notion of a paradox by claiming that there does not have to be any demos to begin with. Most of the debate about the European Union's lacking democratic credentials has revolved around these two antithetic positions, which we label national substantialism and civic voluntarism.[1] As a third possibility, we introduce a sociological approach, labeled bounded institutionalism, that could fill the theoretical void between the other two positions.

National Substantialism

In general terms, national substantialists insist that political life has to be based on the nation defined as a cultural community. The substantialist theme derives from their proclivity to view nations as real and given entities, although their stability may be due to a variety factors (Brubaker 1996; see also Emirbayer 1997). Yet most of these theorists converge on a conception of the nation as a closely knit community held together by a "thick" sense of culture (cf. Walzer 1994). In its critique of liberal individualism, then, substantialist theory emphasizes the social and cultural embeddedness of political actors. Democracy, in particular, requires the presence of a deep sense of shared meaning. Based on the doctrine of popular sovereignty, this perspective stipulates that culturally distinct nations serve as the only viable demoi, and the nation-state is the only acceptable framework for the constitution and reproduction of these demoi. Asserting that the locus of democracy has to remain the nation, national substantialists therefore resist attempts to solve the European Union's legitimacy dilemma by the introduction of supranational arrangements.[2]

By identifying the demos with the nation, substantialist scholars make

[1] Elsewhere, Cederman (2001a) uses the terms "ethno-nationalism" and "post-nationalism" to make a similar distinction. Kraus (2003) refers to "Westphalians" and "cosmopolitans" in a related context. These two camps overlap to some extent with the well-known controversy in political theory between communitarians and liberals (Mulhall and Swift 1992). It should be noted, however, that this literature often draws different conclusions about the role of the nation-state.

[2] It should be noted, however, that in principle, the substantialist logic could be applied to cultural communities at different levels of aggregation. There are also those theorists who believe a full-fledged cultural pan-European identity to be possible (see references in Cederman 2001a). Others favor community building below the level of the nation-state. Yet *national* substantialism refers exclusively to the level of the nation-state. In this chapter, we will have this type of theory in mind even when we drop the national qualification.

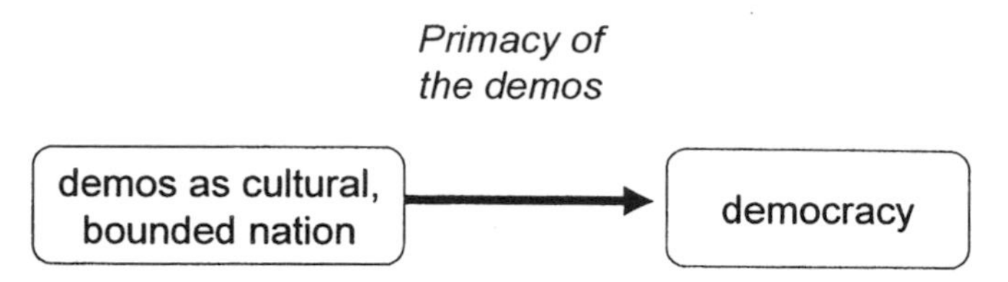

Figure 1. The logic of national substantialism

two basic assumptions. First, they view the national demos as existing logically and historically prior to the democratic process. Without the presence of such a demos, democratic governance has few chances to develop since it relies on common values and profound sense of community. Second, and related to the first point, it is assumed that the demos is culturally constituted in a deep sense. Liberal notions of thinner, multicultural identities are seen as too brittle to carry the weight of democratic decision making. Figure 1 summarizes this logic based on the "primacy of the demos."

There are both applied and theoretical illustrations of this type of reasoning. For example, in an important decision reached in 1993, the German Constitutional Court examined the democratic legitimacy of the European integration process. After ten months of deliberation, the court "acquitted" the Union since it contended that, in the absence of a culturally defined demos, the question of democracy does not even apply: "On this view, a parliament without a demos is conceptually impossible, practically despotic" (Weiler, Haltern, and Mayer 1995: 13). Similar thinking can also be traced in many strands of scholarship including the ethnocultural approach of A. D. Smith (1992; 1995), the intergovernmentalism of Hoffmann (1995), and the constitutional arguments of Grimm (1995) and Kielmansegg (1996).

Civic Voluntarism

Pitting their theory against what they perceive as cultural determinism, civic voluntarists reject national substantialism on the grounds that it reifies the demos and ties democracy to the nation-state, thus ruling out "postnational" options for the future (Habermas 1998).[3] Firmly rooted in a liberal, cosmopolitan tradition dating back to the Enlightenment, these scholars reverse the substantialists' basic assumptions.[4] First, rather

[3] Going even further than that, Brubaker (1998: 15) alleges that civic voluntarists and other critics of nationalism adhere to a substantialist viewpoint in their depiction of the nation as a case of "false consciousness."

[4] Some critics of substantialism attempt to circumvent the cosmopolitan position (see

than following the establishment of a culturally defined, in their view, prepolitical demos, democratic politics should (and typically does) precede it, since it has the potential of bringing people together through deliberation. Thus, it is the conscious act of constitutional choice that sets up the foundations for democratic practice. To circumvent Rousseau's paradox, civic voluntarists typically replace traditional electoral definitions of democracy with deliberative and procedural definitions. According to Jon Elster (1998: 8), deliberative democracy can be defined as collective decision making (1) "with the participation of all who will be affected by the decision or their representatives" and (2) "by means of arguments offered *by* and *to* participants who are committed to the values of rationality." Here the main stress is on the mode by which democratic decisions are reached. While requiring the procedure to be both rational and inclusive, deliberate democracy reduces the criterion of civic membership to those who are affected by particular decisions. This pragmatic solution implies that there does not necessarily have to be a single, stable, ascriptive demos.

Second, once, or if, the demos materializes, its nature differs dramatically from the "thick" notions favored by national substantalism. Explicitly downplaying the cultural contents of political identities, civic voluntarists assert that what holds people together is the commitment to political principles, something that Habermas (1992a; 1992b) has labeled "constitutional patriotism." While culture cannot be entirely eliminated as the foundation of democratic politics, it should be kept at a bare minimum capable of sustaining political communication.

Again, we sum up this reasoning with the help of a diagram. Figure 2 shows how the voluntarist perspective reverses the substantialists' logic. Instead of starting with the demos, democratic practice grows out of constitutional deliberation. The dashed arrow indicates that the step to the demos is optional and less important than the democratic process itself.

It goes without saying that these two assumptions make it easier to envisage a postnational future for democracy. There is no need to wait for a cultural demos to materialize at the European level. In fact, civic voluntarists often view such a high level of ethnic cohesion with suspicion, due to its exclusionary impact on immigrants and internal minorities. Most voluntaristic scholars, including Habermas, call for a constitutional process, which could serve as a social contract and provoke a debate in the wake of which political identity-formation would set in. At a more mundane level, the voluntarist perspective assumes that European public spaces emerge as "resonance structures" in response to supranational institution building (Eder 2000). For example, many of the everyday activities of the European Union, including countless committee meetings that

Sassen in this volume), but this possibility falls outside the purview of civic voluntarism as we have defined it here.

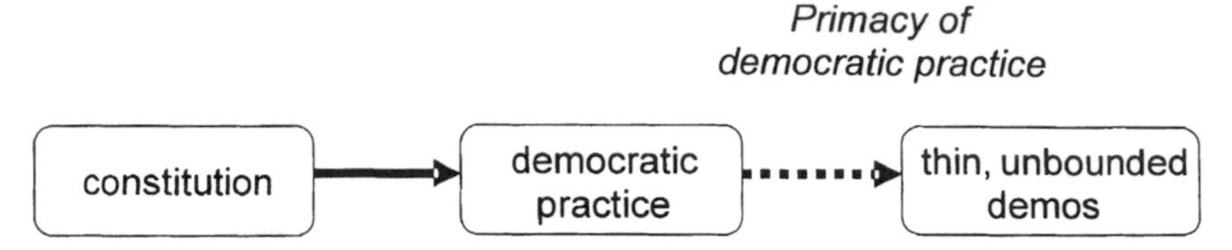

Figure 2. The logic of civic voluntarism

are often referred to as "comitology" (Joerges and Neyer 1998) and a thickening web of nongovernmental organizations (Zürn 1998; 2000), could contribute to identity-building deliberative practices. Other presumably identity-conferring civic measures include referenda (Zürn 1998; Schmitter 2000) and the creation of a truly European party-system (e.g., Börzel and Risse 2001).

Bounded Institutionalism

While sympathizing with the Habermasian project, we contend that it underestimates the infrastructural difficulties implied by Rousseau's paradox. In particular, the postnationalist vision fails to provide a concrete account of how political communication would materialize despite institutional and cultural divergences. The root of the problem relates to Habermas's own framing of politics in discursive and interpersonal terms within an idealized "lifeworld." In an era of mass media and nationalism, however, such a perspective fails to offer a complete rendering of the principles governing modern politics. In fact, to a large extent, political life hinges on culturally and symbolically mediated indirect relationships (Simmel 1971 [1908]; Calhoun 1991; 1992b). In contrast to premodern society, which was based on direct interpersonal relationships, the large scale of the nation requires abstract categorization: "In modern societies, culture does not so much underline structure: rather it replaces it" (Gellner 1964: 155; see also Gellner 1983; Anderson 1991).

In his constructive and sympathetic critique, Craig Calhoun (1991: 101) faults Habermas for partly overlooking the systemic setting within which political communication takes place. What is needed is an explicit theory of infrastructural technologies that together constitutes the "scaffolding of social integration":

> Habermas's failure to develop this sort of foundation for his argument contributes to several problematic aspects of his generally stimulating and powerful theory: its difficulties in achieving cultural and historical specificity; its too-uncritical acceptance of the systems-theoretical description of systemic integration; its tendency to idealize life-world relationships; and its underdevel-

oped account of practical, situated activity that cannot readily be reduced to purely communicative, strategic, or rational action.

We argue that these critical points still hold up well, both with respect to Habermas's own, more recent writings on European postnationalism and especially to the work of his followers in IR. There is still an Enlightenment-inspired undercurrent of technological optimism that can be traced all the way back to the early, heady days of systems theory. It could be argued that the voluntarist project resembles David Mitrany's (1966) technocratic functionalism more than Ernst Haas's (1958) pragmatic and politically grounded neofunctionalism or Karl Deutsch's (1966) communication-based integration theory (for an overview, see Keohane and Nye 1975).

In an attempt to rectify these shortcomings without letting the pendulum swing back to a substantialist position, we propose an alternative perspective that renders categorical identification mechanisms explicit. Such a sociological and cultural strand of theory, which we label bounded institutionalism, conceives of political institutions in broader, cognitive terms than the voluntarists' narrow notions of political culture (Hall and Taylor 1996). Based on such analysis it is possible to reassess the two main assumptions pertaining to the priority of the demos and its nature. With respect to the first assumption, we advocate a bi-directional logic featuring a gradual developmental process that links the demos and democracy. Rather than favoring primacy of either the democratic unit or democratic practice, this "tandem hypothesis" features an ongoing exchange in both directions.[5] In agreement with voluntarist thought, democratic practice is believed to help shape popular cohesion, but unlike in that perspective, the demos can be expected to play an active, constitutive role as democratization proceeds and thus should not be seen as a mere side-effect of democratic practice. In addition, and crucially, peoplehood does not emerge merely as a result of a voluntaristic bargain or everyday politics but may require a considerable degree of cultural standardization: "From an institutional perspective, comprehensive change in a political order involves not only affecting human conduct and formal-legal institutions, but also affecting peoples' inner state of mind, their moral and intellectual qualities, their identities and their sense of belonging" (Olsen 2001: p. 173). Historical examples typically feature explicitly identity-forming mechanisms operating both inside and outside the polity. Whereas educational institutions (Gellner 1983), linguistic policy (Brass 1991), and mass media establishments (Anderson 1991; Schlesinger 1991; Warner 1990) belong to the internal category, wars and other types of exchange fall into the external category (Cederman 2001b).

[5] More recently, Habermas (2001) has come to embrace this hypothesis as well, though he still tends to underestimate the cultural implications of identity formation.

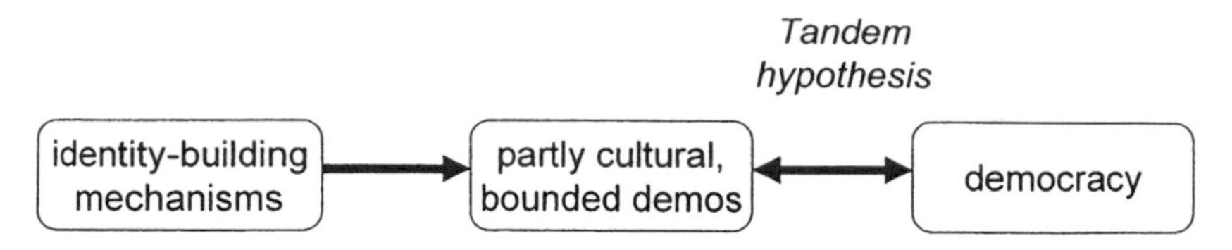

Figure 3. The logic of bounded institutionalism

The second assumption, pertaining to the very nature of the demos, also leaves plenty of room between the other two extreme positions. Compared to the thickly ethnic and reified definition favored by substantialists, we qualify the extent of cultural coordination required by democratic rule, but not to the degree suggested by civic voluntarists. Democracy presupposes a measure of value convergence and cognitive compatibility. Such a motivational and cognitive compatibility rests on shared communicative resources, though not necessarily a single lingua franca. Moreover, as the label would indicate, our institutionalist alternative emphasizes the need for a bounded demos (March and Olsen 1998; Schlesinger 1999). Yet this institutionalist perspective differs from substantialist principles in that it does not insist that the boundaries be inert and ethnic in most respects. Whereas substantialists usually take the popular unit for granted as a cultural fact, we insist that the demos be problematized by uncovering the institutional mechanisms responsible for creating and maintaining it. In fact, boundaries may persist despite significant cultural change and interaction flows across them (Barth 1969).

Figure 3 centers on the "tandem hypothesis" that links a bounded demos together with the practice of democracy. Unlike the two other approaches, our strand of cultural institutionalism brings in a set of specifically identity-conferring institutions. Instead of constitutional bargaining triggering the process, the institutional perspective calls for specific identity-building mechanisms that are at least partly independent of democratic decision making.

How could one tell which of these three perspectives will serve as the most useful guide to transnational democratization? Following the lead of Craig Calhoun, we suggest that a historical evaluation might be helpful in this connection. Robert Dahl's (1989) three "transformations" of democracy represent an appropriate analytical starting point. If the first transformation that led to direct democracy emerged in the Greek city states, it was within the framework of the nation-state that democratic rights for the masses finally developed. Thus, the current issue concerns democracy's transformation to a third, postnational stage (see also Held 1995). To contextualize the issue along these lines, the following section deals with democracy during the second transformation, followed by a section that discusses the third transformation to the supranational level.

The Emergence of National Demoi in Europe

The advances in communication technology that flowed from the industrial revolution prompted an immense increase in the scale of democratic politics. Ushering in a transition from citizens' local assemblies to representative government, the historical changes at work were reflected in vast modifications of the democratic polity's institutional framework. In the course of the second transformation, such specific features of the democratic process as "enlightened understanding" (Dahl 1989: 111) and political participation transcended the narrow settings of traditional civic activities. In this process, they became related to extensive and complex units of government that had to be conceived of in highly abstract ways by their citizens as opposed to the much more palpable world of the ancient polis.

In modern European history, the communicative space of a democratic political unit has largely coincided with the communicative space of the nation-state. The vision that the Enlightenment thinkers had of a cosmopolitan *république de lettres* in Europe never materialized. Reflecting a varying set of politico-cultural background situations, the democratic message was to be delivered in different tongues. Modern European democracies were generally built upon an encompassing context of political communication shared by their citizens; this communicative space constituted the public sphere (Calhoun 1992a).

Nationalism forged the links between the democratic public and the communicative infrastructure that such a public required. Seen in this way, nationalist politics relates the processes of social and political communcation within a population to those of a people. Thus, in political discourse, the nation tends to become coterminous with the demos. One can argue with Pizzorno (1986: 369) that the process of democratic representation serves the purpose of interpreting and expressing the nation's collective identity.

In their classical studies, Deutsch (1966) and Gellner (1983) capture the close bond between nationalism and modernization. The main challenge concerned the bringing together of diverse communities that had previously had few interactions and mostly relied on direct interpersonal contacts for their internal communication. As we have seen, Gellner (1964) suggested that the solution lay in the invention of "high culture," that is, cultural abstractions that created a bond between people who had never met, and most likely never would, a construction that Anderson (1991) refers to as an "imagined community."

Reassessing Rousseau's paradox in the context of the emerging democratic nation-states, we would like to know which of the three analytical perspectives conforms most closely to the historical record. If the national substantialists were right, democracy would virtually always have to be preceded by a culturally constituted demos in the form of a nation. Even a casual glance at the country cases tells us that this perspective cannot be accurate, because even where a stable and ideal-typical nation was supposed to have existed, as in revolutionary France, it took almost an entire century to extend the French national identity beyond the cultural core of the Île de France (Weber 1976). Other cases, such as the German lands in the early nineteenth century, may have constituted a reasonably homogeneous *Kulturnation,* but this did not provide a stable basis for democratization. At least within the context of Dahl's second transformation, successful democratization does not hinge directly on the preexistence of a culturally defined demos.

If substantialist theory can be easily dismissed as a flawed guide to democratization in the age of the nation-state, it is less obvious whether civic voluntarism or bounded institutionalism offers the best account of this historical transition. To find out, it is necessary to return to the core assumptions of both approaches. A voluntarist interpretation would be vindicated if and only if there are historical cases in which (1) democratic decision-making preceded the formation of the demos and (2) it was democracy that bore the main responsibility for the creation of the demos. Our institutionalist case questions both assumptions. For we do not only contend that the timing between democracy and demos was much less of a one-way street than the civic voluntarists would have it. In addition, we suggest that, historically, the reasons for demos formation can be found in processes unrelated to, and sometimes even in contradiction to, democracy.

We start by addressing the question of historical sequencing of democracy and nationalism in Europe. To evaluate the plausibility of the voluntarist account, we turn to three seemingly "easy cases," namely, France, Britain, and Switzerland. France is also often considered to be a classical example of a European "state-nation." If the French Revolution is seen as the introduction of democracy, it could possibly be argued that it preceded the development of a national demos. Following Eugen Weber (1976), we know that it took most of the nineteenth century to turn "peasants into Frenchmen." The problem with this argument, however, is that it exaggerates the degree and depth of democracy in revolutionary and early postrevolutionary France. Indeed, French representative democracy did not mature until well into the twentieth century. After 1789, and especially in the early period of the Third Republic, state elites tended to consider Jacobin nationalization policies to be an indispensable requirement for democratization (Birnbaum 1998; Jacob and Gordon 1985).

Likewise, the British case casts doubt on the presumed temporal precedence of the national demos over democratic institutions. It is true that Britain is often seen as the pioneer of parliamentary democracy in Europe, but it was also one of the forerunners in the process of nation-building (Greenfeld 1992). Also in this case, it seems more reasonable to argue that the development of democracy and nationhood coincided rather followed upon each other.

Against the background sketched out so far, Switzerland represents perhaps the only case of democracy clearly preceding the nation. The process of democratization set in at an early point. Based on a conventional interpretation of the Rütli myth, one could even argue that democratization, at least as far as the male population was concerned, parallels state-formation.[6] At the same time, however, Switzerland's national identity remained weak and did not emerge until the late nineteenth century (Sciarini, Hug, and Dupont 2001).

It is thus not a coincidence that civic voluntarists, such as Habermas (1992b), refer to Switzerland as a paradigmatic example of their theory. Whereas they may be right in terms of the timing, it does not confirm the second assumption regarding the nature of the causal mechanisms. In fact, it is far from clear that democratic politics itself was the prime force driving demos formation. Such a liberal assumption appears to overlook the operation of both internal and external mechanisms that are mostly unrelated to democracy.

To begin with the latter dimension, warfare and external security threats loom large as processes promoting nation-building in all three cases. It is impossible to make any sense of Swiss identity without reference to the institution of neutrality that solved the federation's security dilemma after repeated conflict with the Habsburgs and other neighboring powers (Sciarini, Hug, and Dupont 2001).[7] Likewise, the French and British national identities emerged from a series of wars stretching well into the eighteenth century in and beyond Europe (Tilly 1990). For example, Linda Colley (1992) has persuasively shown that British identity formation owed much to the fights against Catholic France. In all these cases, the expedience of protracted warfare, rather than a conscious contractual choice, shaped the nation as demos. Eventually political and social citizenship was gradually extended to the masses as a recognition of their combat efforts.

Analysis of the internal dimension of demos formation turns the attention to the institutions that create and support a political communicative

[6] A more modern reading dates state formation at the establishment to the Swiss Federation in 1848.

[7] These authors also argue that direct democracy contributed to Swiss nation-building, but it should be recalled that this type of political participation coincides with Dahl's first generation rather than with the second, representative one.

space. Again it would be a mistake to interpret this process exclusively, or even primarily, from a voluntarist standpoint. Quite on the contrary, democratic integration often went hand in hand with coercive cultural homogenization. According to Michael Mann, the push toward cultural uniformity was not an anomalous offshoot of political modernization but was built into the dynamics of democratization itself. Ethnic cleansing and other types of assertive identity formation thus became "the dark side of democracy" (Mann 1999).

In Mann's provocative line of argumentation, this dark side does not only appear in the organic democracies established in Central and Eastern Europe during the peak period of nationalist mobilization. Although with some qualifications, Mann also includes the typically "liberal" democracies of Europe's Northwest (and of North America) in his "dark" account. Such an approach may seem extreme, but Mann offers plenty of historical evidence, reaching from Britain's attitude toward the Irish to the generalized ethnic "readjustments" of the European interwar period, to corroborate the argument that the constitutional affirmation of popular sovereignty ("*we,* the people") did not necessarily clash with the political continuity of *Herrenvolk*-mentality ("we, *the* people"). Accordingly, the present prevalence of culturally homogeneous patterns in a large majority of the European Union's member states should not be interpreted as the result of a spontaneous expression of free constitutional choice; it has to be attributed to the conscious use of power by state elites who claimed to hold a democratic mandate while amalgamating *the* people in opposition to the national "other."

In this connection, it is necessary to consider formative mechanisms that underpin ongoing political practice, such as language policy and educational institutions. Jules Ferry's famous educational reforms are a case in point (Weber 1976). The French Republic was deliberately built on the principle of cultural uniformity, which was interpreted as a necessary condition of civic equality. Secularism and *francophonie* are seen as the safeguards of an integrated system of political communication (Laborde 2001).[8]

Similarly, it should not be forgotten that British identity formation was anything but smooth in the periphery. In fact, national integration and democratization had severe limitations at the Celtic fringe. Michael Hechter (1975) even goes as far as calling this process of cultural homogenization "internal colonialism." This is especially clear in the Irish case: Ireland never became British. Moreover, both the "greatness" and "Britishness" of Great Britain are still being challenged in Wales and Scotland. To the

[8] For an elite-driven perspective on Switzerland, see Altermatt et al. (1998).

extent that this will affect the structural foundation of the British state, it will have an impact on the continuity of its demos too (Nairn 1977).

To conclude this section, it seems that in most historical cases, common cultural identities were not simply forged through the negotiation of civic contracts. Where the demos had not been created under predemocratic conditions, states claiming to pursue democratic goals made deliberate attempts at turning the population into a culturally homogeneous people. Hence, in these cases, popular sovereignty came into being not so much as the manifestation of a collective will articulated from below as through a state-led, top-down process devoted to "people-making." The demos of constitutional theories, then, looks more like the product of institutional fabrication than like the original embodiment of political sovereignty.

Where does this brief account of nation-state formation and democratization in the West European context leave us? On the one hand, the sequencing of political developments seems to matter especially when the beginning of democratic mass politics complicates late state-building and contributes to a short-circuiting of national and democratic development. On the other hand, the relationship of democracy and nationalism is historically contingent. Culture cannot be politically discounted, as civic voluntarists tend to do. However, the existence of a cultural nation should not be considered as a prerequisite of a successful democratization either, as the substantialists suggest. Thus, we conclude that bounded institutionalism provides a more promising account of Dahl's second transformation. Focusing on the new communication technologies, we now turn to an analysis of the third transformation.

A European Demos from Cyberspace?

Will new sociodigital formations open new channels of political participation at the transnational level that can take the place of the previously established structures of representative democracy? The answer to this question hinges on more than technology. Indeed, the sociocultural context of the new mechanisms determines the chances of successful democratic governance. Uneven and insufficient spread of information technology threatens to cause social exclusion both within and among the European Union's member states. In this section we argue that empirical evidence casts doubts on the viability of a digital demos in Europe, at least within the forseeable future.

In terms of normative and policy analysis, there seems to be a consensus on the central role that communication and information will have to play in order to enhance the opportunities for democratic participation in the complex system of European governance. Implicitly or explicitly, the use of new communication technologies is expected to play a crucial role in this context. In the "White Paper on European Governance," published by the Commission of the European Communities in Brussels in 2001, we find a section outlining proposals for reforming Europe's institutional framework with the following statements:

> Democracy depends on people being able to take part in public debate. To do this, they must have access to reliable information on European issues and be able to scrutinise the policy process in its various stages. . . .
>
> Information and communication technologies have an important role. Accordingly, the EU's EUROPA Website . . . , is set to evolve into an inter-active platform for information, feedback and debate, linking parallel networks across the Union.
>
> Providing more information and more effective communication are a precondition for generating a sense of belonging to Europe. The aim should be to create a trans-national "space" where citizens from different countries can discuss what they perceive as being the important challenges for the Union. This should help policy makers to stay in touch with European public opinion, and could guide them in identifying European projects which mobilise public support. (11–12)

Throughout, the document stresses the dissemination of information especially in connection with communication technologies. Since the launching of the Information Society Project in the 1990s, the European Union, and especially the Commission, has indeed put forward several important policy initiatives in order to promote the idea of e-government (Chadwick and May 2003: 272). After the presentation of the Bangemann Report (European Union 1994), with its recommendations concerning "Europe and the global information society," EU institutions have been playing a prominent role as supporters of the infrastructural changes associated with the spread of new information and communication technologies.

Regarding the approach adopted in the "White Paper on European Governance", it would seem that the experts working on behalf of the Commission drew inspiration from the widely acclaimed trilogy Manuel Castells (1996, 1997, 1998) devoted to a thorough analysis of the "Information Age." According to Castells (1998: 318–32), the European Union is the most advanced political response to the globalization process, including the challenges this process involves in the field of communication technology. The Union is interpreted as the foremost manifestation of an emerging new kind of polity: the "network state." Not only does it represent an original form of linking an integrated economic zone to variable and decentralized "nodes" of political authority; it also constitutes a particular communicative space, in which information flows bypass the control of nation-states and cultural identities are transformed (Castells 1998: 324). However, Castells' approach leaves open to which extent, and in what specific ways, information technology could be used to give the communicative space of the European "network state" the shape of a democratic transnational public sphere.

To find a more concrete treatment of the communication theme, we turn to the perspective adopted by Joseph Weiler, one of the main scholarly authorities in matters of European constitutionalism. With the explicit purpose of "discussing some proposals concerning the technology of transnational democracy," Weiler (1999: 349) brings the Internet into focus. The plan, which is presented under the suggestive label Lexcalibur, consists in creating a virtual "European Public Square." This should be an Internet web site covering "the entire decision-making process of the Community, especially but not only comitology" (351). By facilitating the access to important informations and enhancing the transparency of the policy process, Lexcalibur is meant to enrich the trans-European public debates and to serve as a *virtual* resource for strengthening the participatory dimension in the world of *real* politics. The goal is to use the Internet to create more opportunities for both collective actors and individual citizens to get involved in EU politics, thus improving the Union's legitimacy and increasing its democratic potential. As Weiler suggests, in such a scenario the Internet "is to serve as the true starting point for the emergence of a functioning deliberative political community, in other words a European polity-cum-civic-society" (353).

There can hardly be any objections to plans to put the Internet to democratic use. In the current discussions about transnational democracy in Europe, however, there is a strong tendency to blur the line between the normative and the empirical analysis of the contribution that communication technology may make to restructuring the democratic public space. Quite often it is taken for granted that what the Internet *could* and *should* do *is* already becoming an empirical reality. Yet, unfortunately, noble political intentions do not always find immediate reflection in actual trends.

This observation is confirmed by some basic facts that should be taken into account in a provisional assessment of the impact the infrastructural changes in the sphere of information have had on the making of a democratic public sphere in the European Union. Eurobarometer surveys that contain data on media use and access to modern information tools in the Union report that the increase in access to novel forms of information technology in Europe during the last few years has been tremendous. Thus, the household access rate to the Internet at the level of the European Union, which had been 28 percent in October 2000, reached 43 percent in November 2002 (Flash Eurobarometer 2002: 5). Internet connectivity in Europe is being extended swiftly. Nevertheless, it should also be noted that the distribution of Internet resources remains remarkably uneven across EU member states. Whereas the access rate is above 60 percent in Sweden, Denmark, and the Netherlands, it is as low as 14 percent in Greece, with Portugal and Spain at 31 percent, Italy at 35 percent, France at 36 percent, Germany at 46 percebt and the UK at 50 percent (Flash Eurobarometer 2002: 4; for international figures, cf. Choucri 2000).

Moreover, the evidence provided by Eurobarometer data sheds some light on an aspect of Internet use that seems particularly relevant for the context of our discussion, namely, the formation of a European public sphere in cyberspace: Up to now, the Internet has not developed a high profile as a source of information about the European Union. When European citizens look for such information, 59 percent watch television, 35 percent read the daily newspapers, 23 percent turn on the radio, and 19 percent consult information brochures. Only 15 percent of the respondents use the Internet as a source of information on European matters, in spite of the considerable efforts EU institutions have put into developing attractive and highly accessible web sites.[9] At the same time, notwithstanding the sharp increase in the use of new information technology experienced in "e-Europe," there has not been a parallel development in the self-perceived knowledge of EU affairs expressed by the public. While the proportion of people considering they knew "quite a lot to a great deal" about the European Union was 24 percent in spring 1999 (Eurobarometer 1999), the corresponding value for 2001 went down to 21 percent (Eurobarometer 2001b) and increased again to 27 percent in spring 2003 (Eurobaromètre 2003). While based on a short time interval, the trend shows that there is no automatic correspondence between the technological infrastructure and the transnational consciousness of European citizens.

So far, then, the Internet's transformative effects on the European public space have been relatively modest, at least as far as the domain of political mass communication is concerned. If we put it bluntly, in the empire of new information and communication technologies, television is still king. Thus, according to data presented by the European Commission's DG Press in 2003 (4–6), almost all Europeans (98 percent) watch television, with news and current affairs being the most watched kind of program (89 percent). In contrast, a majority of Europeans (53 percent) still do not use a computer, and only one-third (35 percent) surf the Internet. The example of the Internet shows that the use of new information technology does not necessarily widen the communicative range of

[9] For a more detailed listing of other less important sources of information on the EU, see table 4.2 in Eurobaromètre Standard 59 (2003).

Europe's political space; rather, it highlights a more general, elitist tendency that has haunted the formation of a European public sphere since the political integration process began to intensify in the mid-1980s. As Philip Schlesinger (1999) argues, there *is* evidence of the emergence of new communicative spaces in Europe. However, the scope of these spaces is severely constrained. The data collected in Eurobarometer surveys reveal that the groups who allegedly have the highest levels of knowledge on the European Union are managers and those who acquired the best education; at the opposite end, we find women and the unemployed. Schlesinger (1999: 271) maintains that business surveys more and more frequently take the use of the World Wide Web and of e-mail as an indicator of elite status.

Against this background, one has to keep in mind that, as Mansell and Steinmueller (2000: 39–45) have pointed out, the formation of a European cyberspace is closely linked to changes in the citizenship status and the emergence of new forms of social inequality: Internet access and use are related to issues of inclusion and exclusion. Accordingly, advanced information and communication technologies, while offering a potential instrument against exclusion, can also contribute to new types of disadvantages. The extent to which new technologies can be used for overcoming exclusion does not depend on technology per se; it is a matter of institutional provisions that may contribute to reduce the cognitive barriers that restrict the use of new informational assets by socially excluded groups. As a matter of fact, costs and lack of skills turn out to be one of the main reasons Europeans indicate for not using the Internet, according to a survey realized under the auspices of the Commission (Eurobaromètre 2001: 5).

As long as a proper "virtual citizenship" regime is not established by political means, the Internet might even be contributing to sustaining the top-heaviness of the structures of transnational communication in Europe. Both inside and outside the realm of new information technologies, the emerging sphere of interrelated European publics is basically a restricted communicative space occupied by elites (Schlesinger 1999: 276). Public communication flows are typically channeled through specific print media; the *Financial Times* can be considered as the most characteristic transnational press organ for political and economic elites with a strong interest in European affairs.[10] In the context of European elite communication, English has been successively consolidating its position as a nascent lingua franca (de Swaan 1993).

[10] In a report published in the German weekly *Die Zeit,* the *Financial Times Europe* is considered to be the only daily newspaper articulating a European public sphere; cf. *Die Zeit* 29/2001.

Having already spoken of the cognitive barriers that limit a broader access to the emerging European cyberspace, we would like to single out briefly the question of language. As virtual as it may be, political communication in cyberspace still requires a linguistic medium. Irrespective of the communication technology at work, this medium will be related to language communities with palpable cultural identities in the real world. Hence, the issue of using the Internet for the sake of transnational democracy should not be addressed without taking into account its linguistic aspects. The politics of language has often remained a marginal topic in the discourse on transnational democracy. At the same time, EU institutions have been making remarkable efforts to prevent language from becoming the subject of a potentially explosive public debate in Europe (Kraus 2000). It is quite symptomatic that in the whole White Paper on European Governance, the language question receives scant attention beyond a short note on the relevance of guaranteeing the linguistic accessibility of information on the European Union for a broad European public. According to the document, this implies delivering information in all the official European languages (eleven at present), "if the Union is not to exclude a vast proportion of its population—a challenge which will become more acute in the context of enlargement" (Commission of the European Communities 2001: 11).

Nonetheless, the current practice adopted by European institutions when using the Internet for informational purposes does not always match this ambitious objective. On the Commission's web site, many documents are available only in English or French, with German being the most likely third candidate. The limited capacity of the EU translation services is obviously a constraining factor in this respect. It is bizarre, however, that information about a regional development project financed with European funds and located on a North Frisian island (belonging to Germany) is given only in French on the official EU web site. Similarly, it is difficult to understand why an EU project carried out to help homeless children in Palermo gets Internet coverage only in English.[11]

Communication problems of this kind go well beyond the anecdotal level. According to the special Eurobarometer report on *Europeans and Languages* (Eurobarometer 2001), 47 percent of all Europeans declare to know only their mother tongue. According to the survey, which was conducted in December 2000, English is unsurprisingly the leading foreign language in the European Union: 41 percent of Europeans know English in addition to their mother tongue. The corresponding figure for French is 19 percent, with 10 percent for German, 7 percent for Spanish, and 3

[11] The examples are taken from an article "Englisch bevorzugt" published in the *Frankfurter Rundschau* on July 6, 2001.

percent for Italian. It should be noted that these are very rough and approximate findings, based on the self-evaluation of the respondents. In particular, they do not reveal any precise information on levels of linguistic proficiency. Thus, when the respondents were asked if they can take part in a conversation in a language other than their mother tongue, the percentage for English goes down to 32 percent. The proportion given for French is 11 percent, and only German remains relatively stable with 8 percent. In spite of all necessary reservations concerning the reliability of such figures, we conclude that linguistic barriers in Europe's emerging ensemble of communicative spaces remain fairly high. Moreover, foreign language skills are distributed very unevenly, both socially and geographically, displaying a pattern similar to the use of the Internet. In brief, the linguistically and informationally versatile citizen, who is prepared to get actively involved in European public debates, belongs to the upper strata of society and lives disproportionally in northern or central Europe rather than on the Union's Latin rim. All in all, the democratic potential of transnational Internet use will remain limited, as long as "user resource issues, such as ability to receive and interpret information" (Chadwick and May 2003: 272) are neglected.

New information technology itself is unlikely to transform the complicated interplay between the political and the cultural dimensions of European integration. It should be added that the situation observable in the field of other communication technologies does not point toward the quick formation of an integrated transnational public sphere either. Both broadcast media and the press, which appear to play a more important role in the political process than the Internet, still suffer from national fragmentation, in terms of both production conditions and consumption habits (Gerhards 1993; 2000). For example, Gerhards (2000: 294) cites figures showing that the European (as opposed to national and international) coverage in German broadsheets didnot move above 10 percent from 1951 through 1995.

According to a well-known argument, the invention and dissemination of print technologies originated fundamental changes in the patterns of human communication and played a decisive role in the making of the modern nation-state as a historically new form of political organization (Deutsch 1966; Anderson 1991; Warner 1990). At present, it would be premature to try to assess the actual effect of the Internet in fostering the birth of large-scale political communities transcending national borders. So far, however, the European situation does not offer much evidence that new information technology will swiftly lead to more extensive forms of political integration at the transnational level.

We have argued that bounded institutionalism provides the best understanding of how democracy developed during Dahl's second transformation from the city state to the nation-state. In principle, this does not automatically imply that this perspective outperforms national substantialism and civic voluntarism with respect to the transition beyond the nation-state. To be sure, the structural conditions of democratization and demos formation in the latter phase differ dramatically from the former. Whereas the European nation-states, and thus their demoi, formed at least in part as a result of warfare, it seems quite unlikely, and equally undesirable, to rely on such a mechanism in the case of the European Union (Flora 2000). In fact, the European integration process was initiated to rule out this eventuality. While militarized crises cannot be entirely excluded, the most likely external means of identity formation relies on more peaceful exchange processes pertaining to the movement of goods and people, and boundary drawing relating to the European Union's enlargement process (Wæver 1996; Cederman 2001b). In view of the worsening transatlantic relations, it is not surprising that some intellectuals have revived the possibility of building a political platform that challenges the unilateral bias of U.S. foreign policy.[12]

The historical differences do not concern only the external dimension. The challenge of cultural unification was less daunting in nineteenth-century Europe because the units to be integrated were mostly inhabited by unmobilized populations. By contrast, today's situation within the European Union differs from this picture because the member states are all democratic and nationally mobilized. Where political identities are already activated, attempts at assimilation are likely to backfire (Deutsch 1966).

For these reasons, the challenge of continued integration and identity building has become more arduous than during the second transformation, despite the recent revolution in communication technology. Notwithstanding the scarcity of hard evidence confirming our theoretical expectations, we believe that our bounded version of institutionalism should apply even more readily to socio digital formations in the current phase of integration. In this sense, our perspective stresses the dynamics of "citizenization,"[13] and the cognitive processes by which people learn to act as citizens in a digital world.

Conclusion

What does the bounded institutionalist outlook entail in terms of future theorizing and policymaking? We have made two key assumptions that

[12] In a newspaper article cosigned with Jacques Derrida, Habermas interpreted the public outcry triggered by the U.S.-led invasion of Iraq as the expression of a nascent public sphere in continental Europe. See the *Frankfurter Allgemeine Zeitung*, May 31, 2003.

[13] We take this term from Tully (2001: 25).

have important consequences for the debate concerning the European Union's legitimacy crisis. First, we have postulated that the demos and democracy have to develop in tandem. A mere expansion of democratic practice supported by technological innovation, such as the Internet, will not in itself help enlarge the public sphere constituted by the demos. For the second assumption tells us that the nature of the demos is inherently culture-laden, and so its emergence depends on the operation of specifically identity-forming mechanisms that surpass the democratic process itself. Truly democratic deliberation, at least in the Habermasian sense, does not materialize as an automatic response to more intrusive supranational decision making. Moving beyond disparate issues, such as the BSE or corruption scandals, to a more coherent ideological parsing of political conflict will require more than reform proposals based on referenda or "cyber democracy," initiatives that could be interpreted in populist terms (Calhoun 1988).

It may seem that our reasoning exaggerates the "top-down" dimension of EU institutions. Yet our focus on mass-based mechanisms of identity formation does not discount the value of bottom-up initiatives of participation and mobilization. It goes without saying that interest-group politics and spontaneous citizen protest serve essential functions in any democracy. Indeed, Europeans appear to be organizing in increasing numbers across a wide spectrum of policy issues (Imig and Tarrow 2001). However, it should be recalled that, at the national level, similar expressions of such decentralized activities unfold within, and in opposition to, a stable framework of formal electoral and politico-cultural institutions that provide an infrastructure of meaning, adjudicate access, and channel participation. Summing up a volume devoted to the study of contentious politicis within the European Union, Sidney Tarrow (2001: 250) argues that European-level public interest groups "have great difficulty creating and maintaining representative links with their claimed constituencies in the member-states. And without such ties, it would be suprising if such groups gained much political clout either in Brussels or with respect to national governments."

However appealing Habermas's notion of communicative action may be in theory, most of its practical applications to the European Union fail to account for how to undergird the communicative process with a mass-based, participatory infrastructure. Clearly, communication technology itself cannot be the answer. Indeed, the ambiguities that characterize the processes of public communication in the Internet at the national level are likely to become even more pronounced in the transnational context. On the positive side, we find decentralization of expertise and the building of issue networks across borders. Yet on the negative side, there are tendencies toward an ongoing fragmentation of the public and the corresponding loss of concern for the "common good" (Sunstein 2001). At any rate, there is a very real danger that the elitist and populist tendencies of modern democracy will be reinforced in the absence of institutional rules establishing a political framework for the public use of information technology in Europe.

While well-informed democratic decision making at the mass level may not require the cultural cohesion of the nation-state, it would be foolish to throw overboard the achievements of representative democracy in favor of an abstract notion of deliberative democracy that is to a considerable extent unsupported by shared linguistic and educational repertoires, and media institutions at the supranational level. Due to the member states' successful opposition to policies threatening their national identities, the fact remains that these policy dimensions are still notoriously underdeveloped within the European Union (see, e.g., Theiler 1998; Cederman 2001a).

In terms of policy, then, a cultural-institutionalist compromise could be found between the national substantialists' call for "autonomy protection" (Scharpf 1999) and the civic voluntarists' recommendations in favor of accelerated political integration and constitutional reforms. While doubting that the European integration process can be halted, let alone reversed, we believe that there must be a balance between efforts to create a transitional communicative space that is more accessible to a broader public and future steps toward increased supranational authority. Such a gap can be filled only by complementing the thrust toward technological innovation and deepened integration with comprehensive reforms in the areas of education, public communication, and the media that serve to enhance, and distribute more equally, the knowledge and engagement in the European integration process. The possibilities include improved foreign-language training and a strengthened "European dimension" in civic education. Failure to enact such initiatives risks creating "lost generations" of Europeans who lack the capacity or willingness to participate in the democratic process beyond their national communities.

References

Altermatt, Urs, et al. 1998. *Die Konstruktion einer Nation: Nation und Nationalisierung in der Schweiz, 18.–20. Jahrhundert.* Zürich: Chronos Verlag.

Anderson, Benedict. 1991. *Imagined Communities: Reflections on the Origin and Spread of Nationalism.* London: Verso.

Barth, Fredrik. 1969. Introduction. In *Ethnic Groups and Boundaries: The Social Organization of Culture Difference,* edited by Fredrik Barth. Boston: Little, Brown and Company.

Birnbaum, Pierre. 1998. *La France imaginée*. Paris: Fayard.
Börzel, Tanja A., and Thomas Risse. 2001. Who Is Afraid of a European Federation? How to Constitutionalize a Multi-Level Governance System. In *What Kind of Constitution for What Kind of Polity? Responses to Joschka Fischer,* edited by C. Joerges, Y. Mény, and J.H.H. Weiler. San Domenico: Robert Schuman Centre, EUI.
Brass, Paul. 1991. *Ethnicity and Nationalism: Theory and Comparison.* Newbury Park: Sage.
Brubaker, Rogers. 1996. *Nationalism Reframed: Nationhood and the National Question in the New Europe.* Cambridge: Cambridge University Press.
———. 1998. Myths and Misconceptions in the Study of Nationalism. In *The State of the Nation,* edited by John A. Hall Cambridge: Cambridge University Press.
Calhoun, Craig. 1988. Populist Politics, Communication Media and Large Scale Societal Integration. *Sociological Theory* 6:219–41.
———. 1991. Indirect Relationships and Imagined Communities: Large-Scale Social Integration and the Transformation of Everyday Life. In *Social Theory for a Changing Society,* edited by Pierre Bourdieu and James S. Coleman. Boulder: Westview.
———, ed. 1992a. *Habermas and the Public Sphere.* Cambridge: MIT Press.
———, 1992b. The Infrastructure of Modernity: Indirect Social Relationships, Information Technology, and Social Integration. In *Social Change and Modernity,* edited by Hans Haferkamp and Neil J. Smelser. Berkeley: University of California Press.
Castells, M. 1996. *The Rise of the Network Society.* Malden, MA: Blackwell.
———. 1997. *The Power of Identity.* Malden, MA: Blackwell.
———. 1998. *End of Millennium.* Malden, MA: Blackwell.
Cederman, Lars-Erik. 2001a. Nationalism and Bounded Integration: What It Would Take to Create a European Demos. *European Journal of International Relations* 7(2).
———, ed. 2001b. *Constructing Europe's Identity: The External Dimension.* Boulder: Lynne Rienner.
Chadwick, Andrew, and Christopher May. 2003. Interaction between States and Citizens in the Age of the Internet: "e-Government" in the United States, Britain, and the European Union. *Governance* 16:271–300.
Choucri, Nazli. 2000. Introduction: CyberPolitics in International Relations. *International Political Science Review* 21: 243–63.
Colley, L. 1992. *Britons: Forging the Nation, 1707–1837.* London: Pimlico.
Commission of the European Communities. 2001. White Paper on European Governance. Brussels, July 25.
Dahl, Robert A. 1989. *Democracy and Its Critics.* New Haven: Yale University Press.
———. 1994. "A Democratic Dilemma: System Effectiveness versus Citizen Participation." *Political Science Quarterly* 109:23–34.
de Swaan, Abram. 1993. The Evolving European Language System: A Theory of Communication and Language Competition. *International Political Science Review* 14:241–56.

DG Press. 2003. European Citizens and the Media." National Reports. Public Opinion in the European Union. Brussels, May.
Deutsch, Karl W. 1966. *Nationalism and Social Communication.* Cambridge: MIT Press.
Deutsch, Karl W., et al. 1957. *Political Community and the North Atlantic Area.* Princeton: Princeton University Press.
Eder, Klaus. 2000. Zur Transformation nationalstaatlicher Öffentlichkeit in Europa: Von der Sprachgemeinschaft zur issuespezifischen Kommunikationsgesellschaft. *Berliner Journal für Soziologie* 10:167–284.
Elster, J. 1998. Introduction. In *Deliberate Democracy,* edited by J. Elster. Cambridge: Cambridge University Press.
Emirbayer, Mustafa. 1997. Manifesto for a Relational Sociology. *American Journal of Sociology* 103:281–317.
Eurobarometer. 1999. Report Number 51.
Eurobarometer 2001a. Special Report Number 54. *Europeans and Languages.* Brussels, February.
Eurobarometer. 2001b. Report Number 55.
Eurobaromètre. 2003. Eurobaromètre Standard 59.
Eurobaromètre. 2001. Les Européens e la E-INCLUSION. Rapport rédigé par The European Public Research Group pour la DG Emploi. 55(2) (Spring).
European Union. 1994. Recommendations to the European Council: Europe and the Global Information Society (Bangemann Report). Available at http://europa.eu.int/ISPO/infosoc/backg/bangeman.html.
Flash Eurobarometer. 2002. Internet and the Public at Large." Realized by EOS Gallup Europe upon Request of the European Commission (Directorate General Information Society), 135 (November).
Flora, Peter. 2000. Externe Grenzbildung und interne Strukturierung—Europa und seine Nationen: Eine Rokkanische Forschungsperspektive. *Berliner Journal für Soziologie* 2:151–66.
Gellner, Ernest. 1964. *Thought and Change.* London: Weidenfeld & Nicolson.
———. 1983. *Nations and Nationalism.* London: Blackwell.
———. 1997. *Nationalism.* London: Weidenfeld & Nicolson.
Gerhards, Jürgen. 1993. Westeuropäische Integration und die Schwierigkeiten der Entstehung einer europäischen Öffentlichkeit. *Zeitschrift für Soziologie* 22:96–110.
———. 2000. Europäisierung von Ökonomie und Politik und die Trägheit der Entstehung einer europäischen Öffentlichkeit. In *Die Europäisierung nationaler Gesellschaften,* edited by Maurizio Bach. Wiesbaden: Westdeutscher Verlag.
Giddens, Anthony. 1985. *The Nation-State and Violence.* Cambridge: Polity Press.
Greenfeld, Liah. 1992. *Nationalism: Five Roads to Modernity.* Cambridge: Harvard University Press.
Grimm, Dieter. 1995. Does Europe Need a Constitution? *European Law Journal* 1:282–302.
Haas, Ernst. 1958. *The Uniting of Europe: Political, Economic and Social Forces.* Stanford: Stanford University Press.

Habermas, Jürgen 1990 [1960]. *Strukturwandel der Öffentlichkeit.* Frankfurt a.M.: Surhkamp.

———. 1992a. *Faktizität und Geltung: Beiträge zur Diskurstheorie des Rechts und des demokratischen Rechtsstaats.* Frankfurt a. M.: Suhrkamp.

———. 1992b. Citizenship and National Identity: Some Reflections on the Future of Europe. *Praxis International* 12:1–19.

———. 1996. *Die Einbeziehung des Anderen: Studien zur politischen Theorie.* Frankfurt a. M.: Suhrkamp.

———. 1998. *Die postnationale Konstellation: Politische Essays.* Frankfurt a. M.: Suhrkamp.

———. 2001. A Constitution for Europe? *New Left Review* 11:5–26.

Hall, Peter A., and Rosemary C. R. Taylor. 1996. Political Science and the Three New Institutionalisms. *Political Studies* 44:936–57.

Hechter, Michael. 1975. *Internal Colonialism: The Celtic Fringe in British National Development, 1536–1966.* London: Routledge and Kegan Paul.

Held, David. 1995. *Democracy and the Global Order: From the Modern State to Cosmopolitan Governance.* Cambridge: Polity Press.

Hoffmann, Stanley. 1995. *The European Sisyphus: Essays on Europe, 1964–1994.* Boulder: Westview.

Huntington, Samuel P. 1991. Democracy's Third Wave. *Journal of Democracy* 2:12–34.

Imig, Doug, and Sidney Tarrow, eds. 2001. *Contentious Europeans: Protest and Politics in an Emerging Polity.* Oxford: Rowman & Littlefield.

Jacob, James E., and David C. Gordon. 1985. Language Policy in France. In *Language Policy and National Unity,* edited by William R. Beer and James E. Jacob. Totowa, NJ: Rowman and Allanheld.

Joerges, C., Y. Mény, and J.H.H. Weiler, eds. 2001. *What Kind of Constitution for What Kind of Polity? Responses to Joschka Fischer.* San Domenico: Robert Schuman Centre, EUI.

Joerges, C., and J. Neyer. 1998. Von intergouvernementalem Verhandeln zur deliberativen Politik: Gründe und Chancen für eine Konstitutionalisierung der europäischen Komitologie. In *Regieren in entgrentzten Räumen,* edited by B. Kohler-Koch. Opladen: PVS-Sonderheft 29.

Kaiser, Karl. 1971. Transnational Relations as a Threat to the Democratic Process. In *Transnational Relations in World Politics,* edited by Robert O. Keohane and Joseph S. Nye, Jr. Cambridge: Harvard University Press.

Keohane, Robert O., and Joseph S. Nye, Jr. 1975. International Dependence and Integration. In *Handbook of Political Science,* edited by Fred I. Greenstein and Nelson W. Polsby. Reading, MA: Addison-Wesley.

Kielmansegg, Peter. 1996. Integration und Demokratie. In *Europäische Integration,* edited by M. Jachtenfuchs and B. Kohler-Koch. Opladen: Leske + Budrich.

Kraus, Peter A. 2000. Political Unity and Linguistic Diversity in Europe. *Archives européennes de sociologie* 41:138–63.

———. 2003. Cultural Pluralism and European Polity-Building: Neither Westphalia nor Cosmopolis. *Journal of Common Market Studies* 41: 665–86.

Laborde, Cécile. 2001. The Culture(s) of the Republic: Nationalism and Multiculturalism in French Republican Thought. *Political Theory* 29:716–35.

Mann, Michael. 1999. "The Dark Side of Democracy: The Modern Tradition of Ethnic and Political Cleansing." *New Left Review* 235:18–45.

Mansell, Robin, and Steinmueller, W. Edward. 2000. *Mobilizing the Information Society.* Oxford: Oxford University Press.

March, James G., and Johan P. Olsen. 1998. "The Institutional Dynamics of International Political Orders." *International Organization* 52:943–69.

———. 2000. "Democracy and Schooling: An Institutionalist Perspective." In *Rediscovering the Democratic Purposes of Education,* edited by Lorraine M. McDonnell, P. Michael Timpane, and Roger Benjamin. Lawrence: University Press of Kansas.

Mitrany, David. 1966. *A Working Peace System.* Chicago: Quadrangle Books.

Mulhall, Stephen, and Adam Swift. 1992. *Liberals and Communitarians.* Oxford: Blackwell.

Nairn, Tom. 1977. *The Breakup of Britain: Crisis and Neo-Nationalism.* London: New Left Books.

Olsen, Johan P. 2001. How, Then, Does One Get There? In *What Kind of Constitution for What Kind of Polity? Responses to Joschka Fischer,* edited by C. Joerges, Y. Mény, and J.H.H. Weiler. San Domenico: Robert Schuman Centre, EUI.

Pizzorno, Alessandro. 1986. Some Other Kinds of Otherness: A Critique of "Rational Choice" Theories." In: *Development, Democracy, and the Art of Trespassing. Essays in Honor of Albert O. Hirschman,* edited by Alejandro Foxley,, Michael S. McPherson, Guillermo O'Donnell. Notre Dame: University of Notre Dame Press.

Rokkan, Stein. 1975. Dimensions of State-Formation and Nation-Building: A Possible Paradigm for Research on Variations within Europe. In *The Formation of National States in Western Europe,* edited by Charles Tilly. Princeton: Princeton University Press.

Rousseau, Jean-Jacques. 1987 [1762]. *Basic Political Writings,* translated and edited by Donald A. Cress. Indianapolis: Hackett.

Scharpf, Fritz W. 1999. *Governing in Europe: Effective and Democratic?* Oxford: Oxford University Press.

Schlesinger, Philip. 1991. *Media, State and Nation: Political Violence and Collective Identities.* London: Sage.

———. 1999. Changing Spaces of Political Communication: The Case of the European Union. *Political Communication* 16:263–79.

Schmitter, Philippe C. 2000. *How to Democratize the European Union . . . And Why Bother?* Lanham, MD: Rowman & Littlefield.

Sciarini, Pascal, Simon Hug, and Cédric Dupont. 2001. Example, Exception, or Both? Swiss National Identity in Perspective. In *Constructing Europe's Identity: The External Dimension,* edited by Lars-Erik Cederman. Boulder: Lynne Rienner.

Simmel, Georg. 1971 [1908]. How Is Society Possible. In *Georg Simmel: On Individuality and Social Forms,* edited by D. N. Levine. Chicago: University of Chicago Press.

Smith, A. D. 1992. National Identity and the Idea of European Unity. *International Affairs* 68:55–76.
———. 1995. *Nations and Nationalism in a Global Era.* Cambridge: Polity Press.
Stein, Eric. 2001. International Integration and Democracy: No Love at First Sight. *American Journal of International Law* 95:489–534.
Sunstein, C. 2001. *republic.com.* Princeton: Princeton University Press.
Tarrow, Sidney. 2001. Contentious Politics in a Composite Polity. In *Contentious Europeans: Protest and Politics in an Emerging Polity,* edited by Doug Imig and Sidney Tarrow. Oxford: Rowman & Littlefield.
Theiler, Tobias. 1998. The European Union and the "European Dimension" in Schools: Theory and Evidence. *Journal of European Integration* 21.
———. 2001. The "Identity Policies" of the European Union. Ph.D. diss., Oxford University.
Tilly, Charles. 1990. *Coercion, Capital, and European States, AD 990–1990.* Oxford: Basil Blackwell.
Tully, James. 2001. Introduction. In *Multinational Democracies,* edited by. Alain-G. Gagnon and James Tully. Cambridge: Cambridge University Press.
Wæver, Ole. 1996. European Security Identities. *Journal of Common Market Studies* 34:103–32.
Walzer, Michael. 1994. *Thick and Thin: Moral Argument at Home and Abroad.* Notre Dame: University of Notre Dame Press.
Warner, Michael. 1990. *The Letters of the Republic: Publication and the Public Sphere in Eighteenth-Century America.* Cambridge: Harvard University Press.
Weber, Eugen. 1976. *Peasants into Frenchmen: The Modernization of Rural France, 1870–1914.* Stanford: Stanford University Press.
Weiler, Joseph H. H. 1999. *The Constitution of Europe.* Cambridge: Cambridge University Press.
———. 2000. Federalism and Constitutionalism: Europe's Sonderweg. Jean Monnet Chair, Harvard Law School.
Weiler, Joseph H. H., Ulrich R. Haltern, and Franz C. Mayer. 1995. European Democracy and Its Critique. In *The Crisis of Representation in Europe,* edited by J. Hayward. London: Frank Cass.
Williams, Shirley. 1991. Sovereignty and Accountability in the European Community. In *The New European Community: Decisionmaking and Institutional Change,* edited by Robert O. Keohane and Stanley Hoffmann. Boulder: Westview.
Zürn, Michael. 1998. *Regieren jenseits des Nationalstaates.* Frankfurt a. M.: Suhrkamp.
———. 2000. Democratic Governance Beyond the Nation-State: The EU and Other International Institutions. *European Journal of International Relations* 6:147–82.

Information Technology and State Capacity in China

DOUG GUTHRIE

> In the new century, liberty will spread by cell phone and cable modem. . . . We know how much the Internet has changed America, and we are already an open society. Imagine how much it could change China. Now, there's no question China has been trying to crack down on the Internet—good luck. That's sort of like trying to nail Jello to the wall.
>
> —President Bill Clinton, March 8, 2000

INASMUCH AS IT is fashionable to claim that information technology will allow liberal ideals to enter formerly closed societies, thus hastening the fall of authoritarian regimes, China is an interesting case within this discussion. Where the first decade-and-a-half of China's economic reforms have been organized around the development of an export-led economy and the transformation of industrial organizations, since the mid-1990s the focus has shifted to the realm of technology. It is widely believed among experts and Chinese leaders alike that technological development will play a key role in China's continued economic expansion. As China's Minister of Science and Technology Xu Guanhua recently asserted, "Science and technology will play a major role in economic and social development in China. . . . China has achieved remarkable progress in seeking new technology discoveries, which have been important to the national economy over the past few years" (Cui 2001: 1; see also Segal 2002). If China is to develop an indigenous high-tech industry—an essential stage in the push to become a first world economic power—it must follow global trends in technological development and compete with the strongest advanced industrial economies of the world in the realm that is unfolding as one of the crucial battlegrounds of global capitalism: information technology (Cooper 2000). Openly acknowledging this necessity, the Tenth Five-Year Plan calls for information technology to take its place alongside other industrial sectors and take a central place as a "pillar" in the economy; according to planners, the sector is expected to grow 30 percent a year.[1] Further highlighting the importance of information technology in China is the extent to which this sector of the economy is tied to other newly emerging economic sectors. For example, if the country is to develop viable and healthy financial markets, as is its stated goal, it must have in place the information technology to undergird this industry. In many ways, information technologies, and more generally a thriving high-tech economy, hold the key to China's continued economic development and growth into the economic power it aspires to become.

Yet, even as it is widely acknowledged as a central part of China's economic health and development, information technology holds at once promise and peril for the Chinese government. While the evolution of information technology is a necessary step in the continued development of the Chinese economy, it may also provide citizens the tools of privacy and, ultimately, resistance. To the extent that individuals can communicate and gather information beyond the reach of the state, the authoritarian government's power to control its people may be compromised and weakened. Tacitly acknowledging this possibility, Beijing has maintained a tighter control over the emergence of new information technologies in general, and the Internet in particular, than it has over the evolution of any other industry precisely because the free flow of information poses such a threat to the already withering one-party system.

Beijing's tight control over information technology seems to affirm the sentiments of those who believe information technology will bring about democracy. The argument from this camp is that information technology allows for the rapid sharing of information across political and social divides; that information technology (specifically the Internet) is an inexorable force, chipping away at the veneer of authoritarian regimes and laying the seeds of democracy.[2] The view here is first that, through the Internet, exposure to Western liberal ideals of democracy and freedom will help to create an understanding of these ideals and eventually foster a groundswell of popular support for them. The Internet and information technology more generally, the theory goes, cannot be controlled by any government, and authoritarian governments will be overrun by the availability of liberal ideals and the freedom to communicate across boundaries.

[1] For discussion of the role of information technology as articulated by the Tenth Five-Year Plan, see Harner (2000).

[2] See, for example, Freidman (2000, 1999); Hill and Sen (2000).

While few scholars of technological change hold the simplistic view of the technological revolution articulated by President Clinton above, politicians and promoters of technological change often do impute an independent causal role to the technology itself. The problem with this view, at least in the case of China, is that while radical changes are occurring there—changes that amount to significant encroachments on the Chinese government's sovereignty—and while these changes have, to some extent, coincided with the emergence of information technology there, there is not necessarily a causal relationship among these realms. In this chapter, I present a grounded analysis of the relationship between information technology and the ability of China's authoritarian government to control its population. The Chinese government desperately needs to foster a healthy high-tech industry, but is IT the Pandora's box that some Western business leaders and Chinese bureaucrats seem to think it is? To answer this question, I look at four areas where the evolution of IT has had an impact—foreign investment, private social networks, the emergence of newly autonomous sectors of society, and popular resistance—all of which are clearly related to the control exercised by China's authoritarian government and perhaps to democratization. I argue that, first and foremost, the answer to this question depends upon the level of analysis: if we are looking at the macro level, IT does not play a causal role in the changes that are transforming Chinese society—fundamental institutional changes in China originate with governmental reforms, and these institutional reforms have been in motion for many years. However, on the micro level, IT does appear to play a role in the evolution of new types of social networks and in creating opportunities for newly emerging sectors of society. Access to information, the ability to communicate, and individual freedom in the economy and society have changed dramatically in reform-era China, and some of these changes have been fostered by the emergence of new forms of technology. The availability of information technology played an important role in the scale and scope of China's most significant resistance movement of the reform era; and it plays a very particular role in setting the tone of negotiations with foreign investors. Further, the introduction of new technologies in China's private-sector economy and in the academy has created many new opportunities and options for the actors within these sectors of society. In many ways, in each of these areas, the use of information technology by actors in society has shaped the dynamics of governmental-society relations and, by extension, the reform process itself. Information technology has not played a direct causal role in the Chinese government's declining control over its population, but once certain institutional changes were in motion, information technology has interacted with these reforms and perhaps hastened this process of change in significant ways.

Market Reforms and Political Reforms

It is important to acknowledge from the outset the extent to which the economic reforms in China have been driven by a political process. State policies have driven the methodical transformation of old sectors of the economy and the creation of new sectors, the creation of new markets and new economic institutions, and the development of new practices for the actors operating within these new markets. In the two decades of economic reform in China, the state has consistently and methodically guided the reforms, maintaining control over the majority of the industrial economy and tightening fiscal constraints for the inefficient state sector at only a gradual rate. More than this, the state has introduced the policies and laws through which the new markets that increasingly govern economic processes in China have been constructed. Even beyond methodical involvement of the state in shaping China's transition path, the political nature of economic change runs even deeper, as legacies of the former institutions of the state-run economy shape the country's development path in fundamental ways. In recent years, the hands-on policies of the government have extended to the New Economy as well, as the state has sought to regulate and control the emergence of new forms of technology, while at the same time allowing enough freedom for innovation and the emergence of viable free markets. In this section, I will briefly lay out the political aspects of economic reforms in China as they relate to the emergence of an information technology sector. This discussion is relevant, as the political nature of economic reform in China is a necessary starting point for a discussion of the capacity and autonomy of the Chinese state, especially in the realm of new information technologies.

Political Forces in the Emergence of New Sectors

Even as the state has receded from direct control over the economic decisions of individuals and firms, and even as foreign capital and technology have played an increasingly important role in China's emerging markets, the Chinese government continues to play a central role in the reform era. As such, politics have played a fundamental role in China's emerging high-technology sectors. One of the clearest examples of the political role in the creation of the New Economy has to do with the development of China's newest Special Economic Zone (SEZ) in Beijing, *Zhongguancun* (Chinese Silicon Valley) Science and Technology Park.[3] The development of the SEZs has been a purposeful political process, where the government

[3] For discussion of Zhongguancun, see Zhu (2000). See also Cooper (2000: 4). Unless otherwise noted, the discussion here derives from these two sources.

has targeted specific areas and aspects of the reform project to encourage the advancement of a new sector of the economy and a new part of economic reform. There are many SEZs throughout the country, the most famous being Shenzhen in Guangdong and Pudong in Shanghai, which are tied to the development of export industries and attraction of foreign investment into the industrial economy. While the development of Zhongguancun is also tied to issues of export and the continued development of an industrial infrastructure, it is also driven by the government's decision to invest in the high-tech economy. Established in 1998 as a way of encouraging software development, the technology park was taken over by the Beijing municipal government and the Ministry of Science and Technology in 1999. Over the next decade, the park will receive investment from the municipal and national government for several large-scale construction projects to facilitate its development into a global software center. By June 2002 the government was to have built or renovated sixty kilometers of road in the thirty-square-kilometer area and constructed four new sewage plants. The investment yielded immediate dividends: in 1999 the park earned over $10 billion in revenues, surpassing China's other high-tech parks, and by the end of 2000, it had 8,224 high-tech companies. The park is now establishing itself as the economic engine of Beijing: in the first five months of 2001, for example, high-tech exports accounted for half of the city's total and represented an increase of 79 percent over the same period in the previous year.[4]

Another example of the politics surrounding the evolution of information technology in China has to do with the extent to which this sector has remained under tighter control than other rapidly developing industries. This sector is monitored closely by the central government for a variety of reasons. First, it is a sector in which very significant technological transfers are occurring in joint venture deals between foreign and Chinese firms. The Chinese government knows all too well that, as big as the Chinese market for information technology portends to be, it is this market that foreign investors are after. The Chinese government would like to limit the extent to which foreign producers, such as Motorola and Nokia, are able to control that market, as Chinese companies, such as Huasheng, will eventually be able to compete with these companies. Yet the government also knows that it needs the technology that companies like Motorola and Nokia can deliver. As a result, close monitoring has become a central part of the sector's development process. And when it has become apparent that certain companies are doing too well, the government has stepped in and leveled the playing field somewhat.[5] Second, and perhaps

[4] *China Daily,* June 26, 2001: 5.

[5] This was the case with Motorola in 1996. Up until that time, Motorola had only a

more important, because the telecommunications industry provides an infrastructure for the spread of information, the government is clearly afraid of completely losing control over individuals' access to information. Accordingly, telecommunications is the last sector to remain closed to foreign capital, as Chinese law still forbids foreign capital in this sector.[6] It is for this reason that exceedingly complicated deals have been worked out in the establishment of companies in this sector, as in the case of Sina.com.[7] In addition, telecommunications has been the target of the most aggressive regulations.

Access to Information and the Growth of the IT Sector

There has unquestionably been a great deal of activity in the IT sector in recent years. However, before looking at the development of new information technologies per se, let us first take into account the spread of information more generally. Table 1 presents some indicators of the growth in access to information in China over the last two decades. For both newspapers and magazines, the growth has been exponential over the two-decade timeframe, with the number of newspapers expanding from 186 in 1978 to 2,038 in 1999, and magazines expanding from 930 to 8,178 over the same period. Television programs have seen greater than

Wholly Owned Foreign Enterprise (WOFE) in China and a licensing agreement with a variety of factories, including the Hangzhou Telecommunications Factory, to produce their handsets. Motorola made a great deal of money through this arrangement, which allowed them to produce and sell phones without transferring significant technology in the process. Then in 1995 they began negotiating a joint venture with the Hangzhou Telecommunications Factory. In a personal interview with one of the insiders on this deal, I inquired as to what had led to the change of heart. The American manager said, "Let's just say that the [Chinese] government decided it was time for us to share the wealth. And if we were going to keep doing what we are doing in China, we were going to have to set up a joint venture deal with someone."

[6] With China's recent entry into the World Trade Organization, changes in the state's control of this sector are imminent, as the agreement China and the United States reached in the negotiations over China's entry mandates that foreign firms be able to own minority stakes in the telecommunications industry.

[7] It is very likely that the recent strife between the former CEO, Wang Zhidong, and the board of directors was caused by the complex business structure that was required in establishing Sina.com's initial public offering, which was a result of Beijing's prohibitions against foreign ownership in this sector. When Sina.com went public, the company had to give up its control over the Internet within China. Sina.com, which, as an Internet portal company in China, actually can only provide "technical assistance" to the Chinese-based Sina Internet Information Service Co. Ltd, which has an Internet content provider license, of which Wang also owns a majority stake. Thus, we have an American listed company, with an American board of directors, that is purportedly an Internet content provider but does not have an Internet content license in China and has to rely solely on a Chinese-based company for access to the Internet.

TABLE 1
Access to Media of Information in China

	Magazines	*Newspapers*	*Television*
1978	930	186	—
1980	2,191	188	—
1985	4,705	1,445	38,056
1986	5,248	1,574	—
1987	5,687	1,611	—
1988	5,865	1,537	—
1989	6,078	1,576	—
1990	5,751	1,444	91,572
1991	6,056	1,524	—
1992	6,486	1,657	—
1993	6,486	1,788	—
1994	7,011	1,953	—
1995	7,325	2,089	383,513
1996	7,583	2,163	—
1997	7,918	2,149	616,437
1998	7,999	2,053	477,893
1999	8,187	2,038	526,483

Source: Statistical Yearbook of China (2000: 712–14).

exponential growth over this period, with 38,056 programs in 1985 growing to 526,483 programs in 1999. While these media are not typically placed in the category of new information technology, they are indicative of an important trend of growing access to information and thus relevant for any discussion about information and social change.

Table 2 shows the growth in information technology since the economic reforms began two decades ago. Use of pagers, mobile telephones, e-mail and the Internet, and the development of optical and digital cable lines—all-important aspects of a growing IT economy in China—have all expanded dramatically in this period. The growth in pager and mobile phone use has been rapid in the last decade: both of these forms of technology were basically nonexistent in China in the mid-1980s and grew to 46 and 43 million registered users in 1999, respectively. The use of mobile telephones has undergone another period of extreme growth since 1999, growing to approximately 116 million subscribers as of June 2001, according to Lou Qinjian, vice-minister of the information industry.[8] The penetration of these technologies, while dramatic, is not surprising: in developing societies around the world, it has been much faster and easier to implement mobile technology as the primary form of communication

[8] *China Daily*, June 26, 2001: 5.

TABLE 2
Growth of Information Technology in China

	E-mail subscribers	*Internet subscribers*	*Pagers*	*Mobile phones*	*Land-line phones*	*Optical cable lines*	*Digital lines*
1978	0	0	0	0	3,868,200	0	0
1980	0	0	0	0	4,186,400	0	0
1985	0	0	0	0	6,259,800	0	0
1986	0	0	0	0	7,059,100	0	0
1987	0	0	30,900	0	8,057,200	0	0
1988	0	0	97,200	3,200	9,417,900	2,717	0
1989	0	0	237,300	9,800	10,893,300	5,670	0
1990	0	0	437,000	18,300	12,313,300	11,453	0
1991	0	0	873,800	47,500	14,544,300	23,613	0
1992	0	0	2,220,200	176,900	18,459,600	51,352	109,300
1993	0	0	5,614,000	639,300	25,673,500	162,861	298,045
1994	2,329	0	10,330,000	1,567,800	38,018,600	330,359	518,915
1995	6,068	7,213	17,391,500	3,629,400	53,993,200	484,231	677,672
1996	10,107	35,652	23,562,000	6,852,800	70,467,500	754,143	965,263
1997	15,246	160,157	32,546,100	13,232,900	87,878,300	935,835	1,139,476
1998	20,959	676,755	39,081,600	23,862,900	107,371,500	1,351,665	1,560,201
1999	19,855	3,014,518	46,744,700	43,296,000	132,378,400	—	—
June 2001	—	—	—	116,000,000	—	—	—

Sources: Statistical Yearbook of China (2000: 543–546); *China Daily*, June 26, 2001: 5.

than to lay grounded lines. Given the recent introduction of mobile phone technology into China, the growth in this area has been truly dramatic—25 percent of the 175 million phones in China are mobile phones—and virtually all industry experts agree that China will very soon become the largest market in the world for mobile telephones. It is also likely that the figures for mobile telephones are underrepresented here, as the numbers listed here are subscribers to official services, and the unregistered mobile phone market is huge in China. Estimates on just how big this market is do not exist, but one need only go through the process of buying a secondhand phone and setting up an unregistered account to understand just how popular this practice is in China.

With the relatively low penetration of personal computers in China, it is somewhat surprising that there are more than three million registered Internet users in China. Yet, as with the mobile phone reports, it is also likely here that the figures on the Internet are underrepresented, as the most popular web sites in China are those that do not require subscriber registration.[9] Instead, the majority of Chinese gaining access to the Internet today do so through a pay-per-minute service provided by the phone company, in which a user can log on anonymously from any phone and access the Internet or publicly maintained e-mail accounts on one of the main Internet portals. For example, 163, 263, and 169 all allow users to gain access to the Internet without establishing a subscriber account. Table 2 also shows the developmental trends of the infrastructure that supports IT, such as the Internet, optical cable, and digital lines, growing from nothing to more than a million lines each in just over a decade.

Finally, figure 1 compares the activity of foreign investors in a variety of sectors in China. Despite the state's tight control over telecommunications, it is nevertheless one of the sectors most heavily invested in by foreign companies, as measured by the number of foreign joint ventures established in this industry. This reflects both China's need for technology in this sector of the economy but also the foreign perception of great market opportunity.

The bird's-eye view of the information presented above tells us the following: First, the spread of information more generally in China has occurred in dramatic ways over the course of the economic reforms. Second, information technology itself is spreading in significant ways in Chinese society, and this spread includes both individual-user access to IT and the hardware and infrastructure that is necessary for the further development

[9] Industry experts predicted that China could reach an online population of about twenty million by 2002, roughly equal to that of Germany and France. See, e.g., "State of the Internet in China," *Chinaonline,* July 21, 2000. In 2001, predictions suggested that the number would be closer to thirty million by 2002; see *New York Times,* June 12, 2001: W1.

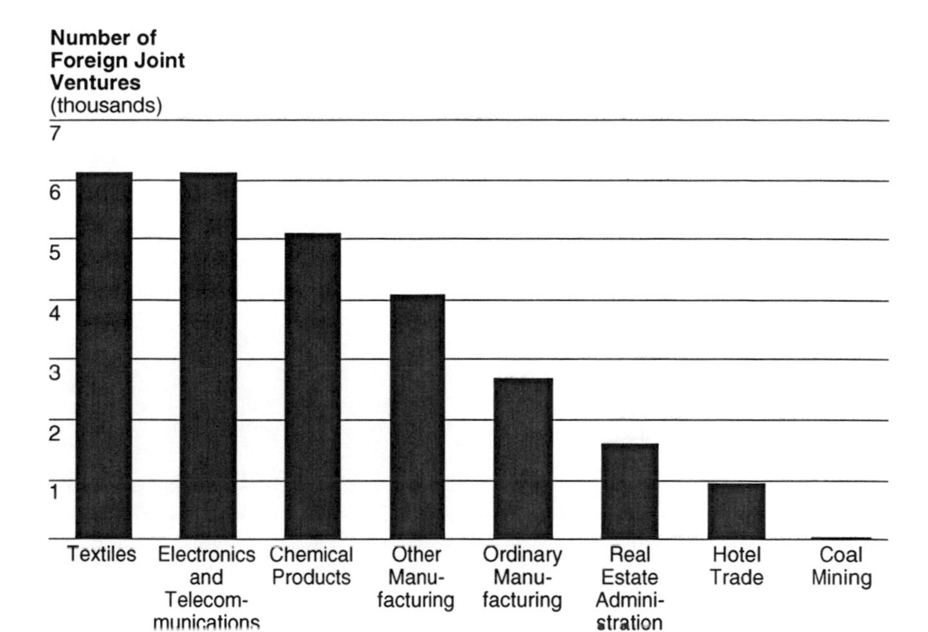

Figure 1. Number of foreign joint ventures in selected industries

of this industry. Taken together, this means that access to information and the high-technology vehicles that facilitate communication about and the sharing of information are significant forces in Chinese society. In addition, the high-technology sectors of the economy, including telecommunications, are among the most active in terms of foreign investment. The question before us now is, what implications do these changes have for Chinese society, for the capacity of the Chinese state to control its population, and for the process of democratization in China?

Information Technology and State Capacity in China

To what extent do these changes, which are driven by macro-level state policies, have an impact on the ability of China's authoritarian government to maintain control over the economy and society? To answer this question, we must approach the issues from the micro level: we must ground the analysis in the ways that actual users are employing the new technologies to their own ends. This grounded analysis will allow us to look more closely at the conditions under which emergent technologies actually have an impact on state capacity. In the following sections I will examine four instances in which information technology appears to be related to incursions against the capacity of the Chinese government to control its population. In the realms of foreign investment, individual privacy, the emergence of newly autonomous sectors of society, and overt resistance to state power, we can find important examples in which information technology seems tied to intrusions against Chinese state capacity and, ultimately, state sovereignty. However, as I will argue below, while information technology has played an important role in the dynamics surrounding each of these areas of inquiry, these dynamics also need to be viewed in the context of the institutional changes already underway in Chinese society.

Foreign Investment

Attracting foreign investment has been a basic part of China's economic development in the reform era. On June 11, 2001, in the City of Shanghai, the Internet giant AOL Time Warner (AOLTW) announced a $200 million joint venture with Legend Holdings, China's largest computer maker.[10] The venture is working toward the development of Internet services that will be bundled with Legend's computers, which currently holds about one-third of the market share for personal computers in China. Despite the fact that foreign companies as of 2001, were still not allowed to own stakes in Internet services or Internet content providers, AOLTW has committed $100 million to the development of a venture that will place the company primarily in a position of consultation and technical support. The reasons the company is willing to accept such a deal likely include the upside potential of Internet development in China and the fact that the prohibitions against foreign ownership in the telecommunications sector are going to change with China's entry into the World Trade Organization. For China, the positive aspects of this deal are many: it entails a large amount of committed capital, even compared to other large-scale joint venture deals;[11] it brings technology to China in an area that is rapidly evolving; and it carries international caché and branding from the largest personal access Internet service provider in the United States.

Yet despite these many advantages for both sides of the partnership, both sides also take on significant risk. The risk for AOL Time Warner is largely economic: given that many of the joint ventures involving multi-

[10] Details of this venture taken from Smith (2001).

[11] Most of the large joint venture deals come in at just under U.S. $30 million, as this is the level at which approvals need not go beyond the municipal government. A joint venture deal the size of the AOL Time Warner–Legend deal must be approved directly by the State Council.

nationals in China have reported losses for the entire time they have been in operation in China, it is unlikely that AOLTW will see a return on its investment anytime soon. This is an investment for the future, and the future is always somewhat unpredictable in developing countries like China. The Chinese risk is, in some ways, more fundamental: when the State Council endorses a deal of this size, it is giving up some amount of control over the development of the sector in which the venture is occurring. In other words, investments such as these pierce the veil of the authoritarian government's sovereign control over the nation and the economy. Presidents and CEOs of multinationals investing large sums of money in China expect to be heard. In Beijing the mayor established an Advisory Council in 1999 made up of presidents and CEOs of companies with significant stakes in China, so that these high-power individuals can have an official forum in which to express their views.[12] Sometimes company executives are afforded even higher access to air concerns. For example, following an incident that involved theft of intellectual property, DuPont used what bargaining power it had to pressure government officials to set forth policies that will safeguard against the recurrence of a similar incident in subsequent investments. In 1994, on the brink of embarking on another joint venture in China, DuPont's chairman, Edgar Woolard, met with Chinese President Jiang Zemin to discuss formal policies that would protect foreign investors. It is unlikely that Woolard was able to elicit any guarantees from President Jiang or that this meeting was a direct precursor to the law protecting intellectual property, which was promulgated in 1995 (the law had been in the works for a long time prior to the meeting). Yet, as China needs foreign investment to develop, such high-stakes negotiations require the Chinese government to create an environment in which investors feel that their assets are somewhat protected. This requires giving up sovereign control over industries and sectors of the economy.

In earlier work I have examined the impact of negotiations over foreign-invested joint venture agreements on Chinese state sovereignty (Guthrie 1999a: chap. 7). That line of research is specifically about the use of arbitration clauses in joint venture contracts, but the issues also apply to joint venture deals more generally. I note in that analysis that "For the first time since the founding of the PRC, foreign parties have input on decisions that affect Chinese internal affairs. Enforcement still lies in the hands of Chinese authorities. But for a country that only a few years ago operated fully on the institution of administrative fiat, turning over power of decision making to a third [party] . . . is somewhat problematic" (163).

[12] The Third Annual Meeting of the Advisory Council of International Business Leaders was held on May 9–10, 2001, in Beijing.

The central point of that analysis is that negotiations with foreign parties require the Chinese government to give up some power and control over Chinese society. The extent to which the Chinese government is forced to give up sovereignty varies with value of the joint venture investment: when the Chinese government is facing a large multinational company that seeks to invest a significant amount of capital and technology, both of which China desperately needs, it must give up control over the venture to a significant extent. And if that company uses arbitration clauses in its joint venture contracts, as most large multinationals operating in China do, the government gives away control of the economic venture in question to a greater extent still.[13]

The AOLTW deal is especially interesting in this vein because it comes in an industry and at a time in which the government seems intent upon maintaining tight control. To argue that IT plays a causal or even central role in diminishing China's sovereign control over economic development or the telecommunications sector specifically would be an exaggeration of IT's role in what is a larger trend. Foreign investment has played an important role in China's reform effort since the country reopened its doors to foreign investment in 1979. From Deng Xiaoping's visit to the United States in 1979 to the Law on Chinese-Foreign Equity Joint Ventures—one of the first laws passed to usher in the economic transition—the attraction of foreign capital and technology has been central to the economic changes occurring in China. Table 3 puts the AOLTW venture into perspective: Despite the size of the venture, this sum of money, while significant, is only one part of an investment trend that has been occurring for the last two decades in China. It is the first venture of its size in the highly guarded telecommunications sector and the first with a major Internet provider, so it will be interesting to see how the Chinese government deals with the inevitable challenges to the economic and social control this venture will bring about. But it is only the most recent in a long line of in-

[13] Arbitration clauses are a particular case in which the government gives away control over joint ventures. If a joint venture contract specifies nothing about how a dispute will be resolved, disputes that arise will be handled by the Chinese courts. This is the best-case scenario for the Chinese government in terms of sovereignty because the courts, at this point, are still an arm of the authoritarian government. However, if a joint venture agreement specifies that disputes will be settled through arbitration, there are two possible venues for this. The first is the Chinese International Economic Trade and Arbitration Commission (CIETAC), an institution of arbitration in Beijing (with branches in Shanghai and Shenzhen). The significant fact about CIETAC in terms of arbitration is that one-third of the arbitrators that sit on any case are from other countries. Thus, once cases go to CIETAC, the Chinese government no longer has control over their outcome. A second possibility is that a joint venture agreement can specify third-country arbitration, in which the dispute will be settled in the arbitration institution of some specified third country. The Chinese government has even less control—if any at all—over the outcome of these cases.

TABLE 3
Foreign Capital Invested in China in the Reform Era

	Foreign capital committed (in billions)	*Number of projects*	*Foreign capital actually used (in billions)*
1985	5.932	3,073	1.661
1990	6.596	7,273	3.487
1995	103.205	37,184	48.133
1999	52.102	17,022	52.009

Source: Statistical Yearbook of China (1999, 2000).

vestments that have placed the Chinese government in a partnership with Western multinationals. So does IT matter for encroachments on state sovereignty in the realm of foreign investment? Yes, but this challenge to the state is only the most recent in a long line of sectoral transformations that have occurred throughout the economy over the last twenty years.

Privacy and Information Technology: Social Networks in China

During a recent research trip in China, a Chinese colleague decided I needed a cell phone, as reaching me was proving more difficult than my collaborator liked or was used to. We fought the traffic across town to Xinshimen, a place famous for, among other things, selling used cell phones. This market, like many of the marketplaces of China, is a bustling scene where the social order seems to border on chaos; it is also a place where many people from Beijing and the surrounding areas go to purchase mobile telephones. Before this excursion, I had not understood how simple it is to have a cell phone in China. Some people have cell phones that operate through a formal telephone service, with an account and a monthly bill, the way that telephone accounts work in many countries throughout the world. However, many people circumvent this system by buying secondhand phones and installing them with miniature phone cards that can be purchased at kiosks and stores throughout the country. Each card has an assigned number, and when a phone is linked with a card, it then responds to that number. Each card is also programmed with a certain amount of money, so that when the money on that card is expired, more money can allow the card to be reactivated or the phone can be fitted with a new card (and new number). It is important to note that setting up such a phone is not only extremely fast and simple, it is also completely anonymous: There is no requirement for a customer's name to be attached to a given card.

With phone in hand, my colleague and I went down the road to a more formal-looking department store, which sold registered mobile phones but also dealt with assigning Internet accounts for cell phones. I had mentioned earlier to my colleague that I was having trouble getting my computer online, and he wanted to see what it would take to solve both of these issues in one outing. After an exchange that evolved too quickly for me to follow, my colleague leaned over to me and said abruptly, "Let's get out of here." As we left, he explained to me that the phone I had just purchased was not a registered phone, and you must have a registered phone to set up an Internet account. Despite the apparent popularity of this practice, I was not too keen on participating in the underground economy without thinking through the implications of such a venture. My friend looked at me quizzically: "Oh no. It's perfectly legal. It's just not formal or registered. The government wants to control the Internet, so you can only set up those accounts on equipment that is registered with the government. Most of us would like to be a little more anonymous, and we would rather use ways that are not so easily monitored by the government."

This comment is reminiscent of a classic book on the role of the gift economy in reform-era China. In 1984 Mayfair Yang published the now classic book on *Guanxi, Gifts, Favors, and Banquets: The Art of Social Relationships in China,* which, as the title implies, examines the resurgence of social relationships in communist China. One of the central issues in the book is what Yang calls rhizomatic networks, the infinitely interconnected social networks that link individuals in the face of overwhelming state power. According to Yang, this phenomenon emerged in response to the Cultural Revolution, when state power was at its peak and also at its most capricious. The point of Yang's treatise is that, in the face of authoritarian power, individuals will find ways to resist the long arm and the watchful eye of the state. In Chinese society, individuals call upon cultural resources for moral authority and the practice of maintaining a bank account of gifts and favors to create a gift economy that operates fully independent of state control. It is Yang's contention that these rhizomatic networks are the ultimate form of resistance in communist China, as they allow individuals to wield power without aligning themselves with the government. In recent years, information technology has become an avenue by which this resistance has been channeled. Inasmuch as information technology allows individuals to operate independent of state control, this form of privacy also becomes a form of resistance.

The use of unregistered cell phones and unregistered Internet services seems, on the surface, to be the technological extension of Yang's treatise of the triumph of the social over the state. The use of unregistered information is, of course, not only about privacy. Indeed, many of the practices in China's IT economy relate as much to the contours of development as anything else. For example, in China, individuals still cannot write checks, so when one wants to pay a phone bill, it requires a trip to the bank to withdraw cash and then a trip to a local bill-paying venue. Yet there is a clear issue of privacy here: individuals are enjoying a freedom of communication, a sharing information, and developing social ties that elude the control by the state that this society has experienced in the past. The technology has become a vehicle for fostering, facilitating, and experiencing this freedom. Here again, it is important to note that these rhizomatic networks existed long before IT was a factor in Chinese society. Information technology like the Internet and mobile telephones are undoubtedly tied to the privacy that individuals seek from an overbearing authoritarian state. IT is not causing this type of privacy or resistance to come about, but it is playing an important part in the evolution of this part of society.

Information Technology and Resistance

A little more than a decade ago, in the spring of 1989, the world watched as Beijing experienced the upheaval of the Tiananmen Movement. Many scholars of this movement have elaborated on the role that information played in the evolution of this movement (Guthrie 1995; Walder 1989; Perry and Wasserstrom 1992). Students used fax machines, telephones, and broadcast equipment to spread their message far beyond the scope of any of the preceding movements.[14] Even beyond the students' resources, other changes in information technology were important in the evolution of this movement: The very fact that the world was able to watch was tied to the changes in information technology that had occurred over the decade prior to the movement. With the major international networks and news media in China for the Gorbachev Summit and the Asia Development Bank meetings, the broadband dissemination of this movement to the rest of China and throughout the world reached unprecedented levels, compared with the popular protests that had occurred in Beijing in 1986, 1979, or 1976. Information technology played a major role in both how much the students were able to broadcast their message and how much of the world was able to hear it.

Again, it is important to place these changes in the context of the insti-

[14] The Stone Corporation, at the time the largest private corporation in China, was the primary provider of many of the tools of information technology to which the students had access. It is also the corporation that founded Legend Computers, the company that is now establishing a joint venture with AOL Time Warner.

tutional changes that were occurring throughout the 1980s. Important as it was in the dissemination of this movement, did information technology play a causal role in this major encroachment on the Chinese government's capacity to control its population? There has been much debate over the causal roots of this movement, with scholars arguing that causal primacy lies with the rise of student networks (Calhoun 1995), the rise of student organizations (Guthrie 1995), students' ability to mobilize cultural resources (Esherick and Wasserstrom 1992), and fundamental institutional change (Walder 1994; Zhao 1997). As Andrew Walder put it, "[W]hat changed in these regimes in the last decade was not their economic difficulties, widespread cynicism, or corruption, but that the institutional mechanisms that served to promote order in the past—despite these longstanding problems—lost their capacity to do so" (1994: 298). Drawing on his earlier work on the communist order in the prereform era (Walder 1986), Walder goes on to specify the mechanisms that were crucial for maintaining order in communist societies as (1) hierarchically organized and grass-roots mobility of the Communist Party and (2) the organized dependence of individuals within social institutions, particularly workplaces. With the beginning of the economic reforms in China, both of these institutional bases of power began to erode. In the first case—the decline of party power—there are two ways this change had fundamental implications for the organization of Chinese society. First, the party no longer had strict control over its own agents. Party cadres operated with an autonomy that increasingly grew in scope throughout the 1980s. This was in large part a direct consequence of the movement away from central planning of the command economy. As the reforms progressed, the new economic policies of the 1980s essentially mandated that local-level bureaucrats assume administrative and economic responsibilities for the firms under their jurisdictions (Walder 1995; Guthrie 1999a). As administrative and economic responsibilities were pushed down the hierarchy of the former command economy, local-level bureaucrats exercised more and more power in the struggle to control resources and survive in the markets of China's transforming economy. Thus, the institutional changes of the reform economy led to the decline of central party control over its own members.

Second, and perhaps more importantly, the party no longer exercised grassroots control over individual citizens. In pre-reform China, the party meticulously exercised such control through local party meetings, usually conducted through an individual's work unit or neighborhood association (Walder 1986; Whyte and Parish 1984). In the reform era, this centrally mandated practice eroded quickly. Managers and administrators no longer required their workers to attend meetings for the dissemination of party ideology. This change is closely related to the institutional changes of the economic transition described above: as economic imperatives replaced strict compliance with detailed directives of the party and the command economy, managers and administrators began to run their organizations less around the dissemination of party ideology and more around the ideals of performance. And the students no longer had to fear that activism would affect their employment prospects, as the emergence of the private economy and the withering of the labor allocation system meant that the state could no longer hold their behavior in check with the threat of consequences in their career placement.

The main point here is that, while information technology may have been important in the evolution of this movement, which posed the greatest challenge yet to the Chinese government's right to rule—a critical issue of state capacity and legitimacy—the causal roots of this movement lay in institutional changes that had been occurring for a decade. It may be the case that information technology affected the scale of this movement and the extent to which the world knew what was occurring, but the roots of the movement itself lay elsewhere. In a certain sense, though, by allowing foreign media to cover the Gorbachev Summit and the Asia Development Bank meetings and by giving the population access to technology—fax machines, telephones, etc.—the Chinese government had inserted itself into a global information network, and its ability to control the flow of information had important consequences for the extent to which it could control the movement. The Chinese government had, in effect, armed its opposition with the tools of resistance, and when the movement occurred, the government could not stop the flow of information beyond its borders, a fact that had profound consequences for the scale and scope of this movement.

More recently, a number of social and political occurrences involving the Internet have illuminated the new role that this form of IT might play in the state's ability to control information. In one incident, when China's top official from the State Administration of Foreign Exchange apparently jumped from his seventh story window on May 12, 2000, government officials were caught completely off guard as the story was posted almost immediately on a bulletin board on the widely visited Sina.com web site. According the Elisabeth Rosenthal's report in the *New York Times,* "The government was clearly not prepared to release the news today, and confusion reigned for much of the day."[15] A similar incident occurred when a story of a Beijing University student who was murdered appeared on a Sohu.com bulletin board on May 19, 2000. In the latter incident, students from all over the country staged a "virtual" protest, forcing officials to allow them to openly mourn and memorialize the student, despite the dis-

[15] *New York Times,* May 13, 2000.

ruptions officials feared the event would cause. In both of these cases, it was clear that the government's mentality regarding control over the flow of information was lagging significantly behind the current reality in this realm. This is a new frontier for outright resistance, and it will be interesting to see over the coming decade what role the Internet plays in the government's ability to control the spread of information and the organization of popular movements.

Newly Independent Sectors

While active resistance has been an important part of the erosion of state capacity and legitimacy in China, the emergence of newly independent sectors has also played a critical role in the political reach of the Chinese state. As in the case of active resistance, new information technology has not caused the emergence of these newly independent sectors, nor has it been fundamental in shaping these sectors of society. However, their emergence and increasingly independent status in the economy have diminished state capacity in significant ways, and the role of information technology has not been insignificant in the evolution of these sectors. Of particular note are the private economy and higher education; the former is a new phenomenon in the reform era with important implications for effective control by the Communist Party; the latter, while predating the reform era, has undergone a radical transformation in the last decade, a transformation that also has important implications for state sovereignty in the age of globalization.

THE PRIVATE ECONOMY

Under Mao Zedong, the private economy was nonexistent. In the 1950s, as the communists sought to overtake and control all sectors of China's economic and social systems, the private economy was all but eliminated. The state controlled all modes of production, private ownership was eliminated, and the freedom to make economic decisions independent of state control became a thing of the past. It was not until the economic reforms began in 1980 that private entrepreneurs were allowed to reemerge in the economy. Since 1980, the number of entrepreneurs in China has grown steadily, as has their legitimacy under the communist mantle. In July 2001, when Jiang Zemin announced that private entrepreneurs would be permitted to join the party, it had become clear that the communist government had finally made its peace with the importance of private entrepreneurs and the private economy in China's transforming economic system. And rightly so: not only has this sector provided an important outlet for the employment overflow that would have otherwise caused a significant strain on the transforming economic system, but, as it has grown, it has also played a crucial role in creating competition for the transforming state sector (Naughton 1995). Figure 2 shows the growth of this sector of the economy since 1980. It is clear from the figure that the sector has grown both in absolute terms and also relative to the other forms of ownership in the economy.

Despite its importance in a system that is still largely owned and controlled by the government, the role of the private sector is limited: even today, as management responsibilities have been passed on to managers and local officials, and as industrial output has shifted to the private sector, the government still owns about 70 percent of the industrial assets (Guthrie 1999a). Yet this sector has played a significant role in encroaching upon state capacity in two ways. First, in times of crisis, it has played a role of outright resistance. During the Tiananmen movement it was the private entrepreneurs of Beijing who provided the students with fax machines, radio equipment, televisions, and other perishable goods that became a staple of the movement (Perry and Wasserstrom 1992). The Stone Corporation was the largest and most famous of these behind-the-scenes participants, but there were many others. It would be a stretch to argue that private businesses in China are predisposed to resistance. However, it is the case that these organizations are structurally the ones that hold

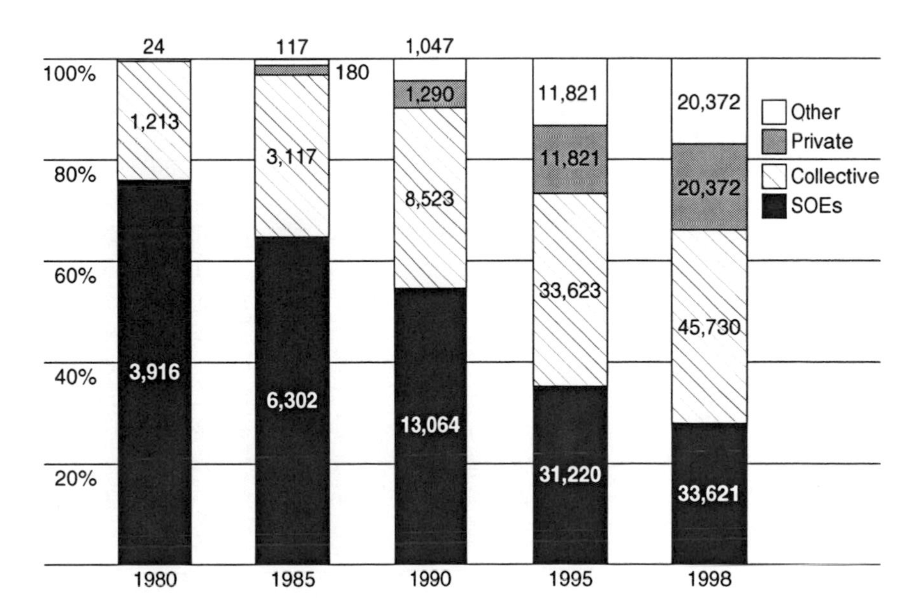

Figure 2. Gross industrial output by ownership type

the greatest degree of independence from the state and therefore have the greatest latitude in protesting when the opportunity presents itself.

Second, and perhaps more importantly, this sector has facilitated and enhanced the institutional changes that have fundamentally shifted incentive structures in China today. In the previous section I briefly introduced Walder's (1994) argument regarding the party's diminished capacity for social control in the 1980s; as I noted above, central to Walder's argument is the fact that China's citizenry was no longer fundamentally dependent upon the state for the allocation of social goods such as jobs. This decline in what Walder (1986) has called "organized dependence" required an opening up of China's labor markets (as opposed to the state's allocation of jobs) and new opportunities for employment. The private sector played a fundamental role on both of these fronts. As Zhao (1997) argues, a fundamental difference in the state's lack of control over the 1989 movement was that students now had options: they were no longer dependent upon the state for the allocation of jobs, so participation in a movement such as this did not pose the career threat it had in the past.

What do these changes have to do with information technology? In both instances, the role of IT lurks in the background: this rapidly growing sector does not play an active role in opposing state power, but it is integrally tied to the evolution of the private sector, and the strength of this sector does diminish the authoritarian government's ability to rule in important ways. In the first case, although the 1989 uprising occurred before the IT revolution, it was information technology that the movement's supporters from the private economy delivered to the students. In the second instance, one of the major sectors that has emerged in recent years to create jobs in the private economy was in the area of information technology, namely, telecommunications. In 1980, when the private economy was first emerging, less than 1 percent of the jobs in this sector were located in telecommunications; by 1999, 5.8 percent of the jobs in the private economy were located in the telecommunications industry—a total of more than two million jobs. The growth in this sector has been driven, in part, by national projects such as development of Zhongguancun in Beijing. It has also been driven by the growth in demand for the products and services this sector provides (see table 2 for examples of growth in demand). So here again, while information technology has not created the private economy, it has played a critical role in the ways this economy has emerged and in its potential for fostering the decline of the Chinese government's control over its population.

HIGHER EDUCATION

As with the private economy, a discussion of higher education in China might require a little background. China has a long tradition of intellectuals advocating political and social change. From the May Fourth Movement of 1919 to the Hundred Flowers Reform of 1956–57 to the Tiananmen Movement of 1989, intellectuals have served as the conscience of the nation in many ways. For more than a century, institutions of higher education have been among the central advocates for social change in China. In the era of economic reforms, however, higher education in China has undergone a radical revolution. In the prereform era, the central government had under its jurisdiction thirty-six universities. During the economic reforms, there has been much discussion in China about the reform of higher education, and in May 1998, after more than a decade of discussion on the topic, a formal reorganization of the higher education system was set in motion. The first change was that ten universities were named as "international level universities" (*guoji yiliu daxue*) and the central government would concentrate its resources on their development. The remaining twenty-six universities under the central government would be gradually turned over to provincial- and municipal-level governments. Within this first-tier group, in addition to the usual funds that the institutions under the central government would receive, the top four of the "international universities" would get extra funds to help them develop as "international level universities."[16] Under this reform, universities were now free to fundraise on their own, develop relations with foreign universities, and generally develop the programs that would make them competitive with top-tier research universities around the world. Table 4 shows the growth and changes in this sector since 1980. Since that time, the number of university students in China has increased by almost 200 percent; the number of faculty has increased by almost 75 percent; the number of student studying abroad has increased by more than 1,000 percent; and the number of study-abroad students who have returned to China has increased by more than 4,500 percent. These changes created a sector that is autonomous from the central government in ways fundamentally different from the situation of the prereform era.

Here again, it should be noted that autonomy does not in and of itself imply an erosion of state power, nor does it imply an influence of information technology within this process of change. However, history teaches that the academy has developed a special relationship with information technology. First of all, while it was the Department of Defense that laid the groundwork for the functional development of the Internet, this medium also found part of its genesis—particularly in the area of content—in the academy (Guthrie 1999b). If it was not for the early adopters

[16] Beijing and Qinghua Universities (widely regarded as the top two universities in China) received an additional 1.8 billion yuan, spread over three years (1999–2001) while Nanjing and Fudan (widely regarded as the third and fourth best universities, respectively) received an additional 1.2 billion yuan.

TABLE 4
Vital Statistics on Higher Education in China

	No. of univ. students per 10,000 pop.	*No. of faculty (higher ed.) per 10,000 pop.*	*No. of students studying abroad*	*No. of students returning from abroad*
1980	11.6	24.7	2,124	162
1985	16.1	34.4	4,888	1,424
1986	17.5	37.2	4,676	1,388
1987	17.9	38.5	4,703	1,605
1988	18.6	39.3	3,786	3,000
1989	18.5	39.7	3,329	1,753
1990	18.0	39.5	2,950	1,593
1991	17.6	39.1	2,900	2,069
1992	18.6	38.8	6,540	3,611
1993	21.4	38.8	10,742	5,128
1994	23.4	39.6	19,071	4,230
1995	24.0	40.1	20,381	5,750
1996	24.7	40.3	20,905	6,570
1997	25.7	40.5	22,410	7,130
1998	27.3	40.7	17,622	7,379
1999	32.8	42.6	23,749	7,748

Source: Statistical Yearbook of China (2000).

of Internet technology in higher education, such as the University of Utah and the University of Southern California, the Internet itself may not have gained the foothold it did before commercialization in 1991; and if it was not for the desire of researchers at CERN to share documents electronically, Tim Berners-Lee might never have had the impetus, much less the insight, to lay down the groundwork for the World Wide Web. As the Internet and Web have become accessible in China, it has been the universities, along with the private economy, that have been at the forefront in the adoption and use of these media. This fact has put these newly autonomous institutions of higher education in much closer touch with the world outside of China. It is because the academy is built on the production, use, and processing of information that access to information technology has such a great potential to speed the evolution of this sector's autonomy from the Chinese government. As scholars from China gain more and more access to different points of view about democracy, economic systems, international politics, social change and many other issues, the state's diminished role in controlling this sector becomes inevitable. Indeed, access to information has always been the impetus for the type of social change advocated by reform-minded intellectuals: from the leader of the Nationalist movement, Sun Yat-sen, and the literati of the May Fourth era, including such notable figures as Lu Xun, to the intellectuals that led the Tiananmen Movement, exposure to foreign models of political and economic systems has been central to the issues these individuals have raised in advocating change.

Conclusions

Many political and business leaders from Western capitalist nations have asserted as President Bill Clinton did on May 8, 2000,[17] that new information technology—particularly the Internet—would play a key role in liberating authoritarian societies. Indeed, in praise of this young technology's potential, we have seen not only bold predictions, but also wild revisionist history. Jim Courter, a former six-term Republican congressman once described the promise of this technology, saying the Internet "has done a lot to bring democratic capitalism to other parts of the world. It was instrumental, I think, in bringing down the Berlin Wall. It was instrumental in having students protest the policies in East Berlin. . . . CNN, the Networks, and the Internet, were instrumental in the demise of the old Soviet Union."[18] Pretty amazing, given that the Internet was not commercialized until 1991 and really had no presence in the world at large before that time. Nevertheless, there are substantively important questions behind these statements, namely, what is the impact of the new information technologies on authoritarian societies? And, more importantly, what causal role do these technologies play in encroaching on the sovereignty of authoritarian governments and in bringing about democratization?

The Chinese government's legitimacy withered significantly over the last two decades in large part because of the institutional changes that were set in motion by the government itself, and there is a dynamic interplay between those institutional changes and the ways that information technology has fostered perhaps greater change than the state originally intended. A number of scholars of China's economic reforms have pointed to the fact that the reform process, which began in 1979 as a set of controlled modifications to the economy, very quickly took on a life of its own, expanding far beyond the purview that the original architects of the reform had in mind (Naughton 1995; Guthrie 1999a). There are many hidden dynamics to this process, which Barry Naughton has called "grow-

[17] Quoted in Drake, Kalathil, and Boas (2000).

[18] Ibid.

ing out of the plan," but one of the dynamics in recent years has been the interaction between state policies and different actors in society. Of particular interest here is how these different actors or groups have employed newly emerging technologies to facilitate the process of change. In each case examined in this chapter, the roots of change lay in state action; yet the state's ability to control these institutional changes, once they emerge into the world of actors in the economy, has been unpredictable at best, and information technology has played a role in that dynamic.

The encroachments on state sovereignty at the hands of foreign investors were just as clear in negotiations over ventures in the industrial economy as they are in the New Economy. However, the state has been more resistant to letting these ventures fully open up to foreign investment, precisely because it fears the potential liberating forces that will emerge with foreign control in this sector, a fact that is tied to the perception that investments in the telecommunications sector specifically may pose a great challenge to the authoritarian government's capacity to control investments in other sectors. Thus, the struggles over sovereignty in the realm of foreign investment are nothing new, but the stakes appear to be higher—as perceived on both sides—when it comes to information technologies such as the Internet. Similarly, with privacy, information technology provides a vehicle for resisting state authority and control, as it provides further outlets to facilitate the growth of these private social networks. But it is only a vehicle, a facilitator, in these processes of change. Yang's rhizomatic networks existed long before cell phones came onto the scene, and if Yang's story is right, these social networks, which became the ultimate form of resistance to capricious authoritarian rule in Mao's China, were playing a role in Chinese society long before information technology existed in any form in China, as the 1970s were a time in China when few people had access to telephones or television, let alone fax machines, cell phones, and the Internet. Yet the social resistance to state control emerged through resources that *were* available. Employing cultural resources, the moral authority of existing social networks, and an accounting system of gifts and favors that defined China's gift economy, people built and assiduously maintained the private systems that could free them, at least to some extent, from monitoring control of an overbearing state. Today, the use of cell phones, e-mail, and the Internet has allowed people to be more connected than ever before, and many people I have spoken with in China see these connections as a form of independence from state control. Not only are unregistered cell phones more convenient than setting up a formal account or a landline phone in their home, but they have the added benefit of being beyond the reach of the state. But note here again that this technology does not create privacy, it only provides new avenues for it.

The same is true for outright defiance and the emergence of new sectors of the economy and society. In the Tiananmen Movement of 1989, fax machines and live feeds to international news programs were certainly important for the evolution of the movement, but it would be a stretch to view information technology as playing any kind of causal role in this moment of collective action. The causal factors driving this movement forward were much deeper, related to fundamental changes in social structure and economic institutions that allowed the state to maintain social control. The role information technology has played in such movements is, once again, a facilitator of processes that are already in play, but it is significant nonetheless. Similarly, with respect to sectors of society that increasingly see themselves as independent from state control, IT also plays a significant role in facilitating this independence—whether through giving intellectuals more access to information or through creating new and thriving sectors of the private economy—speeding up processes that are already in play. In each of the cases discussed in this chapter, actors in society have pushed at the boundaries of state-led change, and newly emerging information technologies are among the tools they have used in this pursuit. In this sense, information technology has played a significant role in the dynamic of change in reform era China.

References

Calhoun, Craig. 1989. Protest in Beijing: The Conditions and Importance of the Chinese Student Movement of 1989. *Partisan Review* 4:563–80.

———. 1991. The Problem of Identity in Collective Action. In *Macro-Micro Linkages in Sociology,* edited by J. Huber, 51–75. Beverly Hills, CA: Sage.

———. 1995. *Neither Gods nor Emperors: Students and the Struggle for Democracy in China.* Berkeley: University of California Press.

China to Become Asia-Pacific Region's Second-Largest IT Market, Study Says." *Chinaonline,* June 22, 2000. Available at http://www.chinaonline.com.

Cooper, Caroline. 2000. Look at India, IT's Where China Wants to Be. *Chinaonline,* June 22. Available at http://www.chinaonline.com.

Cui, Ning. 2001. Technology a Growth Engine to Economy. *China Daily,* June 18:1.

Drake, William J., Shanthi Kalathil, and Taylor C. Boas. 2000. Dictatorships in the Digital Age: Some Considerations on the Internet in China and Cuba. *Information Impacts Magazine* (October).

Esherick, Joseph W., and Jeffery N. Wasserstrom. 1992. Acting Out Democracy: Political Theater in Modern China. In *Popular Protest and Political Culture in Modern China: Learning From 1989,* edited by Elizabeth Perry and Jeffery Wasserstrom, 28–66. Boulder, CO: Westview Press.

Friedman, Thomas. 2000. Censors Beware. *New York Times,* July 25.

———. 1999. *The Lexus and the Olive Tree.* New York: Farrar Straus & Gilroux.

Guthrie, Doug. 1995. Political Theater and Student Organizations in the 1989 Chinese Movement: A Multivariate Analysis of Tiananmen. *Sociological Forum* 10:419–54.

———. 1999a. *Dragon in a Three-Piece Suit: The Emergence of Capitalism in China.* Princeton: Princeton University Press.

———. 1999b. A Sociological Perspective on the Use of Technology: The Adoption of Internet Technology in U.S. Organizations. *Sociological Perspectives* 42(4): 583–603.

Harner, Stephen. 2000. Shanghai's New Five-Year Plan: The Pearl Starts to Shine. *Chinaonline,* December 18. Available at http://www.chinaonline.com.

Hill, David, and Krishna Sen. 2000. The Internet in Indonesia's New Democracy. *Democratization* (Spring).

Naughton, Barry. 1995. *Growing Out of the Plan: Chinese Economic Reform 1978–1993.* New York: Cambridge University Press.

Perry, Elizabeth, and Jeffery Wasserstrom. 1992. *Popular Protest and Political Culture in Modern China: Learning from 1989.* Boulder: Westview Press.

Segal, Adam. 2002. *Digital Dragon.* Ithaca: Cornell University Press.

Shen, Tong. 1990. *Almost a Revolution.* Boston: Houghton Mifflin.

Smith, Craig. 2001. AOL Joins Chinese Venture, Gaining a Crucial Foothold: A Deal to Develop Services for the Internet. *New York Times,* June 12: W1.

Walder, Andrew. 1986. *Communist Neo-Traditionalism: Work and Authority in Chinese Industry.* Berkeley: University of California Press.

———. 1989. The Political Sociology of the Beijing Upheaval of 1989. *Problems of Communism* 38:30–40.

———. 1991. Workers, Managers, and the State: The Reform Era and the Political Crisis of 1989. *China Quarterly* 127:467–92.

———. 1994. The Decline of Communist Power: Elements of a Theory of Institutional Change. *Theory and Society* 23:297–323.

———. 1995. "Local Governments as Industrial Firms: An Organizational Analysis of China's Transitional Economy." *American Journal of Sociology* 101:263–301.

Whyte, Martin, and William Parish. 1984. *Urban Life in Contemporary China.* Chicago: University of Chicago Press.

Yang, Mayfair Mei-Hui. 1984. *Gifts, Favors and Banquets: The Art of Social Relationships in China.* Ithaca: Cornell University Press.

Zhao, Dingxin. 1997. Decline of Political Control in Chinese Universities and the Rise of the 1989 Chinese Student Movement. *Sociological Perspectives* 40:159–82.

Zhu, Enjoyce. 2001. China's Silicon Valley Ready to Take Off. *Beijing Business* (June).

Contributors

HAYWARD R. ALKER, John A. McCone Professor of International Relations, School of International Relations, University of Southern California

JONATHAN BACH, Core Faculty, Graduate Program in International Affairs, New School University

LARS-ERIK CEDERMAN, Professor of International Conflict Research, Center for Comparative and International Studies, Swiss Federal Institute of Technology

DIETER ERNST, Senior Fellow, East-West Center, Honolulu, Hawaii

D. LINDA GARCIA, Director, Communication, Culture and Technology, Georgetown University

DOUG GUTHRIE, Associate Professor, Department of Sociology, New York University

PETER A. KRAUS, Associate Professor, Sociology Department, Humboldt-Universität, Berlin

ROBERT LATHAM, Director, Information Technology and International Cooperation Program, Social Science Research Council

WARREN SACK, Assistant Professor, Film and Digital Media, University of California, Santa Cruz

SASKIA SASSEN, Ralph Lewis Professor of Sociology, University of Chicago

DAVID STARK, Arthur Lehman Professor of Sociology and International Affairs, Columbia University, and External Faculty Member, Santa Fe Institute

STEVEN WEBER, Professor, Department of Political Science, University of California, Berkeley

Index

Abbate, Janet, 148n.7, 154, 156
Abbott, Andrew, 9n.15
Abbott, Bud, 250–251n.8
Abelson, Robert P., 225, 225n.17, 225n.18, 269n.25
Academic economics, growing importance of, 62
Access: organization of, 11; rural-urban discrepancies in, 117–142; simultaneous, 7–8n.12
Accountability: with distributed authority, 45n.10; of networks, 50
Acer group, 104
Actions, organization of, 11
Activism: e-mail mobilization network in, 78–79; global network in, 54; globalization of, 81–82
Activists, digital technologies used by, 75–76
Adams, Paul C., 77–78, 81n.26
Adar, Eytan, 202
Agenda for Peace (Boutros-Ghali), 216–217
Agglomeration economies, 92
Aggregation, 169–172; logic of, 170; patterns of, 174
Agnew, John A., 49
al-Qaeda, organization of, 208
Alexa browser, 42
Alice in Wonderland, interpersonal metafunction in, 249, 251
Alker, Hayward R., 6, 14, 24, 26, 215n.2, 218, 219n.8, 220–221, 222n.14, 223n.15, 225n.18, 228, 230, 232–233
Allen, James F., 258n.15
Alliances, 7; in electronic market infrastructure, 70–71; of NGOs with state and market, 50
Altermatt, Urs, 296n.8
Alternative media network, 79n.24
Alternative media networks, 78n.22
Alternet.org, 42, 43–44
alt.politics.elections, 274, 275
Alumni effect, 183–184
Amazon.com: collaborative filtering technology of, 42–43; recombinant technology of, 41
American Library Association, online partnerships of, 48
Amin, Ash, 125, 129, 130
Amnesty International: information exchange of, 76; Urgent Action Alert system of, 77, 79
Amsterdam's stock exchange, seventeenth century, 62n.8
Analytic categories, 18
Analytical operations, 18–19; destabilizing older hierarchies of scale, 22–25; digital/social imbrications, 19–21; mediating practices and cultures, 21–22
Anderlini, S. N., 219n.8, 227
Anderson, Benedict, 285–286, 290, 291, 293, 303
Anderson, Jon, 21n.33
Anglo-American discourse theorists, 248
Anglo-American media studies, 259–260n.17
Antiglobalization movement, 44; Seattle protests in, 46n.14
Antonelli, C., 95, 98–99
AOL Time Warner: in China, 322–324, 327n.14
AOL Time Warner-Legend deal, 322n.11, 327n.14
Appadurai, Arjun, 46n.12, 50
Apple Computer, Discourse Architecture Laboratory of, 243
Arab banking system, cross-border nature of, 58–59n.5
Archibugi, D., 90
Architectural design research, 223–227
Aristotelian practical reason, 221n.12
Aristotle, 223n.15
Arnold, Erik, 133, 138
Arpa, 169
Arpa networking community, 155, 155n.20
Arpanet (Advanced Research Projects Agency), 12; connection with British computer research networks, 167;

Arpanet (Advanced Research Projects Agency) (*continued*)
development of, 161, 163–164, 170; European nodes of, 147–148n.6; as professional community, 13; restrictions on, 167–168
Arthur, Brian, 132
Artificial, sciences of, 222–223
Artificial intelligence, 222n.14
Asheim, Bjorn T., 130
Asia: cost and time reduction center clusters in, 102; financial crisis of, 72. *See also* China
Asia Development Bank, 327, 329
Asset management funds, 61
Asset specificity, 129n.8
Association: new basis for in knowledge communities, 43–50; technology and new geographies of, 43–50
Association for Progressive Communications (APC), 80
Associative thinking, 50
Asymmetry, 99–100
Asynchronous interactions, 96–97
AT&T, in network development, 164
Atwood, Roy A., 135
Aubert, J. E., 109
Augmented Social Networks (ASN), 43–44
Authority, distributed, 45n.10
Automated early warning systems, 216
Autonomy protection, 306
Avant-garde literature, linguistics of, 250–251
Awareness raising, Internet in, 75n.20
Axelrod, R., 224–225, 237–238
Ayres, R. U., 126n.6

Bach, Jonathan, 6, 10n.17, 10n.21, 13, 15, 24, 26, 27, 46n.14, 48n.17
Baird, Davis, 185n.4
The Bald Soprano (Ionesco), 249–250
Bandwidth, surplus of, 172n.55
Bangemann Report, 298
Bangladesh, Proshika NGO of, 47n.15
Barfield, Woodrow, 10n.20
Barkley, David J., 130
Barley, Stephen R., 39n.2
Barth, Fredrik, 292
Barthes, Roland, 259n.16
Bartlett, C. A., 99, 103
Basel, financial market in, 67n.12
Beaverstock, J. V., 23
Becker, Gary S., 197
Beehive, 42
Beijing University, 333n.16
Bell, Daniel, 5n.10, 163n.40
Bell Telephone Company, 134–135
Bellanet, 75n.19, 80
Bendell, J., 48n.16
Benedikt, Michael, 10n.20
Beniger, James R., 2
Bennett, James, 228n.21, 230
Benyon, David, 243n.3
Berkeley Software Distribution (BSD) model, 198
Berlin Crisis, 168n.48
Berlin Wall, fall of, 335
Berners-Lee, Tim, 334
Bhaskar, 227
Biersteker, Thomas J., 139
Big Brother Inside, 79n.24
Bijker, Wiebe E., 150n.11
Birkinshaw, J., 99
Birnbaum, Pierre, 294
Bitnet, 163; protocols for, 164
Bloomfield, Lincoln, 215n.1
Blue-collar knowledge production, 90
Blumenthal, Marjorie, 137
Boas, Taylor C., 335
Bombay: financial center in, 66; in global financial network, 67; organizing slum dwellers in, 81
Bonds, cross-border transactions in, 63t
Borrus, M., 99, 100
Börzel, Tanja A., 290
Boulding, Elise, 215n.1
Boulding, Kenneth, 222n.14
Bounded institutionalism, 284, 290–293, 294, 303–304; logic of, 292
Boutros-Ghali, B., 216–217
Boyd, D., 105, 106n.8, 107–108
Bradley, S. P., 97
Brand, Stewart, 204
Brass, Paul, 291
Braudel, F., 93
Braverman, Avishay, 119–120, 121, 133
Brenner, Neil, 22, 81
Breschi, S., 90
Britain, national identity of, 295–296, 296–297
British National Physical Laboratories, network development at, 163–164
Brooks, Frederick, 197

Brown, L. David, 45n.11
Brown, Lawrence A., 121, 133
Brubaker, Rogers, 287, 288n.3
Bruzst, Laszlo, 45n.10, 46
Bryan, Cathy, 265
Bryant, Coralie, 39, 48
Brynjolfsson, E., 97–98
Built-to-order production model, 98n.3
Bunn, Julie Ann, 146n.1, 157n.27, 169
Bureaucracy, authority in, 206–207
Burgelman, Jean-Claude, 81–82n.28
Business: computerization of, 163n.40; flexibility of, 126
Business-to-business e-commerce, 127

Cailliau, Robert, 154, 156n.23, 158, 169, 170
Calabrese, Andrew, 81–82n.28
Calhoun, Craig, 13n.27, 266n.19, 290–293, 305, 328
Callon, Michel, 38n.1, 62
Campaign to Ban Landmines, 44
Capacity buffers, 104
Capital: fixity of, 19–20; mobility of, 19–20
Cappella, Joseph N., 269n.24, 272
Carbonell, Jaime G., 258
CARE-Starbucks, 48
Carroll, Lewis, 249
Castells, Manuel, 5n.10, 118, 125, 235n.24, 265, 298–299
Category-bound activities, 257–258
Cederman, Lars-Erik, 7, 15, 26n.43, 28, 287n.2, 291, 304, 306
Center for Victims of Torture, "New Tactics in Human Rights Project" of, 78
Centers, versus peripheries, 15
Centers of excellence, global, 90–91, 101–102, 109
Centralization: of business operations, 122–123; in global financial market, 68–70; in successful NGOs, 48–49
Cerf, Vincent G., 155n.20, 168
CERN research, 334
CEWS. *See* Conflict Early Warning Systems (CEWS)
CEWS Explorer: Chiapas case in, 229; LISP/SCHEME software of, 233
CEWS representational format, 230–231
Chadwick, Andrew, 298, 303
Chandler, A. D., 95, 100
Change, new technologies in, 4
Chen, S. H., 97, 104
Cherny, Lynn, 255
Chiapas crisis, 232; conflict resolution in, 229, 230; tracking phases of, 232–233
Chicago, financial center in, 66
China, 28; access to information and IT sector growth in, 317–321; authoritarian government of, 321–322, 323, 324n.13, 335–336; civil society of, 15; economic development of, 312–313, 322–325; economic transformation of, 335–337; environmental activists in, 76; foreign investment in, 314, 320–321, 322–325, 336; gift economy in, 326–327, 336; governmental reforms in, 314; higher education in, 332–335; industrial output of, 331–332; information technology and state capacity in, 312–337; international financial transactions in, 72; international level universities of, 333; market and political reforms in, 315–321; Nationalist movement in, 334–335; New Economy of, 336; newly independent sectors in, 330–335; private economy in, 330–332; private-sector economy in, 314; R&D budget of, 109; reform economy in, 320–321; social networks in, 325–327; state-controlled IT industrial sector in, 7; Tenth Five-Year Plan of, 312–313; transforming economy of, 328; in World Trade Organization, 317n.6, 322
Chinese International Economic Trade and Arbitration Commission (CIETAC), 324n.13
Chinese Silicon Valley, 315–316
Chomsky, Noam, 225, 225n.17, 257n.14, 258–259
Choucri, Nazli, 300
Christensen, C. M., 95
Ciborra, C. U., 97
Circulation: as function of network performance, 160; of information, 161
Cisco, 107n.10; routers of, 169; system administrators at, 190–191
Citation index, 109n.11
Citizenization, 304
City-based economies, 121
City states, 303–304
Civic voluntarism, 288–290, 294; logic of, 290
Civil rights movement, U.S., 81n.26

Civil society: in cyberspace, 73–74n.17; global, 84
Civil society collaboration networks, 79–80
Cleaver, Harry, 76, 81
Clifford, James, 278
Clinton, Bill, 312, 314, 335
CNN reporting, 235n.23
Coalitions, in infrastructure development, 147–148
Coase-style equilibrium, 205–206
Codding, Jr., George A., 147n.4
Code: cleanness of, 195; gaps between discourse architecture and, 242–243; transparency of, 204. *See also* Source code
Coding, hermeneutic style of, 236–237
Coevolutions, 221
Cognitive knowledge networks, 43
Cognitive social structure, 43
Cohen, M., 224–225, 237–238
Coherence, 9n.15
Coherent configurations, 10
Cold War: international interdependencies in, 218–219; UN peace agenda in, 216–217
Coleman, Scott A., 141
Coleman, Stephen, 266
Collaboration: filtering of, 41, 42–43; of NGOs and state, 47
Colleran, Elisabeth K., 117n.1
Colley, Linda, 295
Collins, Katherine, 140
Comer, Douglas, 168
Comitology, 289–290, 299
Commercialization, 174n.59
Commission of the European Communities, 302; "White Paper on European Governance," 297–299
Commodity futures, 62n.8
Common good, loss of concern for, 305–306
Common sense, media reflecting, 259–260n.17
Commons, global, 84
Commonsense knowledge, 250–251; violation of, 251n.9
Communication: in Internet value, 159–160; problems with in European Union, 302–303. *See also* Language
Communication/information structures, 1; diversity of, 7–8; endogenous capabilities of, 8; versus information technology, 8
Communication systems: coalitions in development of, 148n.8; global networks and internetworks in, 147n.3
Communication technologies: in global flagship network, 96–99; rapid advances in, 117–118. *See also* Information technologies
Communications theory, 160
Communicative action theory, 226, 283
Communicative space, 15; of democracy, 293–297; demos embedded in, 285–286; for European democracy, 300–304; transnational, 298–304
Communist Party, resistance to in China, 328–330
Communitarians, 287n.1
Community, 12–13; of practice, 79–80; sub-nation-state, building of, 287n.2. *See also* Knowledge communities
Community-based information, in rural communities, 131–132
Community informatics, 266
Compaine, Benjamin M., 118
Competition: of global flagship network, 94–96; growing complexity of, 94–95
Complex adaptive systems, 224–225; design of, 222–227; framework of, 237–238
Complexity, management of, 197–199
Computational linguistics-based macroanalysis, 256
Computational modeling, procedurally oriented, 222n.14
Computer-centered information systems, 5
Computer-centered interactive technologies. *See* Information and communications technologies (ICTs)
Computer-centered interactive technology-resources networks, 235–237
Computer-centered networks, 1; in social sciences, 4
Computer-Human Interaction (CHI), 243
Computer-Mediated Communication (CMC), 243
Computer network research, 163–164
Computer networks, 12; Internet as global system of, 150–152; proliferation of, 162–166
Computer programmers: incentives for,

190–192; open source incentives for, 183–186; transboundary cooperation among, 178
Computer science, emergence of, 164
Computer-Supported Cooperative Work (CSCW), 243
Computer technologies: in globalization, 73–74; growth of in China, 320
Computerized case-history analysis, 230–231
Computerized mathematics, 61
Comsat, 147
Concentration process, 27
Conferencing, in global flagship networks, 14
Conflict, 7
Conflict early warning networks, 14
Conflict Early Warning Systems (CEWS), 215–216, 216; design of, 227–233; developing portable information resources, 219–220; ICT-resources networks of, 235–237; modular, networkable information systems of, 235–236; two-stage project design of, 220–221
Conflict grammars, 230
Conflict of interest, between OEMs and contract manufacturers, 107–108
Conflict trajectories, 218–219, 220–221, 228–233
Connections: direct, 171; direct versus indirect, 171–172; indirect, 170–171; social infrastructure for, 69–70. *See also* Interconnectivity
Consensual knowledge, 206
Consolidation, trend toward, 65–68
Constitutional choice, 289
Constitutional patriotism, 289
Constitutionalism, 288; European, 299
Constitutive sociodigitization, 3
Content, in Internet value, 159–160
Contested historicities, 221, 226, 231–233
Continental discourse theorists, 247
Contract manufacturers, 104, 105; limitations of, 105–108; volatility of, 106
Contractor, Noshir, 42–43
Conversation: background knowledge of, 248–249, 251; commonsense knowledge of, 250–251; digitally based large-scale, 6; large-scale, 11; microanalysis of, 257–258; monolinguals and bilinguals in, 252–253; questioning common terms of, 251–253; shaping, 11–12n.23; structures of, 244; systems architecture facilitating, 243–244; very large-scale, 13–14, 15, 25–26. *See also* Very large-scale conversation
Conversation analysis, 247–248; micro-macro divide in, 253–256
Conversation Map, 244; design of, 277–278; system of, 260–265, 266–276
Conversational common sense: dimensions of, 248–253; thesauri and, 256–259
Conversational ontologies, 222n.14
Conversationally oriented philosophy, 220–221
Conversations. *See also* Online conversations
Cooper, Caroline, 312–313, 315n.3
Cooperation, 7; versus co-optation, 48n.17; competitiveness and, 48–49
Cooperative networks, 117–142; replicating U.S. model of, 137–138; in technology diffusion, 133, 134–136
Co-optation, versus cooperation with state and market, 48n.17
Copyright, 205
Cordero-Guzmán, Hector R., 80n.25
Corporate financial services, globalization of, 64
Corporate networks, cross-border, 89–90
Corporate sector, digital information systems in, 89–109
Cosmopolitan globalization, 83n.31
Cosmopolitans, 287n.1
Cost-and-time-reduction centers, 90, 102, 109
Costello, Lou, 250–251n.8
Coulthard, M., 254
Counterfactural histories, 220–221, 231–232
Countergeographies, 78n.22
Counternormative behavior, 196
Courter, Jim, 335
Cowhey, Peter F., 147
Cox, Kevin R., 125
Cox, scaled spaces of engagement of, 80–81
Creech, Heather, 79
Criminal Justice Act (England), protests against, 79n.24
Cronin, Francis J., 117n.1
Cross-border activism, 84

Cross-border cooperation, 22–23
Cross-border dispersion, rapid, 102–103
Cross-border finance, 58–59n.5
Cross-border politics, digital technologies in, 77–78
Cross-border relations, 7; transformations of, 2–3, 4
Cross-paradigm early warning information system prototype, designing, 219–221
Cruikshank, Barbara, 40n.3
Csnet, 168, 170–171
Cui, Ning, 312
Cukier, Neil, 172
Cultural codes, 259n.16
Cultural cohesion, of nation-states, 306
Cultural homogenization, 297; coercive, 296
Cultural-institutionalist compromise, 306
Cultural norms, 194–195
Cultural Revolution, China, 326
Cultural uniformity principle, 296
Culture: high, 293; mediating, 18, 21–22; replacing social structure, 290; standardization of, 291; thick concept of, 289; thick sense of, 287; thin sense of, 288
Curbow, Dave, 243n.2
Currency derivatives, 62n.8
Currency trade, growth of, 58
Customer-relations management, 98n.3
Cyberdemocracy, 265–276
Cybernetic models, second generation, 225n.17
Cyberspace, 10n.20; European demos from, 297–304

Dahl, Robert A., 285, 287, 292–293, 295n.7, 297, 303–304
Dahlman, C. J., 109
Damiras, Niklas, 243n.2
Das, N., 92
Data, versus interpretations, 69
Data stories, 228n.21, 230
Databases: available to locals, 82n.29; sharing of in global flagship networks, 14
David, Paul A., 146n.1, 157, 169
Davis, J., 235n.23
Day, Peter, 266
D'Cruz, J. R., 99, 103
de Saussure, Ferdinand, 258–259
de Sola Pool, Ithiel, 149n.9
de Swaan, Abram, 301
Dean, Jodi, 267n.20
Debian Social Contract, 179n.1
Decentralization, 15; of financial activities, 64; of governing structures, 45; highways and, 124; political authority in, 298–299; of production, 45; shift away from, 45; with simultaneous integration, 83–84; in successful NGOs, 48–49
Decentralized access, 54, 74; for investors, 57
Decentralized conflict early warning systems, 227–233
Deconcentration, 27
Deibert, Ronald, 149n.9
Deliberative association, 13
Democracy: conceptualizing of, 285–293; deliberative, 265–276; etymological meaning of, 285–286; foundations of, 289; postnational, 289; procedural definition of, 286, 289; societal infrastructure of, 285; transformations of, 292–293, 297, 303–304
Democracy in the Digital Age (Wilhelm), 268–269
Democratic decision-making, 294
Democratic integration, 295–296
Democratic process: as enlightened understanding, 293; tandem hypothesis of, 291, 292
Democratic sovereignty, paradox of, 286–287, 289, 290, 294
Democratic state formation trajectories, 28
Democratization: in China, 314, 321; constitutive role of, 291; nation-state formation and, 295–297; Third Wave of, 286–287; transnational, 292–293; universities and in China, 334–335
Demos: bi-directional logic of, 291; conceptualizing of, 285–293; creation of, 294; definition of, 285; embedded in communicative space, 285–286; European, 283–306; identified with nation, 287–288; prepolitical, 288–289; top-down formation of, 297
Denationalization, of financial centers, 71–73
Deng Xiaoping, 324
Denning, Dorothy, 76
Depression, Great, internationalization of financial market and, 59

Der Derian, J., 236n.26
Deregulation: cooperative networks and, 119; in denationalization of financial centers, 71–72; of finance industry, 59; intensification of, 63–64
Derivative sociodigitization, 3
Derivatives: diversification of, 61–62; interest-rate, 62n.8
Derrida, Jacques, 304n.12
Design, 25–26; ethics of for very large-scale conversations, 276–278; shaping technologies, 277
Design firms, 97
Design modules, 97
Design-oriented open systems engineering, 224–227
Destabilization, 22–25
Deutsch, Karl W., 4, 149n.9, 215n.1, 222n.14, 291, 293, 303, 304
Deutsche Borse, foreign listings of, 67t
Developing countries: infrastructure development pattern in, 141–142; telecommunications regime in, 138–139
Development, NGOs fostering from below, 47
Dholakia, Ruby Roy, 117n.1
Dialogue, coherent sequences of, 257–258
Diary, 277
Dicken, P., 93
Differentiation, versus interconnection, 165–166
Diffusion: strategies for in rural communities, 132–134; versus translation, 45–46
Digital divide, 117, 118–119; overcoming, 133–134n.10; rural-urban, 117–142; supply concerns in, 133–134
Digital Equipment Corporation, in network development, 164
Digital formations: definition of, 1; early stages of, 2; institutional and historical trajectories and, 28; limits and logics of, 26–29; platforms for, 5n.11; social logics in structuring of, 6; study of, 8–15
Digital information networks, design of, 14
Digital information systems (DISs), 89–109; in global flagship networks, 91–92, 96–99
Digital networks: private, 5n.11; transformative impact of, 56–57
Digital/social imbrications, 19–21
Digital technologies: capacities of, 5–8; social logics and, 54–84
Digitization, 16; mobility and, 19
Dimaggio, Paul J., 209
Discourse: Anglo-American perspective on, 248; distributional analysis of, 257; micro-macro divide in, 253–256
Discourse analysis: Anglo-American traditions of, 257–258; systems of, 258n.15
Discourse architecture, 242–278; criteria for evaluating, 244; practice of, 243–244
Discourse Architecture Laboratory, 243
Discourse specialists, 242–243
Discussion, themes of, 270
Discussion groups, web-based, 246n.5
Dispersion, 94–95; concentrated, 27, 101–103
Disruptive practices, 17
Distributed outcome, 54
Distributed structures, shift to, 45
Distribution, decentralized, 45
Doctors without Borders, 44
Documents, storage capacity for, 7–8n.12
Domain name system, 151n.12, 169
Dot-corg dual enterprise model, 48
Double movement, 27
Dourish, Paul, 243n.2
Downsizing, 126
Drake, William J., 335
DuBoff, Richard, 123
Dunning, J., 93
Dupont, Cédric, 295
DuPont, in China, 323
Dutch human rights early warning organization, 232n.22

e-business, 92
e-mail: alerts, 78–79; conversation maps of, 260–265; in global flagship networks, 14; transfer programs for, 180
e-mail lists, 246
e-mail readers, 246
Early warning ICTs, conflict-oriented, 235–236
Early warning indicators, empirical validity test of, 232n.22
Early warning networks, design of, 215–238
Early warning systems, 6; for conflict, 219
East Berlin, student protests in, 335

Eccles, R. G., 103
Economic developmental associations, 46
Economic exchange, structural logic of, 181–182
Economic growth: flexibility in, 126n.6; telecommunications investment and, 117n.1
Economic logic, open source, 192–201
Economic reform, China, 330–332
Economides, Nicholas, 162
Economies: financializing of, 58; globalization of, 14–15; globally networked, 125–127; rural, 120–121. *See also* Global economy; Political economy
Eder, Klaus, 289
Edwards, Michael, 48
Edwards, P., 216, 236n.26
Eickelman, Dale, 21n.33
80/20 rule, 190
El Salvador conflict, 230
Electric cooperatives, rural, 136
Electronic activism, expanded options for action of, 78–79
Electronic Disturbance Theater web site, 78
Electronic exchanges, merger of, 70–71
Electronic financial markets: literature on, 56n.1; locational and institutional embeddedness of, 56–73; scaling of, 24
Electronic industry, flagship network model in, 102–103
Electronic markets, 1, 6, 11, 54; rise of, 14–15
Electronic networks, 12
Electronic space, 10, 12n.26; picto-textual dimension of, 11
Ellickson, Robert C., 194n.16
Elster, Jon, 289
EMACS text editor, 188
Embeddedness, of global capital market, 56–73
Emirbayer, Mustafa, 287
Empirical-analytical science, 226–227
Endogenous conditions, 26–27, 27–28
Endurance, 9n.15
Engaged citizenship, 43
Enlightened understanding, 293
Enlightenment thinkers, 293
Enron, 100n.4
Equities, cross-border transactions in, 1975–1988, 63t
Equity funds, growth of, 61
Erikson, Tom, 243n.2, 243n.3, 243n.4
Ernst, Dieter, 6, 14, 24, 27, 28, 90–96, 99, 100–104, 109, 163n.41; global flagship networks of, 16
Esherick, Joseph W., 328
Espinoza, V., 81
Esty, D. C., 232n.22
Ethernets: equipment for, 169; value of, 159
Ethics of discourse architecture, 277–278
Ethnic cleansing, 296
Ethnoculturalism, 288
Ethno-nationalism, 287n.1
E*Trade, centralization of, 68–69
Eudora, 246
Eunet, 153–154, 170–171
Eurobarometer surveys, 299–303
EuroNext: as financial center, 65–66; foreign listings of, 67t
Europe: communications system in, 151; computer networking in, 153–154; emergence of national demoi in, 293–297; emerging networks and interconnections in, 153–154; ethnic cleansing in, 296; Internet connectivity in, 299–300; internetwork development in, 156–157; postnationalism in, 290–291; wired, 7
European Academic and Research Network (EARN), 164
European Commission, DG Press, 300–301
European Community, in network development, 164
European constitutionalism, 299
European demos, 283–306; in cyberspace, 297–304
European financial centers, alliances among, 70
European integration: information technologies in, 303–304; public communication in, 306
European Public Square, virtual, 299
European Union (EU), 93; communication in, 285–306; communication problems in, 302–303; communicative infrastructure of, 284; communicative space of, 15; democratic legitimacy of integration of, 288; enlargement of, 304; everyday activities of, 289–290; in global Inter-

net, 147–148; Internet transformative effects in, 299–304; top-down dimensions of, 305; as transnational communicative space, 283
Evans, Philip B., 120–121
Expert information systems, 14
Extended complex interdependence, 219n.7
Extended enterprises, 101
External economies, in industrial districts, 130
Externalities, maximizing connectivity, 69
Extranet, 96n.2
EZLN guerillas, 232, 233

Fabless design, 97
Face-to-face conversation, micro analyses of, 246
Falk, Richard, 50
Feltham, Jennifer Laura, 140–141
Ferry, Julcs, 296
FEWER. *See* Forum on Early Warning and Early Response (FEWER)
Finance: spatial organization of, 58–59; speed of transactions in, 61
Finance industry, deregulation of, 59
Financial centers: denationalized, 71–73; share of global operations of, 70n.15; trend toward consolidation of, 64–68
Financial crisis of 1997–98, 62–63
Financial flows, short-term, 60–61
Financial instruments: academic economics in development of, 62; digitization of, 56–57, 61–62; versus underlying assets, 62–63
Financial markets: concentration of, 64–66; electronic, 14–15, 19–20, 56–73. *See also* Global financial market
Financial networks, 19–20n.32
Financial services: explosion of innovations in, 61–63; mergers of, 70n.14
Flagship network model: concentrated dispersion in, 101–103; hierarchical networks in, 103–104; network characteristics of, 100–101; theoretical foundations of, 99–100
Flagships, 103–104
Flaherty, T., 95
Flaming, 196
Flamm, K., 96
Flexibility: of Internet protocols, 156–157; in networks, 206–207; in outsourcing, 107–108; of production systems, 126
Flora, Peter, 304
Flores, Fernando, 11–12n.23
Fnet, 170–171
Folktales, 277
Foray, D., 92
Ford Motor Company, electronic market of, 127
Foreign direct investment (FDI), 93–94
Foreign exchange market: concentration of, 65–66; decentralization of, 64; growth of, 58
Formalization level, global market for capital, 56–57
Formations: definition of, 9; delineating categories of, 9; dimensions of, 10, 27
Forrester, Jay, 236n.26
Forum on Early Warning and Early Response (FEWER), 215, 227, 234, 235; regionally oriented teams in, 219–220
Foucault, Michel, 17n.29, 275–276, 276n.26
Fox, Jonathan A., 45n.11
Fox, Vincente, 230
Framing, type of, 9
France: national identity of, 295–296; national network of, 163; network development in, 163–164; Third Republic of, 294
Francophonie, 296
Frankfurt: as financial center, 65–66; financial market in, 67n.12
Fraser, Nancy, 267
Free idea exchange, 181–183
Free-rider trap, 181–182
Free Software Foundation: case study of, 187–188; community of, 189–190; cultural frame of, 195
Freidman, Thomas, 313n.2
French Republic, secularism in, 296
French Revolution, 294
Freud, Sigmund, 251n.9
Frey, E., 105, 106n.8, 107–108
Fudan University, 333n.16
Funds, diversity of, 60
Furness III, Thomas, 10n.20
Futures, development of, 62n.8

"Galactic" city, rise of, 128
Galaskiewicz, Joseph, 255

Galtung, Johan, 228
Game-theoretically oriented international research, 220n.9
Gancarz, Mike, 194–195
Garcia, D. Linda, 5–6, 11, 15, 22, 26–28, 69, 76, 99, 126–127, 131, 134, 136–137n.11, 139
Garcia, E., 219n.8
Gareiss, Robin, 172n.54
Garfield, Eugene, 255
Garfinkel, Harold, 251–252, 253, 257n.14
Garwood, John D., 136
Gateways, administration of, 171
GDP, aggregate, 58
GE Capital-Toho Mutual Insurance Co. joint venture, 72–73n.16
Gellner, Ernest, 290, 291, 293
General Agreement on Tariffs and Trade (GATT), 139–140
General Agreement on Trade in Services (GATS), 140
General Motors: development of corporate network at, 163; electronic market of, 127
General Problem Solvers theory, 225n.17
General Public Licence (GPL), 187–188
Generative resonances, 221
Generative rule systems, 225–226
Genetic system generation, logic of, 170–171
Geneva, financial market in, 67n.12
Geographic location: dispersal versus central coordination of, 68–70; in global economy, 127–128
Geographic region, networked, 130
Geographical scales, social action in, 82–83
Geographies: specialized clusters and, 102–103; types of, 65n.11
George, Alex, 228
Gereffi, Gary, 99, 100, 128
Gerhards, Jürgen, 303
German Constitutional Court, 288
Germany: e-mail alert network in, 78–79; nationalism of, 294; network development in, 164
GFNs. *See* Global flagship networks
Ghosh, Rishab Aiyer, 181–182, 190n.10
Ghoshal, S., 99, 103
Gift culture hypothesis, 185–186
Gift economy: in China, 336; in reform-era China, 326–327
Gillies, James, 154, 156n.23, 158, 169, 170
Ginza, foreign investors in, 72–73n.16
Girard, Monique, 45
Global Business Dialogue, 117
Global business meetings, 73
Global capital market, 54–55, 56–73, 84; distinctiveness of, 59–63; of today versus earlier period, 56–57
Global capitalism, China in struggle for, 312–337
Global circuits, politics of places on, 73–83
Global cities, 19–20n.32, 27n.45, 82n.30
Global city network, 65n.11
Global commodity chains, 100
Global communication flagships, 24
Global communication systems, 5n.11
Global corporations: location decisions of, 95; trade liberalization and, 93–94
Global economy: high-growth components of, 58; information-based, 125
Global Equity Market (GEM), 70
Global financial market, 28; alliances in, 70–71; concentration versus dispersal of, 63–68; incentives in, 62
Global financial products, availability of, 72
Global flagship networks (GFNs), 14, 16, 89–109; concentrated dispersion in, 101–103; conceptual framework of, 91–92; contradictions of, 104–109; forces driving, 93–99; inherent contradictions of, 90–91; integration of, 103–104; network characteristics of, 100–101; rapid expansion of, 106–107; theoretical foundations of, 99–100
Global imaginaries, 75
Global institutions, development of, 73
Global interactive zones, 55
Global media events, 81n.27
Global networks: multiscalar nature of, 55; of resource-poor organizations, 73–74; rise of Internet and, 146–175
Global political space, 37; organization of, 44
Global production system, niches in, 129–131
Global society, basic units of, 217–218
Global South: information network for, 78–79n.23; limited technologies of, 75
Global telecommunications, need for trade regime for, 138–142

Globality, forms of, 54
Globalization: digital information systems and, 89; multiscalar, 23–24; new production techniques of, 127–128; political response to, 298–299; power-related tensions in, 216–217
Globalized industry, regulatory constraints on, 64
Globally administered address space, 169n.50
Globally networked economy, imperatives of, 125–127
Gnutella, 242
Goffman, Erving, 11n.22
Goldstein, Steven N., 168
Google: recombinant technology of, 41; Usenet newsgroup archived by, 268; weighing of sites by, 161n.36
Gorbachev Summit, 327, 329
Gordon, David C., 294
Gorenflo, Neil, 134
Gottman, J., 125
Governance structures, corporate, 101
Government debts, diversity of, 60
Government Open Systems Interconnection Profile (GOSIP), 155n.20
GPL, 196
Grabher, Gernor, 126
Grameen Bank, 47
Grammatical complexities, 225–226
Gramsci, Antonio, 259, 260
Granovetter, Mark, 11–12n.23
Grassroots developmental organizations, 47
Grassroots human rights organizations, 75–76
Greatness, standards of, 186
Greenfield, Liah, 295
Greenpeace, 209; web site of, 77
Grimes, J., 81–82n.28
Grimm, Dieter, 284, 288
Grosfoguel, Ramón, 80n.25
Group Asynchronous Browsing (gab), 42
Group Forming Networks, 43
Group interactions, 6
Grove, A. S., 95
GTE, divestment of, 141
Guangzhou, 316
Guatemalan conflict, 228–230
Guattari, Félix, 278
Guerrieri, P., 100
Guetzkow, Harold, 222n.14
Guice, Jon, 150, 151n.15, 153n.17
Gulia, Milena, 13n.27
Gumperz, John, 252–253
Gurr, T. R., 215n.2, 218n.4, 219n.8, 228, 230, 232–233, 235n.23
Guthrie, Doug, 7, 15, 28, 323–324, 327, 328, 331, 333–334, 335
Guy, Ken, 133, 138

Haas, Ernst, 215n.1, 218–219, 219, 291
Habermas, Jürgen, 216, 225, 226, 227n.19, 266, 267–268, 272, 283, 284, 288–292, 295, 304n.12, 305–306
Habermas-Bhaskar world, 227–228
Habsburgs, 295
Hacker ethic, 187–188
Hackers, 180, 190
Hacking, Ian, 277–278
Hafner, Katie, 170
Haggard, S., 99, 100
Hagstrøm, P., 98–99, 99
Hall, Stuart, 259–260n.17, 260
Halliday, Michael, 249, 251, 257n.14, 260
Haltern, Ulrich R., 288
Hangzhou Telecommunications Factory, 316–317n.5
Hanna, Nagy, 133, 138
Hansard Society, 266
Hardt, Michael, 50
Harlam, Bari, 117n.1
Harner, Stephen, 313n.1
Harré, Rom, 225, 226
Harris, Jed, 243n.2
Harris, Zelig, 256–257
Harrison, B., 130
Hart, John Fraser, 121n.2
Hassard, John, 45, 255
Hawala system, 58–59n.5
Headrick, Daniel R., 147
Hechter, Michael, 296–297
Hedge funds, growth of, 61
Heidegger, Martin, 11–12n.23
Held, David, 125, 285, 292
Henderson, Austin, 243n.2
Henry, Mark S., 130
Herbert, Paul L., 117n.1
Hermeneutic knowledge, 226
Hermeneutics, scientific computational/formal, 225n.18
Herrenvolk-mentality, 296
Herring, Susan, 243n.3, 243n.4

Heterogeneity: in deliberative discussion, 269; of newsgroups, 271–272
Hicks, John R., 159n.30
Hierarchical authority, 197–198
Hierarchical networks, 103–104
Hierarchies: destabilization of, 55; Internet in resistance to, 81–82n.28; networks and, 207–210; shift to networks from, 37–38
Hierarchies of scale: avoiding, 79–80; destabilization of, 22–25
Hierarchy-network interfaces, 207–208
Higher education, China, 332–335
Highway systems, 124
Hill, David, 313n.2
Hill, Will, 255
Hiltz, Starr Roxanne, 265
Himmelberg, Charles, 162
Hintikka, 221n.12
Hirsch, Phil, 154
Hirst, Paul Q., 56n.1
Historical possibilities, characterization of, 231–232
Historico-institutional trajectories, 28
Hitt, L. M., 97–98
Hodgson, M. G. S., 218n.5
Hoff, Karla, 119–120, 121, 133
Hoffman, Stanley, 288
Hong Kong, as financial center, 65–66
Hook, Kristina, 243n.3
Howitt, Richard, 80, 81, 82–83
HR Information and Documentation Systems International (HURIDOCS), 77
Hsinchuh Science Park, 97
HTTP Server Apache, 180
Huberman, Bernardo A., 202
Hudson, Heather, 117, 137
Hug, Simon, 295
Hughes, Thomas P., 146n.1, 169–170
Hugill, Peter J., 147
Hulme, David, 48
Human-centered complex adaptive systems, 216
Human-machine interaction, 38n.1
Human rights: global mobilization for, 81; information dissemination on, 77; movement for, 46n.14; versus sovereignty sanctity, 44
Human rights abuse, as global media event, 81n.27
Human thought, flow of, 223
Humanistic philosophies, 220–221
Humanitarian relief efforts, 44–45
Hundred Flowers Reform, 333
Hungary, Roma Rights organization in, 46n.14
Huntington, Samuel P., 286–287
Hutchins, Edwin, 45
Hypermobility, 19–20

Iammarino, S., 100
IBM: in network development, 164; vertical integration of, 95
ICTs. *See* Computer-based interactive technologies (ICTs); Information and communications technologies (ICTs)
Ideational metafunctions, 249, 251–253
Ideational relations, 258; recurrent patterns of, 255
Identity-building mechanisms, 292
IGOs. *See* Intergovernmental organizations (IGOs)
IKNOW software, 42–43
Imagined community, 285–286, 293
Imbrication process, digital/social, 18, 19–21
Imig, Doug, 305
Immigrants, support networks among, 80n.25
Impacts, search for, 8–9
Incentives: cultural and social norms in, 194–195; macroeconomics, 181–183; microeconomic, 183–186; for open source, 205
Incompatible time sharing system, UNIX replacing, 194n.17
India: development NGOs in, 46–47; grassroots organizations in, 76; Self-Employed Women's Association of, 47n.15
Industrial districts, 130; virtual, 131–132
Industrial organizations, competition and, 94–96
Industrialization, redefining rural communities, 124n.4
Industry, spatial organization of, 64
Indymedia, 79n.24
Information: dissemination of by activist organizations, 76–77; interacting with network purposes, 174; versus knowledge, 40–41; relative worth of, 160–161; social infrastructure in processing, 69–70; types of, 69. *See also* Communi-

cation; Information systems; Information technologies; Knowledge
Information age, access in, 117–142
Information and communications technologies (ICTs): in conflict early warning systems, 227–233; in conflict resolution systems, 221; in globalization, 74–78; services of, 82n.29
Information asymmetries, rural-urban, 121
Information-based global economy, 125
Information-based networking technologies, 126–127
Information broker model, 39–43; verus knowledge facilitation, 40–42
Information ecologies, 8n.13
Information engineering, 222–227
Information exchange, NGO, 75–76
Information gateways, limiting access to, 131n.9
Information networks, 14; rise of global Internet and, 146–175
Information processing, 160n.35
Information resource design, transboundary, 215–238
Information revolution, rapidly accelerating, 284
Information Society Project, 298
Information structures, 1
Information systems: Internet-based, 98–99; social life in, 17
Information technologies, 38; in China, 312–337; in Chinese economic transformation, 331–332; diffusion strategies for, 133–134; in European democratization, 298–304; in fall of Soviet Union, 335; in global flagship network, 96–99; global policy on, 148–149; growth of in China, 318–321; maximizing benefits of, 69; privacy and in China, 325–327; in resistance movement in China, 327–330; in social change, 336–337; state capacity and in China, 321–335; transforming effects of, 303–304
Information technologies (IT): social analysis of, 2
Information Technology Agreement, 140n.12
Information technology industrial sector, state-controlled development of, 7
Information vendors, competition in, 131n.9
Infrastructural logics, 149n.10
Infrastructure deployment, advocates of, 117–118
Infrastructure development: in developing countries, 141–142; global, 146–175; lack of theories of, 146–147; theoretical framework for, 148–149
Infrastructure system formation, 169–172
Innis, Harold, 17n.30, 117–118
Innovation: geographic distribution of, 90; sources of, 109; user-driven, 201
Innovation scores, 109
Instant messaging, 242
Institute for Global Communication (IGC), 77
Institutional change, in global flagship networks, 93–94
Institutional investors: financial asset of in selected countries, 60t; speculative investment strategy of, 61; in today's market, 60
Institutionalism, bounded, 290–293
Institutionalization, global capital market, 56–57
Institutions, 7
Intangible inputs, 98n.3
Intangible outputs, 98n.3
Integration, 94–95; expanded decentralization with, 83–84
Intel, protests against invasion of privacy of, 79n.24
Intellectual property rights, 92
Inter-University Consortium for Political and Social Research, 224
Interactive newsgroups, 271
Interactive technologies (IT), 37; non-governmental organizations and, 38, 39–43
Interactivity, 7–8n.12; definition of, 10; in deliberative discussion, 268–269, 270; new pattern of, 27; shaping spatialization, 11; webs of, 13
Interconnectedness, jump in levels of, 57
Interconnections: development of, 173; versus differentiation, 165–166; economics of, 162n.38; special value of, 166–168; transboundary, 166–169; upper and lower limits on, 172n.54
Interconnectivity, 54; in global market, 127–128
Interfirm transactions, 100–101

Intergovernmental organizations (IGOs), 219; in conflict resolution, 230; in early warning systems, 234, 235n.23
Intergovernmentalism, 288
Intermilitary networking, 151n.13
Internal relations ontological approach, 221n.12
International Alert (IA), 215, 219–220, 227, 234, 235
International Bank for Settlements (Basle), 65n.10
International Campaign to Ban Landmines site, 78
International conflict: contradictory management/resolution practices in, 216–219; databases on, 224
International Connections Program, 168
International interconnections, 151n.13; in internetworking development, 150–152
International Monetary Fund (IMF), 73
International Organization for Standardization (ISO), 149n.10
International organizations: pluralism of, 206; structure and legitimacy of, 207. *See also* Intergovernmental organizations; Multinational organizations; Nongovernmental organizations (NGOs)
International regime analysis, 148
International relations theory: classic, 49; scientific versus classical approaches to, 218–219
International Social Science Council, 215–216
International Standards Organization (ISO), 154
International Telecommunications Union (ITU), 147–148, 149n.10
Internationalization, extent of in capital market, 59
Internet: access to in China, 317–321; backbone capacity of, 141; backbones of, 172n.54; basic model of, 158–162; in China, 313; commercial pursuits of, 174; as community, 12–13; definition of, 152; development of in China, 333–334; discourse architecture for, 242–278; in educational institutions, 333–334; efficiency of, 172; in Europe, 299–300; in European demos formation, 297–304; as global computer communications system, 150–152; global networks and rise of, 146–175; in globalization, 73–74; incentives to join, 171, 172; infrastructure growth of in China, 320; infrastructure system development for, 169–172; large-scale conversations on, 1; lessons about political economy on, 201–210; as network, 12; as new public sphere, 265–276; as one-way publishing network, 266; as public space, 261–265; as resistance movement, 81–82n.28; social purpose in, 173–174; transformative effects of in European public space, 299–304; as transnational institution, 7; U.S. promotion of, 155n.21; uses of by NGOs, 75n.20; virtual private networks on, 165. *See also* Online discussions; Very large-scale conversations; World Wide Web
Internet communities, 43
Internet-constellation networks, 156n.23, 158
Internet media, globalized, 75
Internet protocols, 154–155; conflicting, 156–158; development of, 169
Internet Relay Chat (IRC), 242
Internet Service Providers in Eastern Europe, 48
Internet service providers (ISPs), 152; in Internet backbone, 172n.54
Internet services, business opportunities for in rural communities, 136–137n.11
Internetwork relations, 149
Internetworking: collaboration in, 167; conflicting visions of, 153–158; developmental stages of, 161; infrastructure formation in, 158–162; logics of, 149–175
Interpersonal metafunctions, 249–251
Interpersonal relations, 255; of very large-scale conversation, 260–261
Interpretations, 69
Intrafirm transactions, 100–101
Intranet, 96n.2
Investment, social justice, 207
Investors, decentalized access for, 57
Ionesco, Eugene, 249–250, 251
Iraq invasion, public outcry against, 304n.12
Isenberg, David S., 157n.27
Issue promotion, Internet in, 75n.20
Italian city-state system, interconnections in, 174–175
Ito, Mizuok, 49n.18

Jabil reports, 106
Jacob, James E., 294
Jacobin nationalization, 294
Jacobs, Jane, 120, 125n.5
Jakobsen, Roman, 250–251
Jervis, Robert, 4
Jewish bankers, Venice-based, 58–59n.5
Jiang Zemin, 323, 330
Johnson, Stephen, 44
Joint Network Team (JNT), 156n.23
Jordan, Ken, 43
Jørgensen, H. D., 92, 100
Judd, Dennis R., 80
Just-in-time model, 98n.3

Kaiser, Karl, 285
Kalathil, Shanthi, 335
Kaplan, A., 221n.11
Kaplinsky, R., 100
Keohane, Robert O., 3–4, 218n.6, 291
Kesan, Jay P., 174n.58, 174n.59
Kielmansegg, Peter, 288
Kim, L., 92
King, G., 232n.22
Kirstein, Peter, 167, 168–169
Kissinger, Henry, 207–208
Kluge, Alexander, 267
Knowing subject, 40–41
Knowledge: concentrated dispersion of, 101–103; coordination of, 193–196; digitization of, 92; facilitation of, 40–41; in intergroup cooperation, 6; leveraged across multiple logics, 45–46; mobility of, 17, 89–109; new characteristics of, 44–46. *See also* Information
Knowledge communities, 11, 13; creating new basis for association, 43–50; rise of, 43–44
Knowledge diffusion, 90, 99–100, 108; in global flagship networks, 91–92
Knowledge-intensive support services, outsourcing of, 101, 103–104
Knowledge networks, 14, 79–80; cognitive, 43
Knowledge sharing: in DISs, 96–97; in global corporations, 91–92
Knowledge spaces, 1, 13
Kogut, B., 95
Kollock, Peter, 5n.10, 13n.27, 181
Korzeniewicz, Miguel, 99, 128
Kraus, Peter A., 7, 15, 26n.43, 28, 287n.1, 302
Kravtin, Patricia D., 141
Krippendorf, Klaus, 269n.23
Krogstie, J., 92, 100
Kubatana.net, 82n.29
Kulatilaka, N., 95
Kulturnation, 294
Kuntze, Marco, 79–80
Kuroda, M., 227
Kuwabara, Ko, 185

La Perrière, Guillaume de, 275
Laborde, Cécile, 296
Lakenan, Bill, 105, 106n.8, 107–108
Lamarkian transformations, 225
Land Rights in Africa site, 77
Landmines Ban Treaty campaign, 78
LaNeta, 77
Language: functions of, 251; generative/interpretive power of, 225n.17; ideational metafunctions of, 251–253; interpersonal metafunction of, 255. *See also* Linguistics
Lasswell, H. D., 221n.11
Lateral internetworking, 166
Latham, Robert, 4n.7, 5n.11, 7, 27, 28, 78n.22, 99, 215n.1
Latifee, Dr., 47
Latour, Bruno, 1, 10n.17, 28, 38n.1, 39–40, 45–46, 46n.13, 49, 148n.8, 255, 276
Laurel, Brenda, 11n.22
Law, John, 45, 255
Law on Chinese-Foreign Equity Joint Ventures, 324
Leadership, in open source community, 195–196
Lefebve, Henri, 10n.19
Legend Computers, in China, 322, 327n.14
Legitimacy, 207
Lehman Brothers, 72–73n.16
Leizerov, Sagi, 79n.24
Lerner, Josh, 183–184
Lessig, Lawrence, 174n.59, 204
Levy, Pierre, 44, 49
Levy, Steven, 187
Lewis, Pierce, 128
Lexcalibur, 299
Lexical cohesion, 257
Lexicon, for social groups, 246–247
Liberalization, 28, 93–94
Liberals, versus communitarians, 287n.1

Liberty Telephon Company, 135
Libicki, Martin C., 154
Libraries Online Partnership, 48
Liebowitz, S. J., 157–158n.28
Lifeworld, idealized, 290–291
Lindenberg, Marc, 39, 48
Linguistically encoded meanings, 225–226
Linguistically mediated conduct, 225–226
Linguistics: commonsense knowledge and, 256–259; corpus-based, computational, 255, 257, 258–259
Linus, 181
Linux, 180; archive sites of, 198; authority in development of, 194; case study of, 188–192; community of, 195–196; complexity of, 197–199; decision-making system in, 198; kernel releases of, 192n.12; leadership of, 205; versus Microsoft, 209; modules of, 199; registered users of, 190n.10; Torvald's authority in, 200–201
LISP programs, 222–223, 228–230
Local action, networking of, 83
Local-area networks (LANs), 96n.2, 163
Local communities, reembedding economies in, 125
Local cooperatives, rural, 135–136
Local institutions, as multiscalar, 73–83
Local knowledge, in global market, 127–128
Local organizations: in global capital market, 56–73; in global interactive networks, 54, 55
Local relations, transnationalizing of, 3
Local suppliers, in integrated network, 104
Localized struggles: engaging global actors, 81–82; networking of, 80–81
Locational economies, rural, 129–131
Locational specialization, 94
Lombard system, 58–59n.5
London-based Nongovernmental Organizations, 215
London financial center, 65–66; share of global transactions of, 71
London Stock Exchange: alliances of, 70; foreign listings of, 67t
Lovink, Geert, 82
Lu Xun, 335
Luhmann, Niklas, 160–161
Lukes, Steven, 174n.60
Lyon, Matthew, 170
Lyon, Robert E., 155n.20, 168
Lyon stock market, 67n.12

M-form functional hierarchy, 95
Macher, J. T., 97
MacKenzie, Donald, 62
Macroeconomic incentives: in open source, 181–183; positive network externalities and, 193–194
MAI agreement, 79n.24
Majeski, S., 230
Majordomo listserve, 78–79
Malamud, Carl, 154n.18, 158, 169n.50
Malerba, F., 90
Malmkjaer, Kirsten, 257n.14
Management, digital information systems as tool in, 96–99
Manhattan financial sector, destruction of office space in, 65n.11
Mann, Michael, 286, 296
Manovich, Lev, 10n.18
Mansell, Robin, 140, 301
Manufacturers: global flagship, 104; outsourced, 105–108
Many-to-many exchanges, 244
Mao Zedong, 330, 336
March, James G., 292
Marcus, George E., 278
Margolis, Stephen E., 157–158n.28
Market alliances, 48
Market leadership, destabilization of, 94–95
Market organization, of rural communities, 120–121
Market reform, in China, 315–321
Market segmentation, rapid, 95–96
Markusen, A., 90
Mass communication, age of, 39–40
Mass customization, 97, 98n.3
Mass production, age of, 39–40
Massey, D., 127
Massive interaction dynamics, 255
Mattingly, Garrett, 174–175
May, Christopher, 5n.10, 298, 303
May Fourth Movement, 333, 335
Mayer, Franz C., 288
McClelland, Stephen, 139
McGowan, David, 196
McHoul, A., 247
McKnight, Lee, 266
McSpotlight.org, 79n.24

Meaning: breaks in, 11–12n.23; in open source networks, 22
Media: access to in China, 317–318; architecture of, 243–244; Internet-based, 78–79; as technologies of self, 277
MediaChannel.org, 79n.24
Mediating practices, 18, 21–22
Medicinal Plants Network, 75n.19
Mefford, Dwain, 215n.1
Melbourne, mobilization against World Economic Forum meeting in, 79
Mele, Christopher, 81
Mergers and acquisitions, 61n.7, 106; of global financial firms and markets, 70n.14; stock as currency in, 107
Merges, Robert P., 196
Merrill Lynch, Yamaichi Securities purchase of, 72–73n.16
Message archives, interfaces into, 265
Message threads, 270; structure of, 273
Metafunctions, 249–250; types of, 251
Metcalfe, Robert, 43, 159–161
Metcalfe's Law, 159–161, 166
Mexico, Chiapas case in, 229, 230
Meyer, C. W., 135
Michie, J., 90
Micro-macro divide, 253–256
Microeconomic incentives: cultural and social norms in, 194–195; in open source, 183–186
Microenvironments, 24–25
Microsoft: conflict with open source community, 208–209; Halloween Memo of, 183; online partnerships of, 48; programmers at, 203; R&D budget of, 109
Milan exchange, privatized, 67n.12
Milgrom, P., 98, 100
Millo, Yuval, 62
Mills, Kurt, 76
Minc, Alain, 163n.42
Minitel, 163
Minix, 189
Minsky, Marvin, 215n.1, 225
Mirowski, P., 236n.26
MIT: auto manufacturing study of, 192; programming community at, 187
Mitrany, David, 291
Mobility, digitization and, 19–20
Modernization, nationalism and, 293
Modular design, in open source software, 198–199
Modular information systems, 235–236
Modularization, 92
Moglen, Eben, 181n.181
Monnet, Jean, 283
Monsanto, 207
Montgomery Ward catalogue, 135
Moody's rated bonds, 59
Morrill, Richard, 80
Motorola: in China, 316–317; Wholly Owned Foreign Enterprise of in China, 316–317n.5
Mowery, David C., 97, 151n.14, 151n.15, 169n.49
Mulhall, Stephen, 287n.1
Multiauthored text, cohesion relations in, 258–259
Multi-media hypertext, 235n.24
Multinational corporations: versus global flagship networks, 100–101; legitimacy of, 207; transition from to global flagship networks, 93
Multiscalar transactions, technologies facilitating, 75
Multi-user domains (MUDs), 5n.10
Munro, Alan, 243n.3
Murphy, Kevin M., 197
Mylea, Paul, 42
The Mythical Man-Month (Brooks), 197

Nairn, Tom, 296–297
Nanjing University, 333n.16
Napster, 205, 242
Narration, presentation of, 11
Narratively oriented modeling, 225–226
NASDAQ: Canada, 70; centralization of, 68–69; foreign listings of, 67t; Japan, 70
Nation-building, European, 295–296
Nation-states: community building at level of, 287n.2; cultural cohesion of, 306; democratic, 293–294; democratization and formation of, 295–297; transformation of city state to, 303–304
National demos, emergence of, 293–297
National government power, efforts to bypass, 80
National marketplace, development of, 122–123
National networks, 147–148, 152
National relations, transnationalizing of, 3
National substantialism, 287–288, 306;

National substantialism (*continued*)
logic of, 287n.2, 288; thick cultural concepts in, 289
Nationalism, 293–297
NATO: in conflict management, 237; in Internet development, 168
Naturalistic philosophies, 221
Naughton, Barry, 331, 335–336
Negotiation, across ordering principles and multiple logics, 45–49
Negri, Antonio, 50
Negroponte, Nicholas, 25n.39
Negt, Oskar, 267
Neofunctionalism, politically grounded, 291
Nested hierarchies, ability to escape, 79–80
Netscape Messenger, 246; VLSC view of, 245
Network engineering, 164
Network externalities/effects, 98–99, 158–162; economics of, 148; "small world," 171n.53
Network flagships, framework of, 91–92
Network partners, 103n.7
Network state, European, 298–304
Network subgroups, 167
Network-to-network model, 170
Networking software, 92
Networking technologies: cooperative model for, 137–138; deployment and diffusion strategies for, 132–134; impact of, 121–125; in rural communities, 131–132; two-sided nature of, 117–118; uneven deployment of, 122
Networks, 12; accountability of, 45–46; architecture of, 243; characteristics of in global flagship network, 100–101; definition of, 146n.2; differentiation and interconnection of, 165–166; diversity and proliferation of, 164; dumb, 157n.27; electronic, 12; evolution of, 100; financial, 19–20n.32; flexibility of, 206–207; fragmentation and integration of, 169; global, 55, 73–74, 146–175; global flagship.*See* Global flagship networks; government hierarchies and, 207–210; hierarchical, 103–104, 207–208; highly structured, 11; incentives for investing in, 165; information, 14, 146–175; initial formation of, 161–162; Internet-constellation, 156n.23, 158; intrinsic value of, 171; knowledge, 14, 43, 79–80; metaphorical, 12n.25; national data, 152; proliferation of, 162–166; relations among, 149; shift from hierarchies to, 37–38; social, 12; specialized versus general-purpose, 165; thin and thick, 11–12n.23; transboundary conflict early warning, 215–238; value of, 158–162, 168–169. *See also* Global flagship networks (GFNs); Internet; Internetworking; Networking technologies
Networks of networks, 99, 105, 147n.3, 149
Neuman, W. Russell, 266
New American Model of Industrial Organization, 105
New Economy, 105, 107n.9, 107n.10; China in, 315–321; excessive growth and diversification during, 106
New field of inquiry, locating, 2–8
New technologies, technical properties of, 4–5
New York City financial center, 65; share of global transactions of, 71
New York Stock Exchange, foreign listings of, 67t
Newel, Allen, 225, 225n.17
Newness, social analysis of, 1–2
News media, instantaneous coverage of, 234–235
Newsgroups, heterogeneity of, 271–272
NGOs. *See* Nongovernmental organizations (NGOs)
Nike: online mobilization against, 79; repositioning of, 207
Nikko, 72–73n.16
Nir, Lilach, 269n.24, 272
Nohria, N., 103
Nokia, in China, 316–317
Nolan, Richard L., 97, 163n.40
Nonbusiness infrastructure, 103n.7
Nongovernmental organizations (NGOs), 37–50; in conflict resolution, 219; in early warning systems, 234, 235n.23; expanded role of, 38; growth of in 1980s and 1990s, 47; interactive technologies used by, 38, 39–43, 75–76; international, 3; in international conflict resolution, 218–219; Internet media in, 76–78; knowledge communities around,

13; as knowledge facilitators, 46–47; knowledge networks of, 1; legitimacy of, 207; market alliances of, 48; need to be self-sustaining, 47–48; new collaborative role of, 38, 43–49; new resources and capabilities for, 5–6; in sociodigitization, 17; software for, 75n.19; successful, 48–49; translation versus diffusion model of, 45–46
Nonrivalness concept, 181–182
Nora, Simon, 163n.42
Norberg, Arthur L., 165, 170
Norms, in open source software development, 194–195
North, Douglas C., 93n.1, 204
North, S. N. D., 124
North American Free Trade Agreement (NAFTA), 93
Norway, U.S. cold war internet connection to, 168
NSF: Computer Science and Engineering Advisory Panel of, 168; network development in, 165
NSFnet, 158, 168
Nuclear weapons testing, failure to predict, 234
Nussbaum, Martha, 223n.15
Nye, Joseph S., Jr., 3–4, 218n.6, 291

Obligatory passage points, 39–40
OECD, in Internet development, 168
Off-balance sheet financing techniques, 107
Okabe, Daisuke, 49n.18
Olsen, Johan P., 291, 292
O'Mahoney, Siobhan, 39n.2
One-to-many exchanges, 244
One World International, 80
O'Neill, Judy E., 165, 170
Online conversations: bridge between micro- and macroscale analyses of, 246–247; dynamics of, 255; themes of, 261–264; thesaurus of, 246–247
Online public engagement, limits of, 7
Online workspaces, 79–80
Open/adaptive/complex systems design research, 222–227
Open Development service, Bellanet, 75n.19
Open source: analytic problem of, 179–186; building economic logic of, 192–201; case studies in, 186–192; coordination problems in, 193–196; debugging of, 192; definition of, 178, 179; development of, 17; economic foundations of, 181–186; key features in success of, 190–192; myths about, 201–202; political economy of, 201–210; self-organization of, 203–204; signaling incentive for, 184
Open source community, 6, 13; altruism versus self-interest in, 202–203; battle to consensus in, 200; conflict with hierarchical organizations, 208–210; culture of, 185–186; ego in, 195–196; function of, 186–192; leadership and authority in, 195–196, 200–201; norms in, 194–195; resolving conflicts in, 199–201; shunning in, 196
Open Source Initiative, 179; cultural frame of, 195
Open source licenses, 179n.1
Open source networks, meaning in, 22
Open Source PhP-Nuke software, 75n.19
Open source programmers, as hobbyists, 202
Open source software: complexity of, 197–199; political economy of, 178–210
Open standards, 75n.19, 96
Open systems, biological, 222
Open systems design, 216
Open systems engineering, design-oriented, 224–227
Open Systems Interconnection (OSI) protocols, 154–155, 156–157; development of, 158; flexibility of, 157
Open systems theory, 222n.14
Oral story telling, 277
Organization, 13; complexity of, 69; definition of, 10; shaping spatialization, 11; structure of in rural communities, 120–121
Organization of African Unity, 237
Organizational infrastructure, alliances in, 70–71
Organizational innovation, 38; catalysts for, 97–98
Organizational knowledge, stickiness of, 90
Original equipment manufacturers (OEMs), 104, 105; contract manufacturers and, 107–108; limitations of, 105–108

Ostrom, Elinor, 204n.28
Outsourcing, 101; based on contract manufacturing, 105; for increased flexibility, 126; of knowledge-intensive support services, 103–104; of manufacturing, 105–108; partial, 99; systemic, 99
Oxfam America: Content Management System and Article-Builder of, 77n.21; web site of, 77
Ozawa, T., 99

Padilla, Luis Albert, 230
Parallel problem solving, 191
Paris: as financial center, 65–66; shares of financial market in, 67n.12
Parish, William, 328
Parole, 258–259
Partnerships, 48
Passive economies, 120
Patent protection, 178
Pavitt, K., 90–91, 99
Pax Britannica, 59
Peacemakers, frontline, 227–228
Pécheux, Michel, 257n.14
Pedersen, J. D., 98–99, 100, 101
Peizer, Jonathan, 48
Pennings, J. M., 120
Pennycock, Alastair, 248n.7
Pension funds, growth of, 60
People-making process, 297
Peregrine, Thai operations of, 72–73n.16
Perens, Bruce, 179n.1
Performance, 258–259
Permanently beta organizations, 40–41
Perry, Elizabeth, 327, 331
Persistent online identity, 43
Personal computers: development of, 163; growth of in China, 320
Philippines, democracy movement in, 81n.26
Picto-textual dimension, 11
Picto-textual social artifacts, 10
Pietrobelli, C., 100
Place-boundedness, 20
Pluralism, 206
Political economy: on Internet, 201–210; open source software in, 178–210
Political identity-formation, 289–290
Political projects, large-scale, 15
Political transformation, in China, 315–321
Popular cohesion, 291
Population decentralization, 124
Portals, regional rural, 132
Porter, M., 94
Post, Telegraph, and Telephone (PTT) protocol, 149n.10, 154–156, 157, 158, 164, 173; development of, 170; failure of, 156–157
Post-nationalism, 287n.1
Poster, Mark, 10n.19
Powell, Walter W., 99–100, 209
Power: networked forms of, 54–55; Newtonian image of, 46n.12; technologies of, 276n.26, 277–278
Practical grammars, 225–227
Practical understanding, 223n.15
Practices, organization of, 11
Pragmatic associations, conversational, 258
Prakash, Vipul Ved, 190n.10
Prediction, difficulty of, 8
Price, Vincent, 269n.24, 272
Price breakers, 104
Printing presses, high-speed, 277
Privacy: in China, 325–327; mobile telephones and, 325–327; telecommunications technology and, 336
Private digital space, strengthening of, 73–74n.17
Private sector: in China, 314, 330–332; network development by, 164
Privatization: in denationalization of financial centers, 71–73; global, 137–138; trade liberalization and, 93–94
Process geographies, 46n.12
Procurement costs, 98
Product life cycles, shortening of, 97
Product-specific value chains, 95–96
Production: decentralized, 45; technologies of, 276n.26
Profit margins, eroding, 94–95
Project information, open standard for, 75n.19
Property rights: Lockean, 185; in market economy, 204–205; shift in, 205–206; static conception of, 186
Proprietary code, 188
Proprietary rights, software development and, 57
Protest.net, 79n.24
Protests, antiglobal, 44, 46n.16

Protocols. *See* Internet protocols; Post, Telegraph, and Telephone (PTT) protocol
PTT protocol. *See* Post, Telegraph, and Telephone (PTT) protocol
Public-access Internet, 73–74
Public Citizen, 79n.24
Public spheres: fragmentation of, 305–306; Internet as, 265–276; new types of, 6
Pudong, 316
Pyke, F., 130

Qinghua University, 333n.16
Qinjian, Lou, 318–320
Quarterman, John S., 154n.18, 162n.39, 170–171
QWERTY keyboard, 157–158
QWERTY technology, 157–158n.28

Rabellotti, Roberta, 130
Railroad, advent of, 122
Rapoport, Anatol, 222n.14
Rationality: commitment to, 289; in deliberative discussion, 269; of online discussions, 272–274; in public sphere, 274–275
Raymond, Eric S., 185, 190, 191, 194, 196, 197–200
Reading, cultural codes in, 259n.16
Reciprocation, in deliberative discussion, 268, 270
Reciprocity, technology in, 6
Reclaim the Streets, 79n.24
Recombinant technology, 37–50
Reconstructive research, 226–227
Red Cross, 237
Red Hat, 209
Redden, Guy, 79
Redesign, 26
Reed, David, 43
Referendum: in civic identity-building, 290; reformed based on, 305
Referential contexts, 11–12n.23
Regional rural portals, 132
Regulation, in spatial organization of finance, 58–59
Religion, global networks of, 74n.18
Remote control, new forms of, 92
Renationalization, 284
Representative government, 293
Reputation incentive, 185, 186
Rescaling, 22–25
Resistance movements: information technology and in China, 327–330; Internet in, 81–82n.28
Resonance structures, 289–290
Resources, territorial specific, 129n.8
Rhizomatic networks, 326–327
Richardson, G. B., 94–95
Riemens, Patrice, 82
Rind, D. M., 276–277
Risse, Thomas, 290
RN reader, 246
Roberts, J., 98, 100
Rockerfeller Center, foreign investors in, 72–73n.16
Rogers, Juan D., 165
Roma Rights organization, 46n.14
Ronfeldt, David, 81
Rosenau, James, 50
Rosenthal, Elisabeth, 329–330
Rottmann, Sigrun, 79–80
Rousseau, Jean-Jacques, 286, 287, 289
Rousseau's paradox, 286–287, 289, 290, 294
Royal Commerce One, e-commerce of, 127
Royal Dutch/Shell: e-commerce of, 127; environmentalism of, 207; networks and, 209
Ruggie, John, 22, 37, 204
Rugman, A. M., 99, 103
Rupesinghe, Kumar, 215n.1, 215n.2, 218n.4, 219, 219n.8, 227, 228, 230, 232–233, 237
Rural communities: characteristics of, 119–121; deployment and diffusion strategies in, 132–134; impact of networking technologies in, 121–125; market organization of, 120–121; as self-contained production units, 124n.4; self-sufficiency of, 121–122; social cohesiveness of, 123–124; survival strategies for, 129–131; technology-based strategy of, 131–132; telecommunications regime in, 138–142; telephone technology diffusion in, 134–136
Rural concentration, 27
Rural economies, 28
Rural Electrification Administration (REA), 135–136
Rural out-migration, 124

Rural schools, 124n.4
Rural-urban continuum, 121n.2
Rural-urban divide, cooperative networks and, 117–142
Rural-user-oriented networks, 22
Rütli myth, 295

Sabel, Charles, 46
Sack, Warren, 11, 24, 243n.3, 243n.4; on design, 25–26
Sacks, Harvey, 247n.6, 254, 257–258
Sagawa, Shirley, 48
Sanyal, Bishwapriya, 46–47, 47n.15, 48–49
Sao Paulo, financial center in, 66
Sassen, Saskia, 12n.26, 15, 17, 22, 24, 27, 28, 38, 56n.1, 57n.3, 61n.7, 68n.13, 69, 72, 73–74n.17, 80, 81–82n.28, 83n.31, 92, 99, 118, 125, 127, 128, 215n.1, 288–289n.4; on global cities, 19–20n.32
Satellite networks, global, 147
Saxby, Stephen, 16n.28
Saxenian, Annalee, 206–207
Scalar analytics, 23n.35
Scale-up methodologies, 255–256
Scaled spaces of engagement, 80–81
Scaling, 22–25; questions of, 18–19
Schank, Roger, 225, 225n.18
Scharpf, Fritz, W., 306
Schiffrin, Deborah, 248
Schlesinger, Philip, 291, 292, 301
Schmalberger, Thomas, 215n.1, 230
Schmid, Alex, 232n.22
Schmidt, Susanne, 149n.10, 150n.11
Schmitter, Philippe C., 286, 290
Schuler, Doug, 266
Sciarini, Pascal, 295
Scott, Allen, 128, 132, 139, 141
Scott, James, 17n.29
Search techniques, 41–43
Searle, John, 225
Sears and Roebuck catalogue, 135
Seattle, WTO protests in, 46n.14
Seattle Community Network, 266–267
Secord, P., 226
Securitization, U.S. versus Europe, 62n.9
Securitized debt, distance from underlying assets of, 62–63
Security-sensitive information, 220
Segal, Adam, 312
Segal, Eli, 48
Self, technologies of, 244, 276–278, 276n.26
Self-interaction, 166
Self-interest, in open source software, 202–203
Self-organization, 203–204
Self-reproducing automata, 225n.17
Self-sufficiency principles, 47
Selwyn, Lee, 141
Semantic networks, of online discussions, 262–265
Semantics, 249; in conversation, 258
Sen, Krishna, 313n.2
Sendmail program, 180
Sengenberger, W., 130
Seniority rules, 194
Sennett, Richard, 11n.22
September 11 World Trade Center attacks, 65n.11, 207, 208
Shah, Rajiv C., 174n.58, 174n.59
Shanghai, SEZs in, 316
Shank, Roger, 225n.17
Shannon, Claude, 160
Shapiro, Carl, 159n.31
Sharp, John, 79
Shenzhen, 316
Sichel, D. E., 96
Sierra Leone, conflict in, 219n.8
Sign systems, technologies of, 276n.26
Signaling incentive, 184
Silicon foundry, 97
Silicon Valley, 97
Silverman, Kaja, 259n.16
Silvern, Steven E., 80
Simcoe, Timothy S., 97, 151n.14, 151n.15, 169n.49
Simmel, Georg, 290
Simon, Herbert, 25, 26n.42, 215n.1, 216, 222–223, 224, 225, 225n.17, 236n.26
Simultaneity, 54; of global financial market, 57
Sina.com, 317n.7
Sinclair, J. M., 254
Sinclair, Timothy J., 59
Smith, Anthony D., 284, 288
Smith, Brian Cantwell, 243n.2
Smith, Craig, 322n.10
Smith, Marc, 5n.10, 13n.27, 181
Smith, Michael Peter, 80n.25
Smith, Robert C., 80n.25

Smith-Doerr, L., 99–100
Sobel, R., 95
Social action: cyberspace in, 82–83; geographical scales of, 82–83
Social change, universities and in China, 334–335
Social cohesiveness: of conversational group, 249; networking technologies undermining, 123–124; of very large-scale conversation, 260
Social computing, 243
Social connectivity, in global financial market, 68–70
Social constructivist narratives, 204
Social Contract (Rousseau), 286
Social/digital imbrications, 19–21
Social engineering, piecemeal, 223
Social form, 9
Social groups, lexicon for, 246–247
Social inequality, new forms of, 301
Social informatics, 243
Social integration, scaffolding of, 290–291
Social justice criteria, 207
Social life, *res publica* of, 223
Social logics, 6; in digital formations, 54–84; in electronic markets, 54, 56–73
Social metrics, 277–278
Social movements, 17
Social network-based macroanalysis, 256
Social network theory, 49
Social networks, 12; in China, 325–327; globalizing, 76–77; of online conversations, 263
Social organization, technology and new geographies of, 37–50
Social processes, multiscalar, 23
Social purpose, 173–174
Social relations, rescaling of, 2–3
Social Science Research Council (SSRC) project, 3–4
Social space, 10n.19
Social theory, micro-macro levels in, 254
Social trends, 4
Societé Generale Group, 72–73n.16
Societé Internationale de Telecommunications Aeronautiques (SITA), 164n.43
Society for the Promotion of Area Resources (SPARC), 81
Society-technology divide, artificial, 38
Sociodigitization, 3, 16–18; discontinuity of, 17; multivalence of, 20–21
Socioeconomic change, 5n.10
Sociotechnical logics, 54–55
Software: for disadvantaged NGOs, 75n.19; free, 179; GPLed, 187–188; open source, 178–210; in rural communities, 131; in staging, 11; "word of mouth," 42
Software instruments, financial, 57
Sohu.com, 329–330
Solectron, 107n.9
Solomon, Richard, 266
Source code: debugging of, 192; documentation of, 191–192; modularization of, 198–199; reuse of, 191; standards of, 194–195. *See also* Open source
Sovereignty: democratic, 286–287; in globalized world, 216–217; versus universal norms, 44
Soviet Union breakup, 335; failure to predict, 234
Space: definition of, 10; interaction and, 11
Spatial agglomeration, utility of, 68
Spatial centrality, 65n.11
Spatial mobility, 94
Spatialization, 10–11
Special Economic Zone (SEZ), Beijing, 315–316
Specialization, rapid inorganic growth and, 106
Specialized clusters, 101–103
Specialized external capabilities, 95
Specialized financial markets, multiplication of, 58
Specialized knowledge, coordination of, 193–196
Specialized network suppliers, 92
Specialized networks, 165
Speciation, 193
Specificity, 6
Speculative financial institutions, 60
Speculative investment strategies, 61
Staging, 11
Stallman, Richard, 187–188
Standage, Tom, 147n.4
Standards development, 149n.10
Starbucks, nonprofit partnerships of, 48
Stark, David, 6, 10n.17, 10n.21, 13, 15, 24, 26, 27, 45, 46, 46n.14, 48, 48n.17
Start-up problem, 162
State-building missions, 217
Stavenhagen, Rudolfo, 230, 232

Stedman, Stephen J., 220n.9
Stein, Eric, 285
Steinmueller, W. Edward, 92, 301
Stickiness-of-knowledge proposition, 90, 108
Stiglitz, Joseph E., 119–120, 121, 133
Stinchcombe, A. L., 120
Stock, as currency, 107
Stock exchanges: alliances of, 70–71; foreign listings in, 67t
Stock market capitalization, 62–63
Stock markets, 10 biggest in world, 66t
Stockholm Stock Exchange, mergers of, 70
Stone Corporation, 327n.14
Storper, Michael, 127–128, 129n.8
Strategic interaction, technology in, 6
Student protests, in China, 333–335
Sturgeon, T., 105
Subnational networks, 153
Substantialism, national, 287–288, 306; criticism of, 288–289n.4
Suchman, Lucy, 38n.1, 45
Sun Yat-sen, 335
Sunstein, C., 305–306
Suppliers: higher-tier, 104; lower-tier, 104
Supply chain management, global integrated, 98n.3, 104
Supply-demand mismatches, 107
Surveillance resistance, 278
Sustainability, of rural communities, 129
Sustainable Development Communications Network, 79–80
Sustainable Development Gateway, 80
Sutton, Brent A., 147
Swamintham Research Foundation, M.S., 76
Swann, G. M. Peter, 159n.32, 160n.35
Swanson, Louis, 124n.4
Swift, Adam, 287n.1
Swiss Exchange, foreign listings of, 67t
Swiss nationalism, 295
Switzerland, international banking in, 59n.6
Swyngedouw, Erik, 81
Sydney, financial center in, 66
Sylvan, D., 230
Sylvan-Majeski tradition, 230–231
Symons, Jessica, 79–80
Synchronous interactions, 96–97
System: definition of, 146n.2; formation of, 169–172; scale-up of, 174–175
System-external environments, 223
System generation, logic of, 170n.51
System-internal human environments, 223

Tallinn Stock Exchange, 70
Tambini, Damian, 265
Tanenbaum, Andrew S., 163, 199n.24
Tarrow, Sidney, 305
Task-switching, 188n.9
Tasliki, Liza, 75n.20
Taylor, Peter, 22, 23
TCP/IP protocol. *See* Transmission Control Protocol/Internet Protocol (TCP/IP)
Technocratic functionalism, 291
Technologies: adoption of, 38; contradictory character of, 6–7; deployment and diffusion of, 7; diffusion of in rural communities, 121; rural deployment of, 132–134; as society frozen, 1; types of, 5n.9
Technology-based strategy, 131–132
Technology leapfrogging, 138
Technosocial situations, 49n.18
Teil, Geneviéve, 255
Telecommunications: in China, 336; commodification of, 139–140; in computer networking, 164; foreign investment in, in China, 324–325; growth of in China, 318–320; international systems of, 154; investment in, economic growth and, 117n.1; need for rural regime in, 138–142; privacy and in China, 325–327
Telecommunications Annex, 140–141
Telecommunications businesses, cooperative model for, 137
Telegraph: advent of, 122; in centralized business organization and, 122–123; undermining social cohesiveness, 123–124
Telegraphic network, global, 147
Telenet, development of, 164
Telephones: in centralized business organization and, 122–123; cooperatives in diffusion of, 134–136; global systems of, 147n.3. *See also* Telecommunications
Tensions, between formation dimensions, 27
Territorialized economic development, 129n.8
Terveen, Loren, 255
Text-context conflicts, 248
Textual metafunctions, 249, 251

Textual relations, recurrent patterns of, 255
Thai government, Financial Restructuring Authority of, 72–73n.16
Theater-computer analogy, 11n.22
Theiler, Tobias, 306
Thesaurus: automatic computations of, 262; conversational common sense and, 256–259; of online discussion groups, 246–247; rough-draft, 260–261
Thompson, Grahame, 56n.1
Thomson Financials, on financial market concentration, 71
Threshold effect, 84
Thrift, Nigel, 125, 129
Tiananmen Movement, 327–328, 331–332, 333, 335; information technology in, 337
Tiananmen Square uprisings, 81n.26
Tilly, Charles, 295
Tirole, Jean, 183–184
Toho Mutual Insurance Co.: GE Capital joint venture of, 72–73n.16
Tokyo financial center, 65–66; share of global transactions of, 71
Tokyo stock exchange, foreign listings of, 67t
Tølle, M., 98–99, 101
Top-down transformation, 305
Toronto financial center, 66
Toronto Stock Exchange, alliances of, 70
Torvalds, Linus, 188–191, 194–196, 197–199, 200–201, 205
Toulmin, Stephen, 38, 46n.12, 215n.1, 218n.5, 223n.15
Trade agreement, on global telecommunications, 139–142
Trade barriers, global flagship network growth and, 107
Trade centers, development of, 124
Trade liberalization, 93–94, 119
Trait geographies, 46n.12
Transaction chains, increased length of, 57
Transactions: complex architecture of, 57; digitization of, 61–62; growing range of, 60
Transactivity, capabilities for managing, 57n.3
Transboundary conflict early warning networks, 215–238
Transboundary digital formations, 2–3
Transboundary interconnections, 166–169
Transboundary internetworking, 147–148; approaches to, 152–158
Transboundary processes, 22–24
Transformation, 15
Transformative capacities: of global capital market, 56–73; of global network, 54
Transformative sociodigitization, 3
Transforming structures, 5n.10
Translation: definition of, 45n.11; versus diffusion model, 45–46; Latour's description of, 46n.13; NGOs engaged in, 45–46
Transmission Control Protocol/Internet Protocol (TCP/IP), 152, 154–155, 156, 169; efficiency of, 172; support of, 156–157
Transnational communication: density of in Europe, 285; in European community, 300–304; European demos and, 283–306
Transnational networks, in conflict early warners, 227–233
Transnationalization, 3
Travelers Group, acquisition of Nikko by, 72–73n.16
"True believers," 180
Tsagarousianou, Roza, 265
Tully, James, 304n.13
Tuomi, Ilka, 195
Turkle, Sherry, 5n.10
Turner, Bryan S., 254
Turoff, Murray, 265
Tuthill, W. C., 136

Uncertainty, competition and, 94–95
United Nations (UN): in conflict management, 237; in conflict prevention, 217–218; diplomatic system of, 233–234; High Commissioner for Refugees of, 233; Security Council decisions of, 216–217
United States: elections in, online discussions of, 269–271, 273–276; government interfacing network organizations in, 207–209; Internet anchored in, 151; Internet protocol in, 155n.20, 155n.21
Universal norms, concerns about, 44
Universities, intracampus communication networks on, 163
University College, London (UCL), early international network at, 167

UNIX Network, 153–154
UNIX operating system, 188; development of, 194n.17; programming standards of, 194–195; simplified, 189
UNIX philosophy, 194–195, 198
UNIX software, 188
Urban centers: in global economy, 128; impact of networking technologies in, 121–125; rise of, 122–123
U.S.-European networks, early development of, 167–168
U.S. Patent Office, 109
U.S. telephone cooperatives, 134–136
Use, mediating cultures and practices in, 21–22
Usenet newsgroups, 161n.37, 246, 268; conversation maps of discussions in, 261–265; discussions on, 268–269; index of, 246n.5
User networks, emerging, 161–162

Vallee, Jacques, 168n.48, 170
Value chain, stages of, 99–100
Variability, 6–8
Varian, Hal, 159, 159n.31, 160, 162n.38
VAX, 187
Venn diagram, 12n.25
Vertical bureaucracies, 126
Vertical disintegration, 100
Vertical integration, 95, 126–127; benefits of, 130
Vertical specialization, 94, 95–96, 101, 105; information requirements in, 97
Vertically integrated mutlinational corporations, transition from, 93
Very large-scale conversations (VLSCs), 244–248; maps of, 260–265; micro-macro divide in, 254–256; politics of, 265–276; technologies of self and design ethics for, 276–278; typical contemporary view of, 245
Very small aperture terminals (VSATs), 137–138
Vesterager, J., 98–99, 101
Video-conferencing, 97
Village Knowledge Centers, 76
Viner, Katharine, 44
Viral clause, General Public Licence, 187–188
Virtual citizenship, 301
Virtual communities, 5n.10, 167
Virtual corporations, 5n.10, 16
Virtual enterprises, 100, 101
Virtual environment, rural, 131–132
Virtual industrial districts, 15
Virtual private networks (VPNs), 165
Virtual space, 10n.20
Vixie, Paul, 191
VLSCs. *See* Very large-scale conversations (VLSCs)
Voluntarism, civic, 288–290, 294
von Neumann, John, 236n.26

Wade, Robert, 56n.1
Waever, Ole, 304
Walder, Andrew, 327, 328, 332
Walker, D. R. F., 23
Wallenstein, P., 227n.19
Walzer, Michael, 11n.23, 287
Wang Zhidong, 317n.7
Warf, B., 81–82n.28
Warkentin, Craig, 50, 77
Warner, Michael, 291, 303
Warschauer, Mark, 133–134n.10
Wasserman, Stanley, 255
Wasserstrom, Jeffery, 327, 328, 331
Weaver, Warren, 160n.35
Weber, Eugen, 13, 22, 24, 27, 28, 159n.30, 294, 296
Weber, Max, 206–207
Weblogs, 242; index of, 246n.5
Webs of interaction, 13
Webster, Frank, 5n.10
Wehn, Uta, 140
Wei, Sha Xin, 243n.2
Weiler, Joseph H. H., 285, 286, 288, 299
Weiner, Norbert, 236n.26
Wellman, B., 13n.27
Werle, Raymund, 149n.10, 150n.11
West, Paige, 40
Western European banking system, cross-border nature of, 58–59n.5
Westphalians, 287n.1
Whittaker, Steve, 255
"Who's on first?" skit, 250–251n.8
Whyte, Martin, 328
Wild, Helga, 243n.2
Wilhelm, Anthony G., 268–274, 274, 275–276
Wilkins, M., 93
Wilks, Alex, 82n.29

Willard, Terry, 79
Williams, Shirley, 285
Williamson, O. E., 95, 100
Windrum, Paul, 159n.32, 160n.35
Winnograd, Terry, 11–12n.23
Wittenburg, Kent, 42
Woolard, Edgar, 323
World Bank, Knowledge Bank of, 82n.29
World Economic Forum, 117; online mobilization against, 79
World Telecommunications Organization (WTO), 140
World Trade Center, September 11 attacks on, 65n.11, 207, 208
World Trade Organization (WTO), 73; China's entry into, 317n.6, 322; online mobilization against, 79; protests against, 44
World Wide Web Consortium (W3C), 147–148n.6
World Wide Web (WWW): advent of, 151–152; groundwork for, 334
Wright, Stephen, 219
Wurster, Thomas S., 120–121

Xerox, in network development, 164
Xu Guanhua, 312

Yamaichi International Capital Management, 72–73n.16
Yang, Guobin, 76
Yang, Mayfair, 326–327, 336
Yau, P., 92
Yenta, 42
Yunus, Dr., 47

Zacher, Mark W., 147
Zaloom, Caitlin, 24
Zander, U., 95
Zanfei, A., 95
Zapatista movement, 76–77, 81n.27, 233
Zeng, L., 232n.22
Zhang, Yibin, 130
Zhao, Dingxin, 328, 332
Zhongguancun, 332
Zhongguancun Science and Technology Park, 315–316
Zhu, Enjoyce, 315n.3
Ziff, P., 250n.8
Ziff's Law, 250n.8
Zimbabwe, database in, 82n.29
Zmag.org, 79n.24
Zook, Matthew, 151n.12
Zurich, financial market in, 67n.12
Zürn, Michael, 290

Lightning Source UK Ltd.
Milton Keynes UK
UKOW04f2146070715

254726UK00006B/156/P